Effective Audio-Visua

Effective Audio-Visual

A User's Handbook

Third edition

Robert S. Simpson

FOCAL PRESS

Focal Press
An imprint of Butterworth-Heinemann
Linacre House, Jordan Hill, Oxford OX2 8DP
A division of Reed Educational & Professional Publishing Ltd

A member of the Reed Elsevier plc group

OXFORD BOSTON JOHANNESBURG
MELBOURNE NEW DELHI SINGAPORE

First published 1987
Second edition 1992
Paperback edition 1994
Third edition 1996

British Library Cataloguing in Publication Data
A catalogue record for this book is available from the British Library

ISBN 0 2405 1416 5

Library of Congress Cataloguing in Publication Data
A catalogue record for this book is available from the Library of Congress

Typeset by Datix International Limited, Bungay, Suffolk
Printed in Great Britain by Bath Press plc, Bath, Avon

Contents

Preface

Audio-visual has come to mean many different things to different people. To some the use of a single slide is audio-visual, to others only a program using the latest video technology qualifies for the epithet. Technology itself has made the whole field more intimidating, and some potential users could be forgiven for either giving up the idea of using audio-visual at all because it seems so complicated, or of worrying whether they have chosen the right medium.

This book is written for the audio-visual user. This is a wide public, because audio-visual media can be used by any organization. Usually the person who has to use or commission AV methods has no formal training in their use; his or her job is often some functional role within the organization for which AV is seen as useful support.

The aim of the book is to help these people in two ways. Firstly to convey *advice* on the choice of AV medium and how to prepare programs and materials, and secondly to convey *facts*, especially about equipment, presentation methods and environments that will help in the practical implementation.

It might be argued that someone who has to commission AV only needs advice on choice of medium, and that technical matters can safely be left to technicians. Unfortunately, many good audio-visual shows have not realized their full potential because the sponsor has failed to take any account of the environmental needs of AV. A technician cannot reverse a management decision to give presentations in a totally unsuitable room. On the other hand the manager with responsibility for the success of a presentation who also understands the basic principles involved can ensure that the problem never arises in the first place.

Because the book includes factual information it will also be of use to those already in the audio-visual business, whether as program producers, staging technicians or professionals in one branch of AV who need basic information about another.

The chapters are, as far as possible, complete in themselves. Each is a review of a particular subject, so can be read on its own.

Audio-visual methods have a valuable contribution to make in education, training, government, art, culture, commerce and industry. Their use can be stimulating, effective and fun. This book is intended to help users make the most of them.

In the few years since the first edition of *Effective Audio-Visual* there have been no changes in the basic rules about how to use AV. However there have been some shifts of emphasis, so the opportunity has been taken to update the illustrations and examples.

High definition television, multi-media interactive and sophisticated computer displays with ever more confusing acronyms have appeared on, and sometimes disappeared from, the market. This third edition introduces many of them, but does keep them in their place. AV must not be allowed to be dominated by its own technology.

The second edition of *Effective Audio-Visual* introduced a number of new techniques for interactive and computer-based AV, but made the point that many of these were, at the time, not quite ready for practical application. Without doubt the major change since then has been the coming to maturity of some of these techniques.

Because of its review nature, this book does not pretend to be a detailed textbook; indeed each chapter could support a book of its own. The aim remains that of helping users and practitioners to make the right choice of presentation method.

R.S.S.

Acknowledgements

Parts of this book have previously been published in a house publication of Electrosonic Ltd entitled *What is Audio-Visual?*

In the text there are necessarily references to company trademarks. When known these have been indicated as trademarks: for example, Carousel™, a trademark of Kodak Ltd and Eastman Kodak Co.

Illustrations of audio-visual equipment are representative. Inclusion or exclusion of particular products does not imply any special endorsement or adverse comment. The author acknowledges help with illustrations from many colleagues in the audio-visual business, and the source of illustrations is indicated in the captions. Unattributed photographs and drawings are by Electrosonic Ltd, David Nuttall and the author.

Chapter 1

The choice of audio-visual media

The first confusion that must be faced is the distinction between 'Audio-Visual' and 'Visual Aids'. This book covers both, because they are generally used together, and often use the same equipment. For the purposes of definition *audio-visual* is taken to mean a medium using recorded sound, whereas *visual aids* accompany a live presenter.

A summary of visual aids is shown in Table 1.1.

Table 1.1 Visual aids

Display material, including product display
Flipchart
Overhead projector
Writing board
Magnetic and other display boards
Slides
Filmstrip
Video display of single images
Video display of computer output (graphics and text)

It is clear that some of the aids require more preparation than others, and that some permit writing on by the presenter. A similar summary for audio-visual is shown in Table 1.2.

Table 1.2 Audio-visual

Super 8 mm movie
16 mm movie
35 mm movie
70 mm movie
Filmstrip with sound
Single slide with sound
Dissolve slide with sound
Multi-image slides with sound
Videotape videodisc/displayed by monitors
Videotape/videodisc displayed by projection
Multi-screen video
Computer output display linked to sound
Sound and light technique
Mixed or multi-media

The problem is deciding which medium to use. This chapter reviews the priorities related to business and training presentations. The special considerations required for exhibitions, big audiences, and permanent installations are more fully dealt with under separate headings.

When choosing a presentation medium it is important to concentrate on the particular presentation objectives. Effectiveness of a presentation is more important than the means employed. The specialist needs of a limited audience are very different from those of a publisher of video programs or a producer of public entertainment films, which means there is often an argument for using the simplest possible medium. However, the discipline that a medium imposes can also greatly improve the effectiveness of a presentation.

Often a user finds that more than one medium is needed – usually a simple visual aids system and a separate recorded program system. The choice of medium for a particular application depends on the answer to a series of questions. Unfortunately, there is not always a clear-cut answer to each, but there is often a 'favorite' answer. The actual application then points to one medium as the best single answer to the problem. This is illustrated in Table 1.3.

Table 1.3 reviews the choice of visual aids for groups, but does not answer all questions. The occasions when visual aids are used vary greatly in formality and importance. A group of engineers having a discussion in a laboratory can manage very well with a writing board as their main visual aid; a monthly board meeting of a major corporation may in some ways resemble a discussion meeting, but the speed at

Table 1.3 Choice of visual aids for groups

What is best if . . .	*Favorite answer*
'Once only' material is needed for an informal audience?	Flipchart/overhead projector (OHP)
A 'teaching' session is needed?	Writing board/OHP
A mixture of frequently used and specially prepared material is needed	OHP
The visual must be changed while being displayed?	OHP/magnetic board
The visual must match AV material in impact and quality?	Slides
It must be easy to prepare material 'in house'?	OHP
Daylight is present?	Writing board/OHP/ flipchart
On-line computer data must be displayed?	Video or data projector
A prestige presentation is to be given?	Slides

which discussions can be conducted depends greatly on how well information is presented to the group.

At this point a 'law' applies; the effectiveness of the visual aid can be proportional to the time taken to prepare it. A well-prepared set of slides presenting financial data to a board meeting may take longer to make than a set of figures hastily duplicated and handed round. However, the presentation of financial data to a group by slides is far more satisfactory than the use of handouts because:

☐ Everyone is looking at the same data and is not reading ahead.

☐ The discipline of preparing the data in a form that can be read by a group ensures that it is easier to assimilate.

☐ The formality imposed by a standard medium speeds up the presentation and discussion process.

Chapter 2, on the preparation of visuals, demonstrates that production of good visuals is not necessarily more time-consuming than production of bad ones. In recent years many difficulties have been removed. Lettering machines that are quick and easy to operate mean that it is no longer necessary to choose between typewriting, which is difficult to read, and the delay and expense of typesetting. For those with personal computers and laser printers, the illegible overhead transparency is a thing of the past. Introduction of the high-quality instant slide allows a uniform presentation quality, even when material must be prepared at the last minute.

Figure 1.1 The writing board goes electronic. Anything written on the Quartet Ovonics Electronic Copyboard can be reduced and distributed as a paper copy

When it comes to audio-visual it is necessary to make a distinction between individual and group communication, because the equipment used for the different applications gives a different emphasis to both the questions and the answers. Tables 1.4 and 1.5 review the two fields.

Table 1.4 Choice of AV medium for individual communication

What is best if . . .	*Favorite answer*
A lot of copies are needed?	Video
Only a few copies are needed?	Slides
Easy program making and editing are required?	Slides
Movement is required?	Video
Good image quality is required?	Slides
Portability is required?	Video/slides/ computer
Low production costs are essential?	Slides/computer
Interactive programs are required?	Videodisc/computer

The tables raise a few questions. For example, video is suggested less often than might be expected. This is because most serious business presentations depend heavily on good-quality *still* images (such as charts, graphics and product photographs) that must be presented on a large scale. The normal TV screen can show only a limited amount of information when viewed by an audience. If video projection is used the results are acceptable for moving images in the dark, but are not *always* good enough for large still images with some room lighting.

Starting in business can be easy ...

— or a trifle difficult..

... But whatever your business problems..

... The Bank are always there to Help.

Figure 1.2 The storyboard is the key to any audio-visual program. One image, one idea

Table 1.5 Choice of AV medium for group communication

What is best if . . .	*Favorite answer*
Wide distribution of the program is required?	Video/16 mm
Good image quality is required?	Slides
Good sound quality is required?	Slides/video
Easy program editing is required?	Slides/computer
Movement is required?	35 mm/16 mm/video
A large audience is to be motivated?	Multi-image
A live presenter also takes part?	Slides/computer
Production costs must be kept reasonable?	Slides/computer

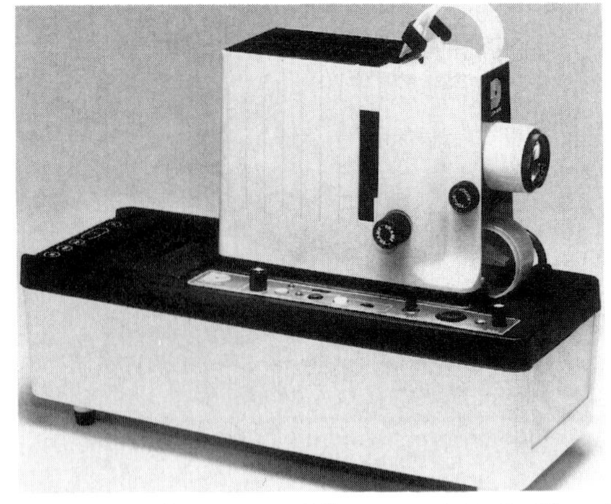

Figure 1.3 The economical filmstrip projector is still used in education, but not in commercial AV. This is the Dukane Micromatic II sound/filmstrip projector

Figure 1.4 The boardroom at Bergesen DY A/S in Oslo is traditionally styled . . .

Figure 1.5 . . . but as the world's largest independent owner of LPG shipping tonnage it needs full AV facilities, which are neatly concealed in the furniture

Figure 1.6 Both the screen and the video projector retract into the ceiling when not needed. (Installation by Audio Grafisk)

The personal computer has become a highly satisfactory way of making legible presentation support material. This in turn has created a big demand for methods of showing the material to groups, which is being met with new kinds of video projectors. Video projectors themselves are now much easier to use, so for many users the computer/video projector combination is becoming *the* standard method of presentation support. However, this should not blind users to the merits of alternative and, possibly, simpler methods of presentation when they are more appropriate.

As users of audio-visual methods gain experience, they may well find it necessary to use a variety of media and presentation methods. So, if much use is made of video sequences with moving images for training, use of slides or overhead transparencies for still images (as opposed, for example, to computer-generated images presented on a video system) can be a positive advantage because this helps vary the pace of the presentation.

All AV methods require some preparation, not only in their initial production, but also in their actual use. Obviously a special occasion justifies special effort and perhaps extra expenditure. But it is in day-to-day use that AV can be most effective, and it is therefore important to plan from the outset how material is to be prepared. It has become increasingly possible to use one medium for production and initial showing with another medium for distribution. The next four chapters develop this theme.

Chapter 2

The preparation of visuals

Some years ago Kodak Ltd in the UK put out a leaflet entitled *Let's stamp out awful lecture slides.* It started with the following:

> We are organizing a crusade. Nothing controversial, of course. We are not party politically minded, and we don't feel too strongly about the Corn Laws. However, what we do get rather steamed up about are *awful lecture slides.*
>
> You know what we mean. An awful slide is one which contains approximately a million numbers (and we've left our opera glasses behind). An awful lecture slide is one which shows a complete set of engineering drawings and specifications for a super-tanker. An awful lecture slide is one which shows about two-dozen dials when only one is necessary.
>
> That's what we're crusading against, and we would like you to join us. Of course we know *you* don't produce that sort of slide, but you probably know somebody who does. Do him a favor and pass on these tips . . .

For 'lecture slides' the leaflet could just as well have read 'overhead transparencies' or even 'flipcharts'; the problem is the same for any text-based visual presented to an audience. So before deciding on which medium to use it is important to understand the ground rules that apply to the preparation of *all* kinds of visual.

Legibility

The biggest single error made in preparing visuals is the idea that legibility in one form means legibility in another. A printed page is read at a distance of 30–50 cm (12–20 in). In a lecture theater a slide or overhead transparency projected onto a 2 m (6 ft) wide screen may have to be viewed from a distance of 20 m (60 ft). Reading text in this way is like reading this book from a distance of 3 m (10 ft).

Thus in AV terms it is just as much a crime to make an overhead transparency directly from a typewritten page, as it is to photograph that page and make a slide from it. To produce a legible visual requires preparation, and in principle there is little difference in the time needed to produce a usable slide, overhead transparency, flipchart page or computer text page. The decisions as to which to use depends, therefore, on other factors.

Figures 2.1–2.4 demonstrate how lettering should be sized for the different methods of presentation. It is no coincidence that they all end up by limiting any one visual to about the same amount of text.

Clutter

The booklet *Slide Rules* by Antony Jay, that accompanies the splendid Video Arts and Kodak Ltd co-production *Can we please have that the right way round?* suggests that a good rule is never to put more text on a slide than would be printed on a T-shirt. This may be taking matters a little further than is practical, but does make the point that any visual should be kept simple. It is not just a matter of limiting the amount of

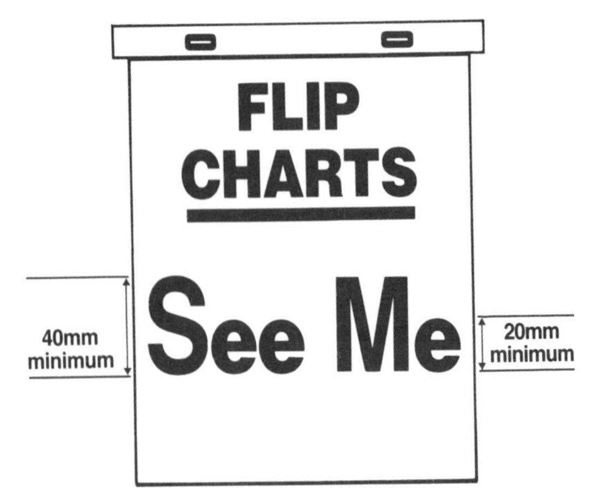

Figure 2.1 On an A2 flipchart letters should be not less than 20 mm (¾ in) high for a maximum viewing distance of 8 m (26 ft). Longer viewing distances mean a bigger chart and larger letters

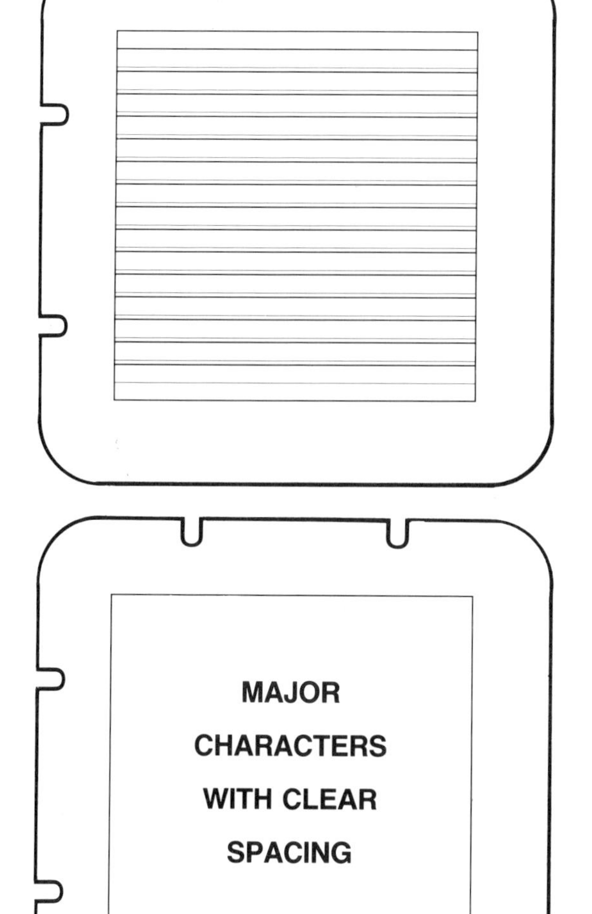

Figure 2.2 On an overhead transparency there should be a maximum of fifteen lines. Major characters should be one space high, and there should be a clear space between each line

Figure 2.3 The rules for preparing text slides are the same as for overhead transparencies . . .

Figure 2.4 . . . therefore it is not surprising that the same rule applies for text on a video screen

text, but of considering *any* image in terms of how it stands up to presentation.

Thus, an engineering drawing or circuit diagram may be ideal for printing in a book for individual study, but is quite useless as a slide or overhead transparency. This is not just because the audience has difficulty in deciphering the details; while they are busy screwing up their eyes to examine the details of the visual, they are not listening to the presenter.

So the next rule is never to put more on a visual than is necessary to make the point being discussed. In fact, it is possible to go further and suggest the best visuals are those that are not quite complete – relying on the presenter to complete the picture.

The specific advice which emerges from these general observations applies to all kinds of prepared visuals, whether slides, overhead transparencies, flipchart pages, or computer pages intended for video projection. Some of the points demonstrated in the

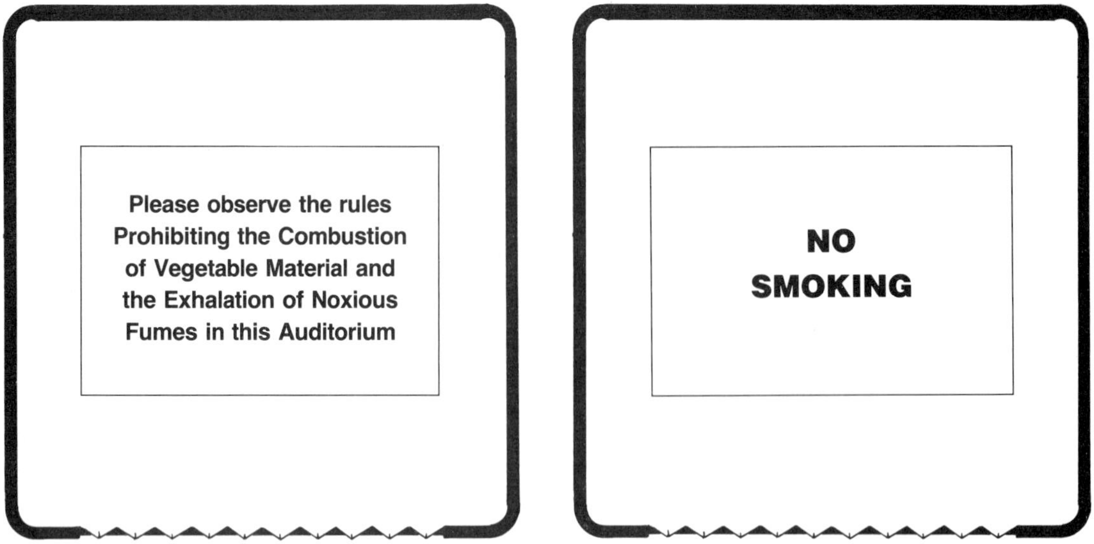

Please observe the rules Prohibiting the Combustion of Vegetable Material and the Exhalation of Noxious Fumes in this Auditorium

Figure 2.5 Keep the number of words on a visual to a minimum

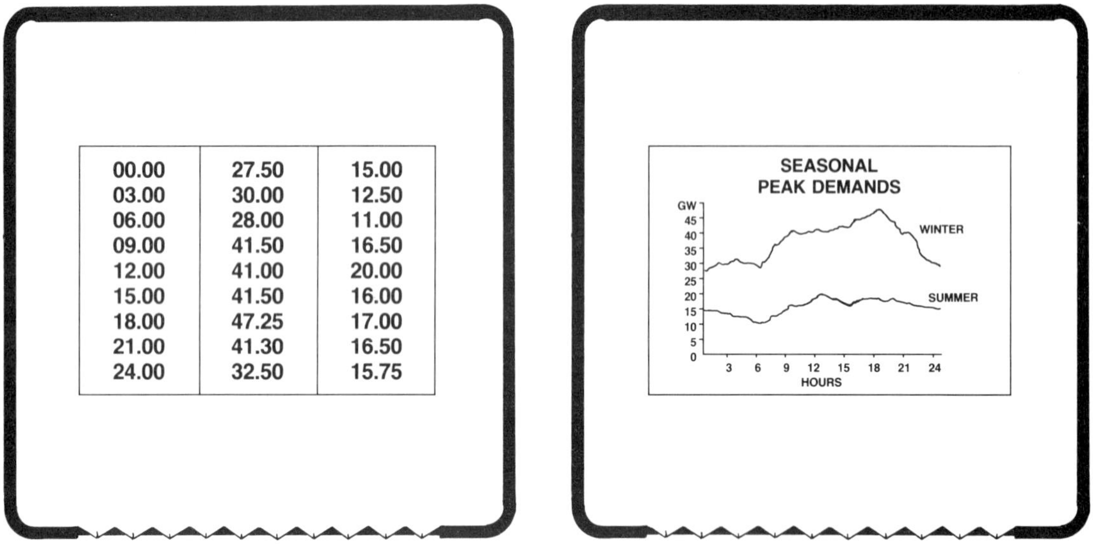

00.00	27.50	15.00
03.00	30.00	12.50
06.00	28.00	11.00
09.00	41.50	16.50
12.00	41.00	20.00
15.00	41.50	16.00
18.00	47.25	17.00
21.00	41.30	16.50
24.00	32.50	15.75

SEASONAL PEAK DEMANDS

GW: 0, 5, 10, 15, 20, 25, 30, 35, 40, 45

WINTER

SUMMER

3 6 9 12 15 18 21 24

HOURS

Figure 2.6 Avoid tables of figures. Use a chart or graph

accompanying illustrations, in no particular order, include:

☐ Keep the number of words to a minimum.

☐ Never show large amounts of text which allow the audience to 'read ahead' of the presenter.

☐ Remember that a visual is a support, not a substitute, for the presenter.

☐ Always simplify diagrams to the essentials.

☐ Avoid big tables of figures. Either break them down into smaller tables, or, better still, convert them to a graph, bar chart or pie chart.

☐ When labelling diagrams or charts use only horizontal lettering – *never* angled lettering.

☐ Use color wherever possible to clearly distinguish between sections of a diagram or sectors of a chart.

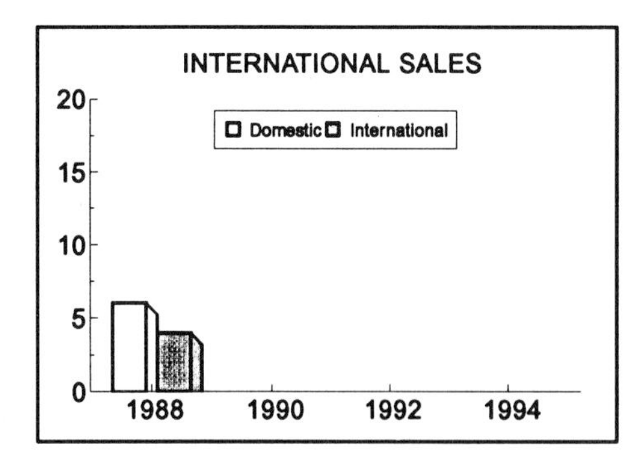

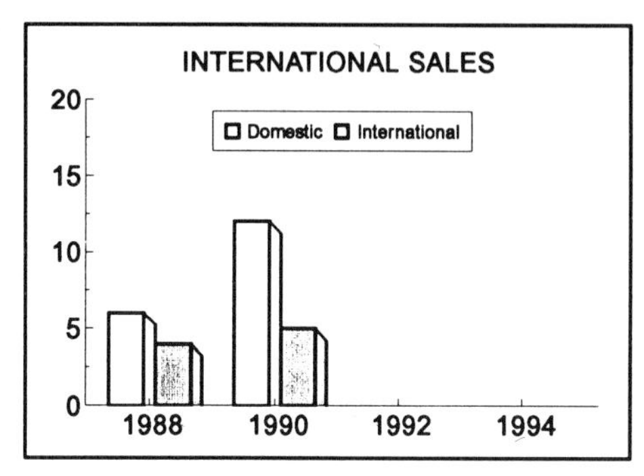

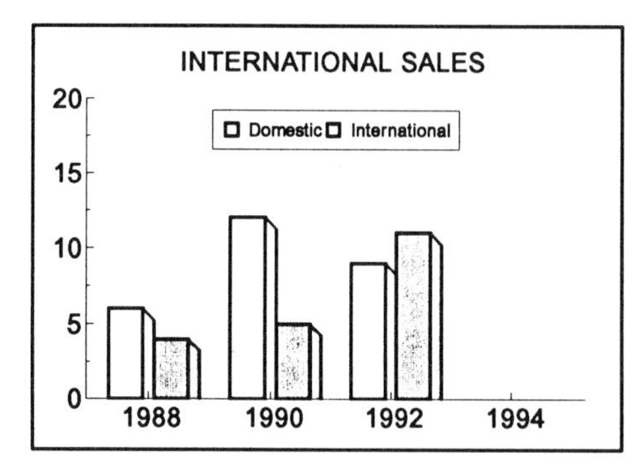

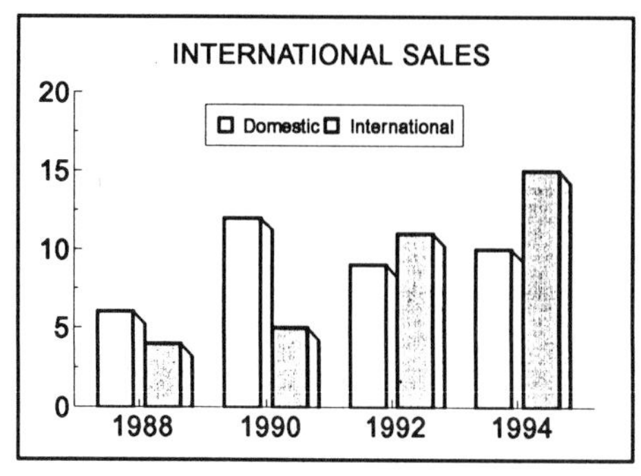

Figure 2.7 Use a build-up sequence if the final chart is complex, or if different parts of the chart need a separate explanation

☐ When using color, be careful to use combinations that complement each other. Some background/foreground combinations, e.g. pink letters on a red background, can be illegible when projected.

☐ When photographs are used, show only the essential. This may mean reproducing only part of an existing picture to eliminate irrelevant background.

☐ Make sure that all visuals for a presentation are the same format. Do *not* mix portrait and landscape formats.

☐ Never put an image in front of an audience that is not directly relevant to what is being said by the presenter. Do not bring on an image too early, and, most importantly, get it off the screen as soon as it has been discussed.

☐ Wherever a complex image is required, try to break it down into several simple images. If this is not possible, show it as a build-up sequence.

The last point is the most important of all. The technique of *'successive reveal'* or *'build-up'* helps communication because it ensures that the audience is only presented with one new piece of information at a time. The presenter is forced to keep to a logical order of presentation, and neither he nor the audience can jump ahead, because they do not see visual information on the screen before it is required. Although principles of visual design are the same for all the standard media, it is the way in which image *sequences* can be presented that affects the choice of medium.

People who give or organize presentations on a regular basis need to develop methods of making or commissioning visuals. It is important to understand that even if users do not intend to make the visuals themselves, they must be aware of the principles of good visual design. A bureau produces the slide asked for – if the slide has too much text, it is the fault of the commissioner, not the bureau.

Making visuals

Now it is time to review different ways in which visuals can be made. At this stage the methods are described without relation to their cost or suitability for in-house use. In many cases there are equally effective alternatives, and it can simply be a matter of personal preference as to which is used. For example, people with an aptitude for using personal computers are likely to want to use them to make visuals; others, who hate the sight of computers, are probably much happier with simple lettering machines.

Common sense must be applied to some of the methods listed. For example, it may seem strange to suggest that one of the ways in which overhead transparencies can be made is directly from a slide. Why not use the slide itself and project that? If a presentation consists mainly of photographic images the best way to present them *is* by slides; but if it is given in a daylit room and most of the material is conventional

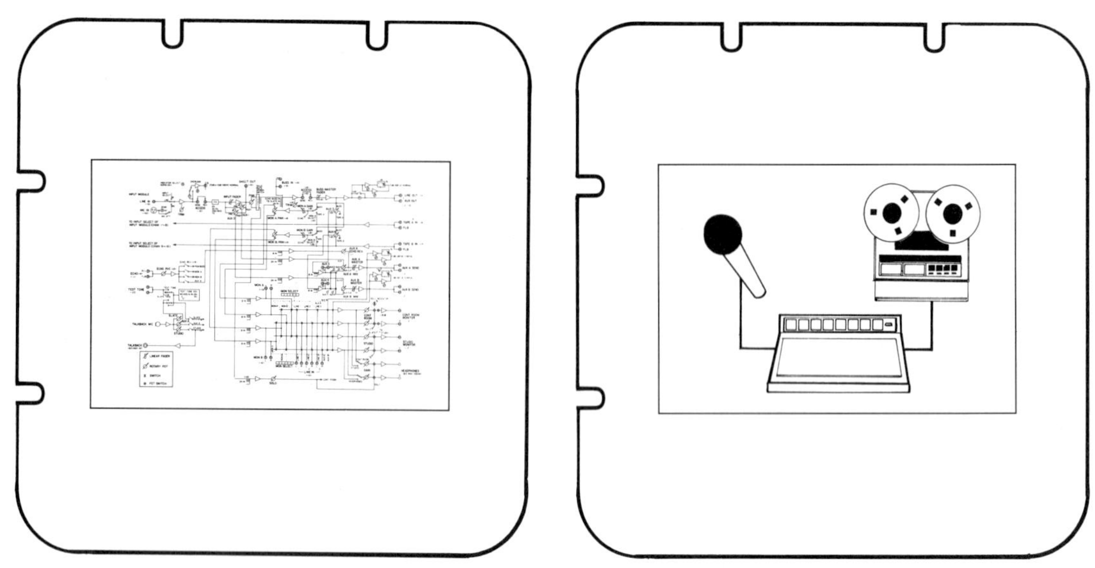

Figure 2.8 Diagrams intended for textbooks or manuals are too complicated for group presentation. They should be simplified

overhead transparencies, it can make sense to convert the one or two photographic images required into big overhead transparencies.

Overhead transparencies

The 'fuel' for the overhead projector (OHP) is the overhead transparency. This is a piece of transparent film, with or without a card border, usually 25 cm by 25 cm (10 in by 10 in) or A4 size. Although such a transparency is of sufficient size that it is possible to prepare a visual direct on the material, a copying process is usually included as part of the preparation.

It is possible to write directly on a transparent film using a felt-tipped marker pen. Usually this is done only when a presenter uses the OHP as a writing board, but sometimes it is appropriate to prepare a sequence of hand-written transparencies in advance.

Better-looking transparencies use proper lettering and even artwork. Although it is possible to apply this directly to a transparent film, it is more usual to prepare the initial material as artwork on plain paper, then make a transparency from it.

Such a transparency is made in one of two ways. Best results are obtained using thermal or infra-red films. The original is placed against the film and the two are passed through a thermal copier. The resulting transparency can then be:

- □ black on clear;
- □ black on color;
- □ color on clear;
- □ clear on color,

depending on the type of film used. The range of colors is typically limited to blue, red, green and yellow.

An alternative method, limited to giving black on clear or black on color tint, is to use a standard plain paper copier. These can either be the dry toner type (e.g. Xerox™) or the liquid toner type (e.g. Nashua™). Here plain paper copier (PPC) film is placed in the copier's feed tray while the original is placed on the machine exactly as if an ordinary copy is to be made. An ordinary copy should be made first – if the original makes a good paper copy, then it should also make a good transparency.

Table 2.1 Some methods of making OHP transparencies

Write or draw directly on to transparent film
Copy original on to clear or colored thermal film
Copy original on to transparent film using plain paper copier
Enlarge a slide to OHP transparency size
Use color computer printer with transparent film
Use color copier on to transparent film
Photograph original on to Polaroid instant transparency film

Lettering on the original can be prepared in several different ways. Typewriting is often used, but it should definitely be avoided unless a display typewriter is available. Ordinary typewritten characters do *not* meet requirements for legibility, however an enlarging photocopier can be used to bring them up to the correct size. Rub-on lettering such as Letraset™ gives good results, especially when additional standard symbols are needed. Much quicker to use, however, are lettering machines that prepare the lettering in aligned strips.

The advent of the personal computer has introduced another highly satisfactory way of preparing OHP transparencies. However, it is not good enough to use just standard letter printing because this would be no better than typewriting. What is needed is a computer with a graphics printer, and a program that allows creation of page layouts using large fonts and, possibly, diagrams and charts.

Figure 2.9 Good quality lettering for the in-house production of visuals can be produced with a lettering machine. (Photo Kroy Europe Ltd)

Figure 2.10 It is easy to prepare an OHP transparency using a simple graphics program on a personal computer

Figure 2.11 The OHP transparency can be printed in seconds on a laser printer

The Apple Macintosh™ computer is very easy to use for visuals creation, and is widely used in the design professions for this purpose. The IBM personal computer and its successors and clones now have a vast range of programs of varying complexity suitable for this work. Most commercial users' needs are met by low-cost programs such as *Harvard Graphics*™, ideal for creating organizational charts, graphs and other business graphics.

Although best results are obtained using laser printers, perfectly acceptable transparencies can be made using dot matrix printers.

Color can be added to overhead transparencies in several ways. Apart from the options offered by basic film, more elaborate coloring can be achieved by:

□ hand tinting with marker pen;

□ hand coloring by cutting and pasting on color adhesive films;

□ mounting several separate color-on-clear originals together in the same mount.

□ using a color graphics printer to print directly onto a transparent film.

The last method is the most satisfactory for the business user. Color printers are now available at sensible prices, and they allow presentation graphics generated on personal computers (for example using Microsoft Powerpoint™) to be directly converted into color OHP transparencies. However, there is an important point to remember when converting color images from one medium to another. Colors will not always match the original seen on the computer's monitor screen, so it is important to make a test copy first and, if necessary, adjust the colors to give the best result on the final transparency.

If more elaborate coloring is required it is best to

use standard photographic methods. Full-color transparencies for OHP use can be made from any color original, whether flat artwork, a color transparency or a photograph in another size. For those in a hurry, Polaroid make an instant OHP transparency film.

Figure 2.12 Tektronix make a low cost thermal printer ideal for making color OHP transparencies from personal computers. (Photo from Autographix Ltd)

Slides

Although there is no reason why an OHP transparency should not look professional, they are normally used for less formal occasions. However, if complete flexibility in image make-up, a full range of colors and the highest possible image quality are required, then slides are called for.

Lettering for slides is prepared in the same way as for overhead transparencies. Typing can be used because it is possible to blow the lettering up to a legible size. However, because this then reveals a poor letter quality, typing tends to be used only for effect. The user requiring only a few slides may use rub-on lettering or a lettering machine. The professional slide maker uses properly typeset lettering, either made on his own machine or by a local printer. It is only necessary to have the letters set with the correct relative size, and in high-contrast black and white, as coloring and positioning is all done by the camera.

The rostrum camera

The enormous flexibility of slides, both as visual aids and as components of AV shows, is due to the way photographic images can be manipulated. For AV work the secret behind this manipulation is the rostrum camera. Many users do not need the full facilities of a professional rostrum camera; their needs can be adequately met by a simple copy stand. Others do not need all the features of large machines. However, a brief description of a typical camera serves to illustrate both what it is and what it can do.

The main component of a rostrum camera is a rigid table. Flat artwork, which can be any size from about the size of a slide up to about 1.2 m (4 ft) wide, can be placed on the table. The artwork can be top-lit by suitably placed lighting, which usually has polarizing filters to eliminate glare. As an alternative, the table can be underlit to allow reproduction of transparencies, which can be any size from about half that of a slide up to about 25 cm by 20 cm (10 in by 8 in). The illuminated stage for transparencies is lit by a special three-color lighting unit, used both for adding color to black and white originals and for color matching or correction of color transparencies.

With a simple table the camera operator must position the item to be reproduced. When, as is often the case, precision positioning is needed, the artwork or transparency is placed on an animation compound. This device allows the original to be moved under the camera by turning handles, one for the X-axis (east-west), one for the Y-axis (north-south) and one to rotate the original through 360 degrees. Each movement is precise to a fraction of a millimeter and can be read off from a display counter. Repeat accuracy is essential.

The camera itself rides on a rigid vertical column that may be 3 m (10 ft) high or more. Although manual focusing is available, the camera usually has an automatic focusing arrangement that ensures lens focus adjustment is moved exactly as required as the camera rides up and down the column. This is located at the top of the column when photographing large pieces of flat art, and at the bottom when copying from another slide.

The camera itself can also be changed. The standard camera movement is for 35 mm slides, but this can be exchanged for a camera designed to take superslides on 46 mm film, or even for a 16 mm camera for filmstrip or movie making. Whichever film is used, the camera accepts bulk-loaded film in lengths up to 120 m (400 ft). The main feature that sets the camera apart from a normal 35 mm camera is pin registration.

Many of the effects that can be achieved by the rostrum camera depend on precise positioning of the image relative to the film. For example, one of the most common techniques used in slide making is double exposure. A number of frames may be exposed for one piece of text, then some of them exposed again to add further text as part of a build-up. If each exposure is not perfectly in register, the result is a mess. Thus, the camera is equipped with registration pins that precisely locate the film in the camera for each exposure.

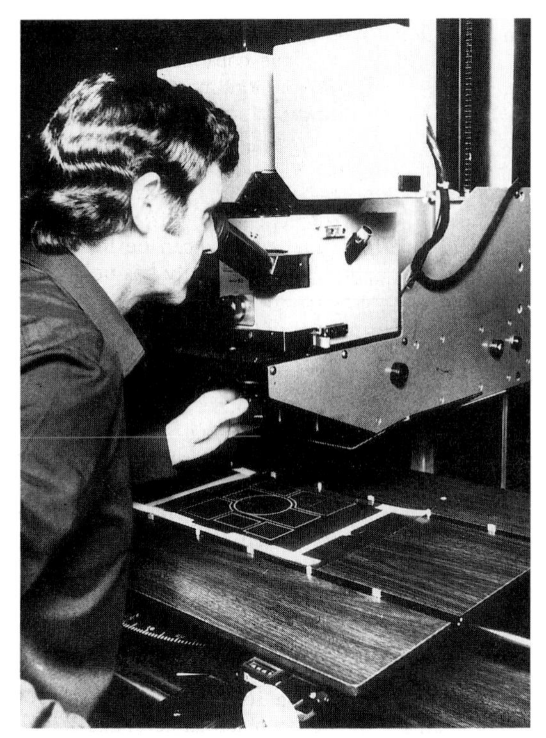

Figure 2.13 A Forox rostrum camera in use. The artwork or transparency to be copied is carried on an animation compound which allows precise location and planned movement. (Photo Prater Audio Visual)

The camera also has controls that allow for multiple exposure, automatic taking of a preset number of exposures, and automatic film movement both forwards and backwards. Camera movement up and down the column is motorized, and the whole ensemble is under either manual or computer control. Viewfinding is done either with an eyepiece viewfinder, or by a system called reticle projection – where an image corresponding to the exact slide format is projected back through the camera on to the artwork or slide stage. This allows artwork and camera to be precisely positioned so only wanted material is copied.

What can a rostrum camera do?

A rostrum camera can copy any piece of artwork, photograph and color transparency on to a slide. It can select any part of the image, or can make up a series of slides from one original (e.g. for multi-image work). It can also do several less obvious things.

It can add color. Most professionally-made text slide start life as a typeset item in black and white. This is usually transferred to high-contrast black and white film known as *ortho film* (e.g. Kodalith™). For some jobs this slide may be used directly or else be sandwiched with a colored gel; but the more usual requirement is for colored lettering on a different colored background. This is done by exposing a positive (black lettering on clear background) version of the slide, illuminated by the required background color. The same piece of color film is exposed again, this time to a negative (clear lettering on a dense black back-ground) version of the slide illuminated by the required letter color. The final result is a two-color slide.

This shows why registration is vital. Any discrepancy between positive and negative images would show up badly. The process is not limited to two colors; further multiple exposures allow as many different colors as required. In the case of text and graphics slides an extension of the process allows production of a series of slides constituting a build-up, each component of which is in precise registration with its predecessor.

The rostrum camera can be used to produce many trick effects in the hands of a skilled operator. These include:

- ☐ Streaking, achieved by a combination of many successive exposures and zooming of the camera lens.
- ☐ Neon glow, achieved by several methods including use of a diffuser glass placed at a short distance in front of the text being copied.
- ☐ Spinning text, achieved by multiple exposure and rotation of the object text.
- ☐ Masking the image with a photographically produced mask of any shape. Masks can be hard or soft edged. By extension of this technique photo montages can be made up, with images merging into each other or occupying defined areas.
- ☐ Posterization, and other tricks with color. An image may be rephotographed to black and white through different color filters, then reconstituted as a color picture with completely different colors.

The modern rostrum camera is a highly efficient means of producing visuals, with enormous scope for creative use. Not everyone needs all facilities of the big rostrum camera, so users with less exacting requirements use simpler copy stands (see Figure 2.14) with cameras that have the vital attribute of pin registration even if they do not have a big film capacity.

Some slide reproduction requirements can be met by other methods. Splitting of one image into several slides for multi-image and panorama-projection purposes is usually done by the rostrum or copy stand camera. However for simple splitting there is another method. This is to use a photographic enlarger to project the image directly on to suitably positioned and pin registered 35 mm film (see Figure 2.15).

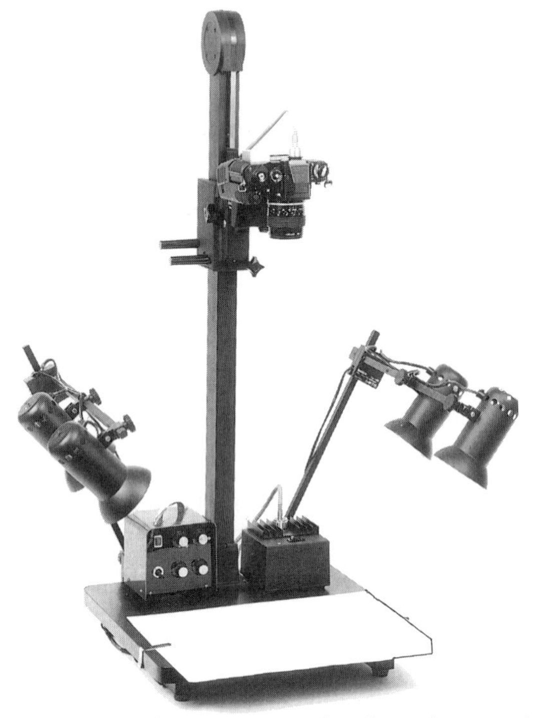

Figure 2.14 Those users not needing the productivity of the full-size rostrum camera, but still needing many of the effects, can use a simpler copy stand with a pin-registered camera. (Photo of the MRX camera from Image Ltd)

Figure 2.15 The Wess Splitter is a device that allows one original image to be split into several slides, for example to produce a panorama for projection. An enlarger projects the original onto the precision held film

In-house or out?

Most OHP transparencies are made at short notice within the organization that is going to use them, but most slides are made by specialist service companies. The exception is the large organization able to justify its own fully equipped AV department.

The description of the rostrum camera makes it clear why this is so. A skilled camera operator is able to produce high-quality slides in large quantities at reasonable price. Most service companies quote standard prices for commonly required slide formats, and always pre-quote for special work.

Slides can be professionally made at short notice. Service companies typically offer a three day standard turnround, and 24-hour turnround at a premium. Even so, last minute slides may be required. This need can

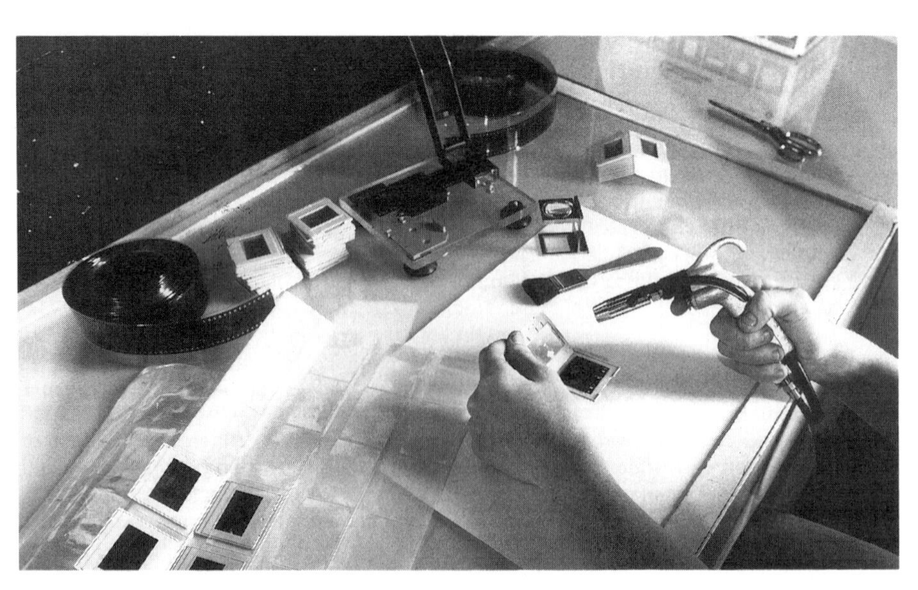

Figure 2.16 Care must be taken when mounting slides to ensure that no dust is trapped in the mount. (Photo Tony Gidley Productions Ltd)

now be met in-house by the use of instant slides. Film for instant slides is available from Polaroid in 35 mm cassettes, which can be used in any normal 35 mm camera.

Slides can also be created with a computer, using a suitable computer graphics program. This method of producing visuals is so important that it warrants a separate chapter of its own. Computer-generated images are taking an increasing share of an increasing market, however, there remain many types of image for which the photographic method is either the only way or the best way. Most slide service bureaux offer both types of slide production, and match the method to the client's requirements.

Increasingly, the 'photographic' and 'electronic' methods of image creation and manipulation are merging. Most 'single screen' business slides are made using computer methods, but 'multi-image' slides and product pictures are mainly photographic. All the techniques described in 'photographic' terms in relation to the rostrum camera have their 'electronic' equivalents.

Chapter 3

The role of computer graphics

If this chapter had been written in 1980 it would have been of limited interest to the average AV user, because the relevant computer graphics equipment available at the time was extremely expensive and able to serve only a limited market. Now the situation is completely different, and anyone who has to commission visuals must be aware of what computer graphics can do. In the early 1980s the computer graphics industry enjoyed explosive growth, and there were hundreds of companies offering equipment and services. It was difficult to determine why apparently similar products could vary so much in price and performance. However, the situation has now settled down, and while there is still amazing progress in graphics systems, the choice for the industrial user is now easier.

This is because of the universal acceptance of the personal computer as a standard item in the office. A combination of the dramatic fall in computer and computer memory prices and the need for computer manufacturers to endow their products with more perceived value, has led to a situation where graphics capabilities which fifteen years ago would have cost $110,000 (£75,000) or more can now be obtained as a commodity item for less than $2,000 (£1,350).

The basics

A brief description of how computer graphics work helps put the subject into perspective, and explains why there has been a sudden take-off in activity. One of the components necessary for practical graphics is *memory*, huge quantities of it. It is the amazing drop in price and increase in density of semiconductor memories (and similar components) that has made high-quality computer graphics available to all.

The way in which computer graphics are seen is, first of all, on a cathode-ray tube – the display component of all TV sets and video monitors. An electron beam is deflected by electrodes so that it 'writes' on the tube surface; when the beam hits the phosphor on the inside of the tube, the phosphor fluoresces. The beam can be modulated so that the resulting spot of light varies in brightness.

An image can be built up on the tube face in one of

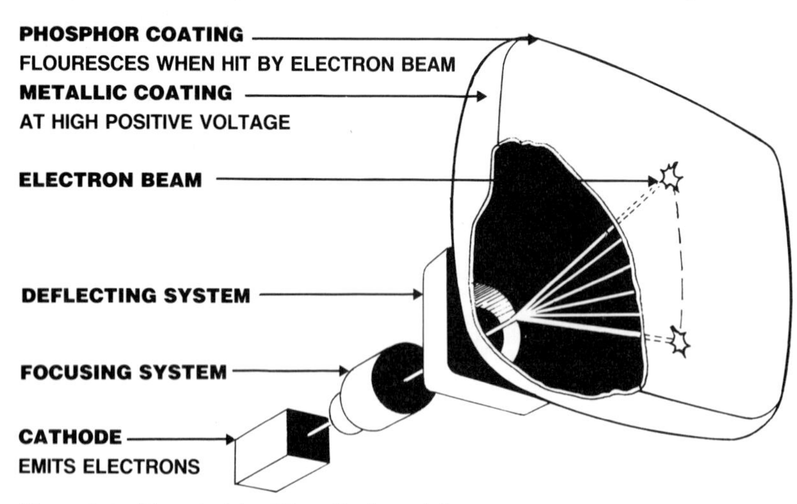

Figure 3.1 The principle of the cathode-ray tube

two ways. In computer graphics they are referred to as vector and raster graphics. Vector graphics can be considered as a set of instructions describing how to draw a particular picture. If the device drawing the picture can think in terms of coordinates on the screen, it can store the instruction set in the form of:

- ☐ Start at 100x, 100y
- ☐ Move to 245x, 100y
- ☐ Move to 100x, 320y
- ☐ Move to 100x, 100y

This succession of moves draws a triangle. If the movement or 'vector scanning' is done fast enough, and the phosphor of the tube has some persistence, the triangle is seen as a still image. A process of continually redrawing the picture is referred to as refreshing the image, and it is usually done about 30 times a second to give the effect of permanence.

A system for vector graphics therefore consists of a means of entering coordinate information, and a memory to store the long sequence of instructions needed to draw the image. Vector graphics can be extremely precise, the accuracy is limited by how precisely the coordinates are specified. Graphics can include arc drawing routines, so that circles and curves look smooth. Vector graphics are widely used in computer-aided design and drafting, where the accuracy of the line drawing is essential, and where mechanical plotters can produce a hard-copy drawing which follows exactly the same vector instructions used to compose the picture on the cathode-ray tube.

Raster graphics

Although vector graphics are accurate, and are excellent for applications like architectural and engineering design, they are not so flexible when large masses of color are required. There is also a problem in displaying vector graphics on a standard video monitor or TV set.

In a monochrome display of a standard television set the cathode-ray tube is used in a different way than in vector display. Instead of moving in a way determined by the image; the electron beam moves in exactly the same way whatever image is to be displayed. It scans the tube face from the top to the bottom, building up a raster of lines.

In the USA the image usually consists of 525 lines, with a complete image being scanned 30 times a second. In Europe the image usually consists of 625 lines, and a complete image is scanned 25 times a second. In fact TV images are interlaced with two 'fields' of 262.5 or 312.5 lines. One field takes one sixtieth or one fiftieth of a second to scan, and the next field places its lines between those of the first field.

Whether an image is seen or not depends on modulation of the electron beam current. No beam current means no picture; maximum beam current causes a bright line to be drawn.

A color tube is just three times as complicated. Subjectively, any color can be built-up by mixing appropriate proportions of red, green and blue. The face of a color TV tube consists of thousands of phosphor dots arranged in sets of three; one dot in each set fluoresces red, one blue and one green. Usually

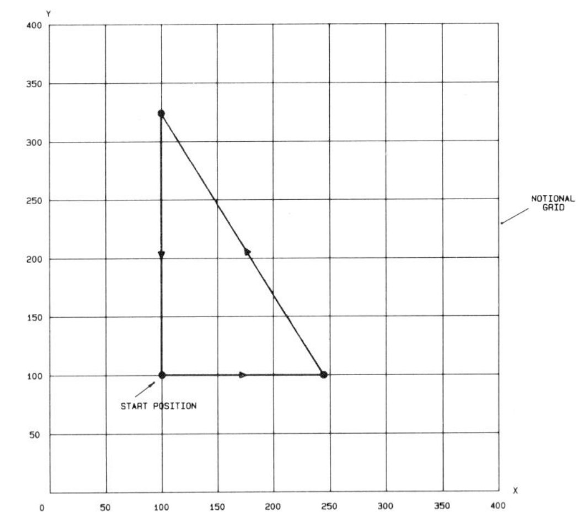

Figure 3.2 Drawing a triangle with vector graphics

Figure 3.3 Computer graphics are widely used in engineering design and drafting. Many of the diagrams and flow charts in this book were prepared using a simple CAD (Computer Aided Design) program

three electron guns are used in the tube to produce three electron beams, one for each color. A clever masking arrangement ensures that the red gun can only hit red dots, and likewise for blue and green guns.

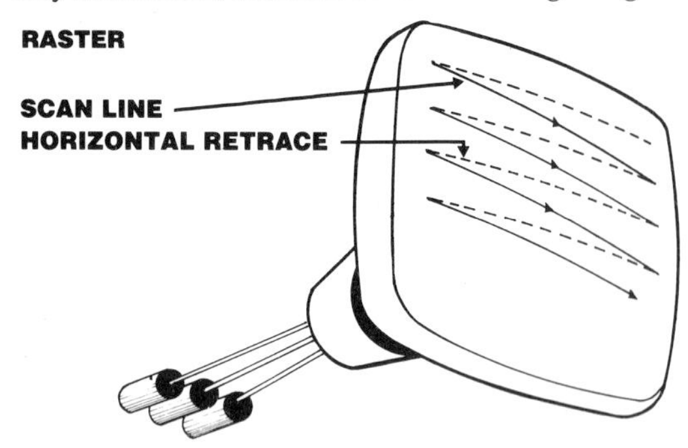

Figure 3.4 Raster display by a cathode-ray tube

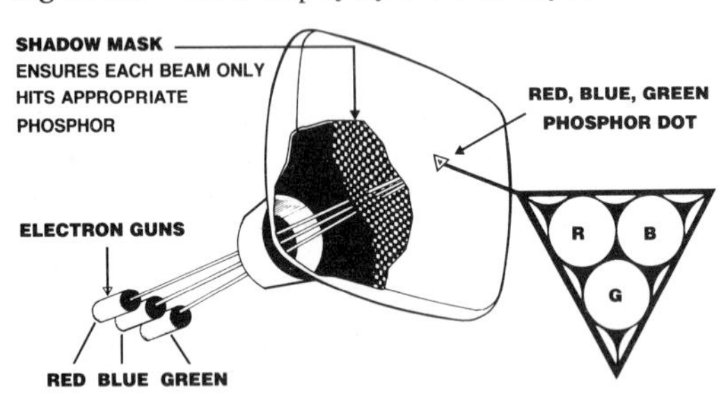

Figure 3.5 The color TV tube. In this illustration the dot pattern shown uses round dots. Now many tubes use rectangular or stripe dots

Resolution of a particular color display therefore depends on two factors: how finely the dot or stripe pattern is printed; and the quality of the electrical signal controlling the electron beams. The normal TV picture is made up from an analog signal. This means the electrical voltage that modulates the beam current moves *continuously* over a range of values to change the momentary brightness of the picture. Picture quality, in principle, depends on how fast the signal changes values, and on the precision with which a particular value is maintained. While this method of image construction is very satisfactory for live picture sequences created by a video camera (whose internal action is a scanning action which neatly matches that of the cathode-ray tube display), it does not lend itself to computer synthesis.

Raster graphics notionally divides the screen into a large number of picture elements, abbreviated *pixels.* An obvious possibility is to match the capability of the graphics system to that of a TV image. In a 525-line system only 480 lines are actually displayed, the time which would be taken by the other lines being required for the 'vertical interval' between fields. So a popular raster display is based on a pixel array of 480 (vertical) and 640 (horizontal). This is the VGA format. In order to eliminate the flicker effect of an interlaced video still image, computer graphics are usually shown by progressive scan with 60 full frames per second – roughly equivalent to twice the 'speed' of NTSC. This requires monitors of higher performance than normal 'video' monitors.

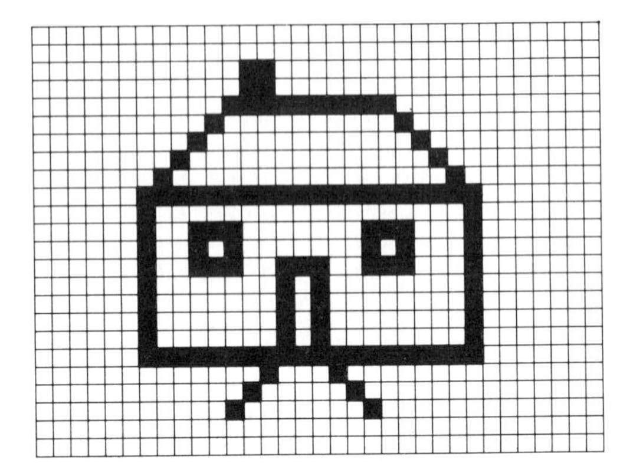

Figure 3.6 The pixel concept

A computer works entirely on a digital basis: that is, it can only think in terms of discrete values. It might seem reasonable to ascribe to each pixel a range of values for each of the red, blue and green components. Reasonable, yes, but quite demanding on the computer memory because, in the example given, values for 307,200 pixels need to be stored.

Computers store numbers as binary digits or *bits,* each of which can only have the value 0 or 1 (on or off). A convenient grouping is a *byte,* which consists of eight bits and is thus able to represent 256 different values. A typical store for a raster graphics display allows one byte per pixel, and it becomes a matter of user choice as to how values are ascribed.

In theory, the user could choose from a range of options, for example:

- 256 different colors, each created by a preset mix of red, green and blue;
- 256 different levels of white;
- 8 levels of red, 8 levels of green and 4 levels of blue.

In practice the first option is the one often used, and the user is able to define which colors will be used. The acronym of CLUT does not describe a clumsy clot who produces mis-registered graphics, but stands for *color look-up table,* used by the computer to match the remembered numerical value to a desired color.

It is already clear that the quality of raster graphics is limited in two ways. *Resolution* is limited by the number of pixels in the display, while *color* and *intensity*

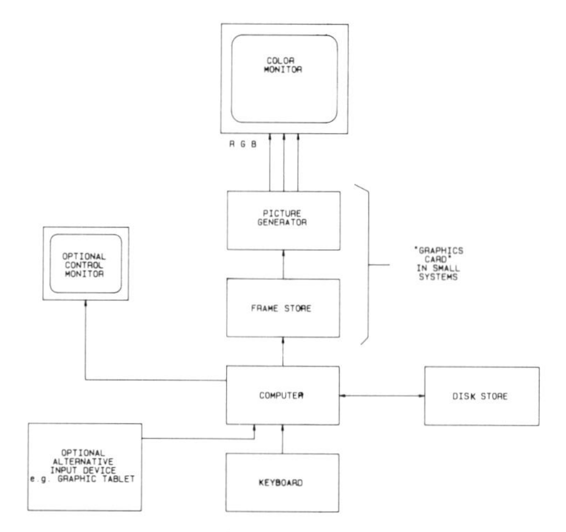

Figure 3.7 Block diagram of a computer graphics system

precision are limited by the number of memory bits allocated to each pixel. As memory prices have come down, so users have begun to expect better performance and even quite modest systems now allow full 24-bit color, i.e. 256 levels of each of red, blue and green by using one 8-bit byte for each color. Precision graphics systems and computer-based film post-production systems commonly use 10 or 12 bits per color.

A typical system

A typical computer graphics system consists of:

☐ A display, which is usually a color monitor. Depending on precision of work to be undertaken this can be either an ordinary commercial monitor or a precision device with high resolution, which is much more expensive.

☐ A frame store. This stores instaneous values of every pixel in the image being presented.

☐ A picture generator. This converts the information held in the store to the separate red, green and blue signals needed by the display.

☐ A central processor (or computer) that controls the whole system and, in particular, allows the user to easily change what is in the memory.

☐ A keyboard, for entering instructions and text.

☐ A long-term memory, usually a floppy disk or hard disk store for the long-term storage of completed images.

☐ Sometimes a second monochrome monitor for displaying instructions, program options etc.

In reality, a complete computer graphics system consists either of a purpose-built system or, more usually, a standard computer to which a graphics package has been added. The graphics package consists of a special graphics controller that carries out the picture memory and picture generation functions, together with computer software allowing easy access to it. Graphics packages are available for popular computers such as the Apple™ and IBM™.

Table 3.1 lists some of the graphics standards that evolved during the 1980s as the price of computer memory came down, and users' quality expectations went up. Table 3.2 lists a few representative examples of low-cost computer programs that can be purchased for use on personal computers.

What can a complete computer graphics system be expected to do? In the past there tended to be a distinction between graphics programs based on the setting-out of text and those intended for artwork, sometimes referred to as paintbox programs. Customer demand and technological advance have now brought both together, so a reasonable power graphics program would be expected to allow:

☐ Setting-out of text anywhere on the screen. Early programs only allowed lettering used in normal data displays. This is virtually useless for display and presentation work. Therefore, the program should allow a reasonable choice of fonts and character sizes. It should be possible to position text anywhere on the screen.

Table 3.1 Examples of computer graphics display standards

Name	*Full name*	*Pixel array*	*Color?*
CGA	Color Graphics Adaptor	640 × 480	2
HGC	Hercules Graphics Card	720 × 348	Monochrome
EGA	Enhanced Graphics Adaptor	640 × 350	16
VGA	Video Graphics Array	640 × 480	16
Super VGA	Super Video Graphics Array	800 × 600	256
8514/A	IBM Graphics Adaptor	1024 × 768	256
X-VGA	X Windows Video Graphics Array	1280 × 1024	256

Note that some of these standards can show alternative resolution and color combinations e.g. VGA can show 256 colors at resolution 320 × 200. These examples are all for the IBM series of personal computers and their clones. Apple Macintosh computers have equivalent displays, but starting at the VGA level. Very high resolution displays using 1600 × 1200 and 2048 × 2048 pixel arrays are also available.

Table 3.2 Examples of graphics programs suitable for visuals preparation

Publisher	*Program title*
Ashton Tate	Applause II and Draw Applause
Business & Professional Software/Polaroid	Presentation Express
Lotus	Freelance Plus
Micrographics	Graph Plus
Corel	Corel Draw
Microsoft	Powerpoint
SAS	Graph PC
Software Publishing	Harvard Graphics
Zenographics	Mirage

Figure 3.8 Making slides on a standard personal computer. (Photo courtesy of Slidework)

☐ Selection of any required color. Computer graphics people advertise somewhat meaningless claims such *as a palette of 16.7 million colors.* In practice, a more realistic choice is given at any one time – for example a choice of 256, but with the possibility that the user can re-define a color. There is sometimes a restriction on the number of different colors that can be used in a single image e.g. sixteen.

☐ The drawing of any outline. This can be on a freehand or a construction basis. In the case of construction the user defines beginning and end of lines and the computer draws the straight line between, or center and radius of a circle are defined and the computer draws the circle. Here the computer is exercising the principles of vector graphics but translating them into raster form.

☐ Filling of a shape with color. This is picture-book painting made easy. Any outline defined by an unbroken line can be filled with a selected color.

☐ Storing and manipulation of a shape, image or pattern. For example, a company logo can be lifted from back-up memory and placed anywhere on the screen. It can be enlarged or reduced and multiple copies made.

Image entry

In practice any program has other special features – for example the ability to modify lettering by rotation, tilting or adding drop shadow, or the ability to draw or shade pictures using a selection of 'brushes' with fine or broad brush strokes. An 'airbrush' facility is often included. It is impractical for the user or artist to enter all painting instructions by a keyboard, so the keyboard is typically used for just text entry and system instruction, while everything else is entered using a graphics tablet, 'mouse' or other special input device.

A graphics tablet contains a coordinate grid so that, if a stylus or cursor is placed at a particular point, the position of that point can be recorded. At first it may seem difficult to both look at the monitor and draw on the graphics tablet, but artists quickly get used to looking only at the monitor and their painting hand follows the activity on the screen. They only look at the graphics tablet if, for example, an overlay is used and an existing image, such as a logo, is being digitized for subsequent manipulation.

Digitizing an existing image by hand is tedious. Again electronics comes to the rescue. A video camera can be used for 'frame grabbing'. Any piece of existing artwork can be viewed by a video camera and transferred into the appropriate memorized pattern of pixels. Often only a simple monochrome camera is used, and color is added by the artist once the frame has been grabbed.

Full-color frame grabbing is also possible using a color camera. The problem here is that the original video image usually has far more color information in it than the graphics system can handle. Either some

Figure 3.9 The Sony Pro-Mavica, shown in the foreground, is a still electronic camera that records high resolution images on a magnetic disc. These can then be read into computers or into image transmission systems

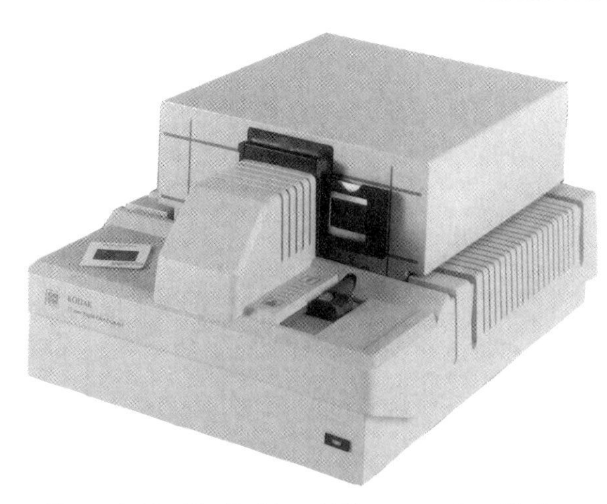

Figure 3.10 The Kodak rapid film scanner is an ideal way to import standard 35 mm slide images into the computer graphics environment

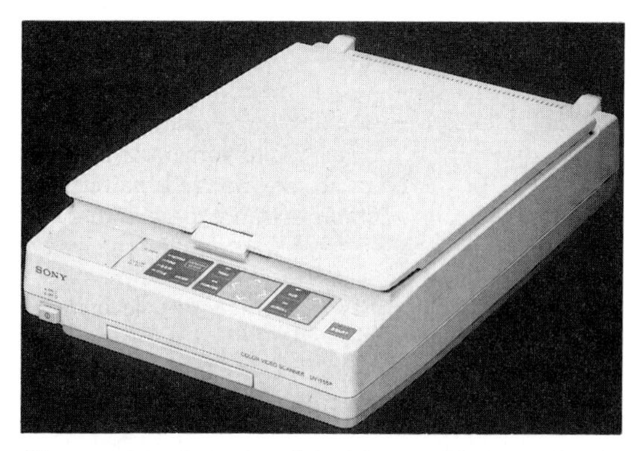

Figure 3.11 Importing flat art images into computers is made easy with the Sony UY-T55 scanner

arbitrary color selection must be made, or the frame-store memory greatly increased, usually to a 24-bit per pixel system which allows 256 levels of each of red, green and blue. This gives an image which, in color terms, is indistinguishable from the original.

Recently low cost 'scanners' have become available that allow any hard copy image to be scanned and copied into computer memory. These work on a digital basis, storing a high resolution image, which then allows the user to select only part of a document and still retain an image with a resolution similar to a TV picture.

Once an image is in a computer memory it can be modified in many ways. Because the entire image is recorded in numerical form, it is easy to apply mathematics to some or all of the numbers. In simple graphics systems this may have the effect of changing a color. In advanced graphics systems used, for example, in the processing of images received from satellite cameras or thermal imaging cameras, the computer can be used for image enhancement. This process of making the invisible visible depends on the detection of very small changes in the nature of an image which are then amplified.

Image files

Images are stored in a computer in the form of 'files'. These can be vector files, raster files or 'metafiles'. Metafiles can hold both raster and vector information. The best known vector files are HPGL (Hewlett Packard Graphics Language) and DXF (Document eXchange Format, used for AutoCAD™ drawings). Vector files are best for line drawing work, and for outputting on line drawing devices. The size of the image is easily changed to match the display or printing device resolution.

Raster files are the most widely used for presentation images, because they are best for scanned images, 'paint' packages, and photo-realistic images. They are best displayed on monitors and printed on dot printing devices. The scaling or re-sizing of images can affect image quality. Most graphics files work on a maximum notional pixel array of 64K × 64K (65,536 × 65,536) so when displayed on a monitor, the image either has to be rescaled or only a part of the pixel array is selected.

Well known raster files include:

□ PCX, the PC Paintbrush Format from Z-Soft Corporation, which was one of the earliest file formats. Its latest version supports 256 colors.

□ TIFF, Tagged Image File Format, a general purpose, multi-platform format developed by Aldus Corporation and Microsoft. It was specifically developed as a basis for importing scanned images into desk-top publishing packages.

□ GIF is the file format used by Compuserve, and is widely used for exchanging images on the Internet.

It is quite possible to convert from one file format to another. The most common requirement is to convert from a vector format to a raster format, and this can be done by a program such as 'Graphics Converter' from IMSI.

Image generation

It is increasingly the case that organizations using computers for design wish to use the images for presentation. For example, architectural or product images may be stored as three-dimensional images in vector format with the associated 'rendering' information for each polygon making up the image. Rendering gives the image the required colors and texture.

It can immediately be seen that, in principle, the display of an image stored in 'bitmapped' or raster format can be quite fast, because all that is required is to read out a memory containing the raster information. But the process of converting a vector image to a form suitable for display requires very fast computer processing. Each image has to be constructed from the vector instructions, and each polygon within the image must be colored and textured; the whole lot then being turned into the equivalent bitmapped image for raster display.

Devices which do this are called 'image generators'. At the top end of the market, companies like Evans and Sutherland and Silicon Graphics, with their 'Reality Engine™', make dedicated hardware to carry out the process. For the business market, slower graphics cards are available for both PC and workstation environments.

Image generators allow you to 'walk round' a three-dimensional environment, or see a possible new product from all directions – you simply tell the computer from which direction you wish to 'see' the object. At present there is a problem with update speed, so while still images can be created to very high resolution, 'film' type animation can only be achieved at lower resolutions and/or with simple images with a limited number of polygons.

Image output

The foregoing brief description of how an image is assembled in a computer is relatively easy to understand. In practice graphics programs vary enormously in complexity, and those at the top of the range employ considerable sophistication in providing routines that speed up image creation. Some programs are application-specific. For example, business graphics programs include automatic chart and graph drawing routines, where it is only necessary to feed in the numerical values to be represented.

Having created the image on the computer screen, what next? Increasingly business and professional users keep the image in the computer, in which case there must be some easy way of storing, organizing and retrieving images. The simple business presentation programs like Harvard Graphics, Applause™, and Powerpoint™ all allow the creation of 'slide shows' complete with transition effects. There can be more of a problem with complex and high resolution images, because generally the programs used for creating these do not have fast sequential presentation in mind. Often the user will need images in another, non-computer, form.

The business user will be using the output to support some kind of presentation. For a written presentation, a printed version of the graphic can be provided by a suitable plotter, either in monochrome or color. This is also satisfactory for simple graph and chart work but, because the plotter is usually plotting an exact replica of the screen memory, pixel structure is obvious. Furthermore, this type of output can only provide a limited choice of color.

If an overhead transparency is required as the final output it can be easily made on suitable plotters or color printers, or can be reproduced from the hard

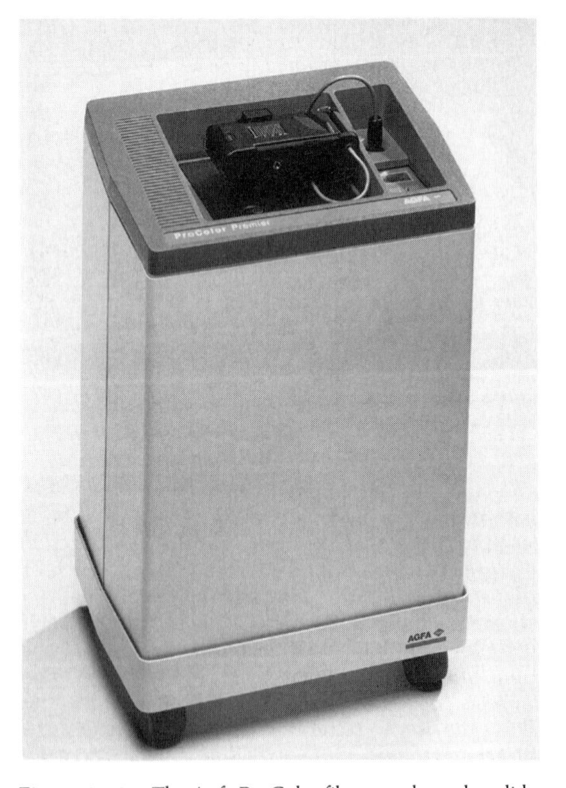

Figure 3.12 The Agfa ProColor film recorder makes slides from computer output in either of two resolutions, 2048 × 1366 or 4096 × 2732 pixels. It is suitable for PC, Macintosh or Postscript environments and can correct for different kinds of film. It takes less than 90 seconds to output the average business 35mm slide

Figure 3.13 Polaroid make two film recorders, the C13000 with 2000 line resolution and the C15000 with 4000 line resolution. They are especially suitable for use with Polaroid's instant transparency film, but can also be used with conventional film

copy. It is also possible to make transparencies photographically, direct from the computer screen.

What is the best way to get the output in photographic form? The obvious method is simply to point a camera at the color monitor. If the correct shutter speed and suitable film are used, the results are remarkably good. This is now made very easy with low-cost cameras using instant film, although ordinary single-lens reflex (SLR) cameras are also quite suitable. A hood arrangement ensures the screen image is not affected by ambient light and that the camera is exactly the right distance from the screen.

While this method is appropriate for production of business graphics slides for in-house use, such as a weekly or monthly management meeting, results are not really good enough for professional presentation or for print origination. The first major limitation is the color monitor tube, which introduces both distortion and a textured look to the image arising from the tube's three-color dot pattern.

This can be avoided by using a specially designed film recorder. These devices vary greatly in complexity, but in principle they consist of a high-definition *black and white* monitor with a flat face plate. The camera, which can range from a simple SLR camera to a sophisticated camera with pin registration, is in a fixed position directed at the screen. In front of the camera lens there is a color filter wheel, which can successively interpose a red, green and blue filter between the screen and the camera lens. The whole assembly is in a light-tight box.

The film recorder receives the image in the form of its separate red, green and blue components. It displays each in turn, for the appropriate time, as a precision black and white image. Because of the insertion of the associated color filters, the camera records the full color image. Slides produced in this way are of high quality and have excellent color. However, because of this high quality, defects of the image creation process are emphasized and the pixel construction of the image is extremely obvious when blown up by slide projection. Circles and diagonal lines suffer from the 'jaggies', which may be acceptable for informal presentation work, but are not acceptable for professional use.

Image enhancement comes to the rescue. Because the image is stored in numerical form, it can be processed to yield a higher-definition picture. The main method is to greatly increase the number of lines in the image in order to eliminate the obvious raster construction, and to increase the number of pixels in the horizontal line. Medium-resolution systems offer between 1300 and 2000 lines, high-definition systems between 4000 and 8000 lines.

Needless to say, high-definition systems are expensive. The business user planning to use computer-generated high-definition graphics has a range of choices.

The choice

Choosing the method of making computer-generated slides (or OHP transparencies, artwork etc.) depends both on personal inclination and on objectively answering a number of questions:

- ☐ How many slides are required annually?
- ☐ Is it a *requirement* that slides be made in-house (e.g. for security reasons)?
- ☐ What level of quality is required?
- ☐ What sophistication is required in preparation of artwork (as opposed to text only)?
- ☐ Who is going to operate the system?

This last question is probably the most important. If all that is required are simple text and chart slides, anyone with rudimentary keyboard skills can produce acceptable results. However, as soon as any element of design or creativity is required, good results are only obtained when an artist or person with graphics training is at the controls.

In ascending order of cost the choices are:

- ☐ Use a low-definition system based on a personal computer making slides directly from the screen. This is only suitable for non-critical once-only work for small meetings.
- ☐ Buy a computer program which allows creation of slides on a low-cost computer. Data held in the computer is then transmitted to a service bureau, either directly by telephone line, or in the form of a floppy disk, the bureau makes a high-definition slide for about £7 ($10) a shot. Results are excellent and the bureau can usually help enhance slides with additional artwork, such as logos, or can produce the more difficult material as an additional service.
- ☐ Buy a complete stand-alone system. If, for reasons of security, it is essential to have the entire system in-house, there is no choice but to have a complete package including the film recorder. Systems using standard SLR cameras vary in price from around £5,000 to £50,000 ($7,000 to $80,000). The price difference is accounted for mainly in definition, which ranges from around 1000 lines to 4000 lines, and in the sophistication of the graphics software.
- ☐ If very high definition is required and a large throughput is expected, then a top-of-the-line system is warranted. These are used either by bureaux or by very large users.

Only the first three choices are of interest to the business user. In Europe there is now a much greater tendency to use the bureau approach; usually the slide-making bureau is asked to do the whole job, but sometimes initial creation is done in-house. In the USA more complete systems are purchased by end-users, although whether systems are used enough to make them a worthwhile investment is open to doubt.

In all the enthusiasm for computer-generated material, it must not be forgotten that there are many cases where traditional all-photographic methods produce better results at lower cost. Thus most slide-making bureaux are beginning to offer a combination of resources, so that the end-user does not have to worry about the actual method used.

Formats

One cautionary word: slides are correctly presented in the 3:2 format and video images in the 4:3 format. Professional computer graphics slide-making systems have as their end product a properly proportioned slide. Low-cost systems sometimes limit the user to a slide in the video format, which may be unacceptable if it is to be used with other photographically produced material.

Figure 3.14 Slidework of Manchester are a bureau specializing in business graphics. Nearly all their output is computer created, using a variety of computer programs and high resolution film recorders

Video output

Can a computer graphics display be directly transferred to video tape? It might seem easy, because the original image is shown on a color monitor. However, in computer graphics the image is stored in its component red, green and blue parts. A normal video system needs a composite image. Some slide-making systems are able to offer a composite (NTSC or PAL) output, and these images can be transferred one at a time onto video tape, using a suitable editing system. Alternatively a device called a scan rate converter can be used (see Chapter 20).

More usually, a graphics system designed specifically for video output will be used. These work within the limitations of the standard video image, and make no attempt at a high-definition output (unless they are specifically designed to work in the field of high-definition television, see Chapter 19). They do, however, make much more of the image sequencing possibilities, and ultimately allow complete animated sequences to be produced.

These sequences are normally produced in slow motion, because the task of replacing one complete image with another involves the movement of a massive amount of information. If this has to be done 30 times a second, a very high speed and large capacity computer is needed.

Chapter 4

Commissioning an audio-visual program

Who decides that an audio-visual program should be made and, having made the decision, how do they go about getting it done?

Within a corporate organization the decision may be made by executives as part of their job. For example, a training officer commissions a product training video, or a marketing director commissions a company profile film. Alternatively an organization might have outside advisers who do the commissioning for them, for example the company staging a product launch may commission multi-image programs, or a public relations firm may recommend the sponsoring of a program suitable for network television.

Justification for making an audio-visual program will sometimes be pragmatic, but should always be based on the conviction that the AV program is doing something that cannot easily be done any other way. It is easy to understand that AV programs can support events like product launches, where attention of an audience must be gained then held. It is more difficult to analyze the comparative advantages and disadvantages of issuing a company report in the form of a videocassette for home viewing in place of, or to accompany, the traditional printed document.

Define objectives

Therefore the first phase in the development of an AV program is the definition of objectives. The first stage must be to examine thoroughly why it is being made in the first place. The sponsor and producer must be able to define clearly:

- ☐ Benefit(s) that will accrue to the sponsor as a result of the show being made.
- ☐ Action expected from the audience as a result of having seen the show.

It is best not to start by saying that a certain piece of information must be communicated. The danger of doing this is that the obvious route is taken, which usually results in far too much being put into the program. By being absolutely clear about objectives, only relevant items are included.

For industrial, commercial and training AV shows there is a very simple rule: *one show, one message*. The temptation is always to try and do too much. The process of defining objectives may well reveal that there are multiple objectives that are best dealt with separately. Sponsors should not be worried about how short a program is, because it is generally better to make two or more short programs than one long one. Often the group of programs can be treated as a single project, at no significant extra cost.

Choose medium

Choice of medium must be made on the basis of the considerations discussed in Chapter 1. The definition of objectives reveals the kind of message that is to be put across which, in turn determines, for example, whether any moving picture sequences are required. More importantly this part of the production process thoroughly analyzes how the final program is to be shown. For example; whether it is only to be shown to group audiences or to individuals; whether it is for a fixed single installation, or should be made in multiple copies; and whether it is for a single special occasion, or for continuous daily use over several months or years.

Often a hybrid solution is favored. A multi-image show may be ideal for the big audience expected at the annual sales meeting, but the same show may be required in a portable version for later use. In principle it can easily be transferred to video, but *only* if planned in advance, because this imposes restrictions on format and content of the original multi-image version.

Some productions are not complete without supporting written material. In this case it is important to agree whether the additional material is prepared within the same budget, and the extent to which it will be necessary for the program producer to be involved in its preparation.

Figure 4.1 A major product launch may not only involve AV programs on a screen . . .

Figure 4.2 but also actors . . .

Figure 4.3 . . . and a complex set. Only the larger production company can handle all aspects of this kind of event. (Photos of the ICL DRS6000 launch from Spectrum Communications Ltd)

Agree budget

Objectives of an AV program should determine its budget. This may seem an obvious statement, but it is amazing how often, in practice, sponsors of AV programs take no account of it.

For many programs this simple statement can be literally applied. If the audience responds in the required manner (by buying the product, by reducing absenteeism, by eliminating waste) a direct cash benefit can be seen. The question then simply becomes how much of this benefit can or should be applied to the show budget. If the planned show is going to cost more than the benefit, it is doubtful whether it should be made.

It is more difficult, but still possible, to apply the same criterion to motivational, public relations, and public entertainment shows. Often it can be a matter of cost per head. Sometimes a show is made to make a small group of people, or even one person, feel good. A senior manager may want a program made that makes the chairman of the board look good, and to keep the chairman off the back of his subordinates. If he is a good manager he is not only honest with himself, and knows that this is the *real* objective of the show, but can actually ascribe a value to it (e.g. cost of recruitment and disruption if he loses some of his subordinates).

It is important to include in show budgeting the realistic cost of delivering the show to the intended audience. It is no good making a superb show which in practice costs thousands of unbudgeted dollars to stage each time it is shown. It is equally no good making a video for use by salesmen if the cost of equipping each member of the salesforce with a video set has not been taken into account.

Prepare first treatment

This is the stage at which the show producer prepares an outline of the show. In the case of a major production this stage includes a draft script and storyboard but, for a small industrial AV, it is simply a summary of how the objectives are to be met. The summary need only to be worked out in sufficient detail to quantify the resources needed for the production, and to provide a production timetable. Where location photography or filming is expected to be required, the production timetable must include a full shooting schedule, so that necessary access can be arranged. The first treatment serves as a 'client approval point' when an outside producer is making a show for a customer.

Confirm likely costs

Work done at the first treatment must be sufficient to allow all likely costs to be confirmed, and to ensure that the show is going to be delivered within budget. The rule is that if costs are going to exceed budget, the first treatment must be done again, and again, until the two match.

Rules are sometimes made to be broken. It can be that the first treatment itself will reveal further benefits that could result from a more expansive approach. The show sponsor must decide if the extra benefit is worthwhile. If not, he must insist that the show sticks to budget.

Producers

Who are AV producers? The term now covers a wide range of companies and individuals. A large public relations agency may have AV production as one of its skills, then there are companies whose sole job is producing AV programs, and finally there are individuals providing either production management or a specialist part of the production cycle.

In practice an AV production, whether it is a documentary-style film or a multi-image show for a product launch, is made by a small team of creative people. Big business and creativity do not seem to go well together. Some years ago it was stated that the largest company making TV commercials in New York City was five years old, and that the oldest company making them was only eleven years old. This situation is a little extreme, and says as much about New York as about AV production, but it does illustrate the point that there is considerable mobility in the production business.

Production companies often only provide production administration. They then contract individuals and other specialist companies to carry out different parts of the production process. This makes choosing a producer a more complicated process than it might at first seem, but clearly the essential element is to find out exactly who is going to be working on the production.

The best way to start looking for a producer is by seeing as wide a range of relevant work as possible. Thus if you are a museum curator, for example, and wish to commission a multi-image show, it is clearly sensible to see some similar productions *in situ*; and find out who made the programs that appeal to you. But then you must be sure that, if you choose a particular company on the strength of its previous work, you are going to get the same creative team, or that you are satisfied with proposed changes in the team.

Production companies will be pleased to show you their 'show reel'. This may demonstrate more than some nice programs. It may show:

☐ Consistency of style, demonstrating that the company is based on a single creative team. If you like the style, then this type of producer is likely to prove reliable.

☐ A surprising variety of styles, almost certainly because the company does not do its own 'creative' work. The questions then are, how good is the production company's management, and how good are they at correctly matching production style to the client's need?

☐ A specific ability. For example the ability to make a documentary program, or the ability to make motivational programs.

When you see a show reel, or when you review work done for other organizations, remember to ask the producer or end-user what the productions that interest you actually cost to produce.

If a production company cannot be found by reviewing existing work, and, in particular, by asking the sponsors of such work how successful the program has been in meeting its objectives, then it is possible to obtain lists of production companies from various trade associations. Some of these are listed in Table 4.1.

Table 4.1 Some AV trade associations, craft associations and technical societies

United States of America

Association for Multi-Media International Inc (AMI)
10008 N. Dale Maybry Highway, Suite 113,
Tampa, Florida 33618–4424
Tel (813) 960 1692

International Communications Industries Association (ICIA)
3150 Spring Street,
Fairfax, Virginia 22031–2399
Tel (703) 273 7200

International Television Association (ITVA)
6311 North O'Conner Road, Suite 230,
Irving, Texas 75039–3510
Tel (214) 869 1112

Society of Motion Picture and Television Engineers (SMPTE)
595 West Hartsdale Avenue,
White Plains, NY 10607
Tel (914) 761 1100

United Kingdom

Association of Business Communicators (ABC)
1 West Ruislip Station,
Ruislip, Middlesex HA4 7DW
Tel (01895) 622 401

Audio Visual Association (AVA)
Herkomer House,
156 High Street,
Bushey, Hertfordshire WD2 3DD
Tel (0181) 950 5959

British Kinematograph Sound and Television Society (BKSTS)
M6–M14 Victoria House,
Vernon Place,
London WC1B 4DF
Tel (0171) 242 8400

International Visual Communications Association (IVCA)
Bolsover House,
5/6 Clipstone Street,
London W1P 7EB
Tel (0171) 580 0962

Australia

Australian Presentation and Multimedia Association (APMA)
Suite F7, 1–15 Barr Street,
Balmain, NSW 2041
Tel (02) 9953 6768

The brief

First stage in commissioning a producer is to get to know the people. At first this is best done informally, because personal relationships are an important factor in ensuring that program objectives are met. It is almost certainly a waste of time briefing people you are not comfortable with.

It is then the sponsor's responsibility to properly brief the producer. While the producer may be asked for advice, it cannot be too strongly emphasized that the brief is the sponsor's responsibility. The brief must show:

☐ Objective(s) of the production(s).

☐ Any background material essential to an understanding of the objective(s).

☐ Intended audience.

☐ Date by which the production must be complete.

☐ Place(s) where the show will be seen.

☐ Method of distribution (if relevant).

☐ Time validity of the program (i.e. is it for a single special event, or is it supposed to be effective for a period of years?).

☐ What action or reaction is expected from the audience as a result of having seen the show.

☐ The budget.

☐ Precisely who has responsibility on behalf of the sponsor.

Competition

Unless the project is a small one, or unless there is already an established sponsor/producer relationship, it is unwise to give a project to the first producer seen. Equally it is inefficient to solicit proposals from a large number of producers.

When a serious project is involved, it is best to seek initial proposals from at least three, but no more than six, producers. They should be asked for a short response to the brief, outlining how they would propose to do the job, and who would be involved as the creative team. Their response should indicate how keen they are to do the job, and whether they can do it within the timescale and budget required.

From the initial response a shortlist can be drawn up. This should consist of no more than three companies. Now the stage has been reached when, in order to get a better idea of whether the producer will fulfil the brief, some creative work must be done.

In the advertising world there is a tendency to the 'free pitch', often involving expenditure of many thousands of dollars to fail to get the contract. This can only result in clients of that agency or producer paying more than they otherwise would. When commissioning AV programs the free pitch should be avoided, especially as a sponsor may be asking small firms or individuals to do creative work.

It is also the case that free pitching is not actually less expensive for the sponsor. It is best to devote part of the overall budget to establishing which is the right company for the job. By paying for the proposal the sponsor gets accurately costed work and, very likely, actually gets better value-for-money as the proposers try to ensure most effective use of the budget. Most important of all, the sponsor has early re-assurance that the approach is correct.

Figure 4.4 At the World Financial Center in New York the Siteguide™ provides a touchscreen directory to the building and its environs. Here the program design, and the design of the display, have been tailored to match the building in which they are installed. (Photo Insightguide)

So the suggestion is that the three shortlisted companies are each invited to present detailed proposals, corresponding at least to the 'first treatment' referred to earlier. The amount of detail depends on the scale of the production, but the proposal should include a first draft script and an outline storyboard. It should then be easy to see whether the producer is likely to fulfil the brief, and to choose between the three competing proposals. Often one will be obviously superior, sometimes it will be a close competition.

As part of the selection process it is important that the competing producers be allowed to present their proposals in person. While this should in no way reduce the importance of a properly prepared written and visualized report, it gives the sponsor a better understanding of the proposed approach. It also allows the producer to give his or her 'best shot', and prevents any possibility of producers feeling they were rejected because their proposals were not understood.

From a simple ethical point of view the idea of paying for proposals is clearly fair. You have chosen three contenders on the basis that you are sure any one of them could do the job, and you are now asking them to do the vital initial creative work – there is no reason why you should expect to get this free of charge.

The amount to be paid for a proposal varies, but a typical arrangement is that the two unsuccessful bidders each get between 3 and 5 percent of the budget, while the winner gets the job for the balance. The idea is to ensure that everyone agrees the whole arrangement is fair, and that the unsuccessful bidders cover their net costs.

An important point here is that, unless specifically allowed for under the terms of the invitation to propose, the sponsor should not assume that he has any rights in the unsuccessful material. A beneficial result of this method of selecting producers can be that new ideas are proposed which, while not being thought suitable for the immediate project, spark new ideas for further programs.

The paid proposal stage of commissioning a production should not be entered into unless there is a definite intention to proceed with it. The result of the competition should be made known as soon as possible, by a date specified in the request for proposal document.

Figure 4.5 The BMW business television project transmits a live program by satellite every week to 160 main dealer sites. Videocassette copies are used for other sites and later viewing. AV production on this scale requires close co-operation between producer and client. (Photo courtesy BMW (GB) and Visage Ltd)

Contracts

Once a producer has been chosen it is important to agree the commercial basis under which the work is carried out. Long standing sponsor/producer relationships often allow work to be carried out on a simple purchase order basis, but commissioning of a special program, or working with a producer for the first time, does call for a formal understanding between the parties.

Some trade associations (see Table 4.1) have model forms of contract which are easy to understand, and which cover all the main points of importance.

Most AV productions can and should be delivered for a fixed price. The only justifiable variations are of the contingency kind: for example, when filming exterior shots weather may be a factor, but where this can be foreseen there should be a defined contingency sum in the budget, only to be used if justified.

However, the sponsor also has responsibilities. Clearly if the sponsor fails to give approvals on time, or fails to make available products for photography when agreed, then it is reasonable that the sponsor is charged for any wasted time. The original briefing document should have specified who, on the sponsor's side, is responsible for approvals. While, at the original commissioning stage, it is acceptable to have a committee approach to the matter with all interested parties having their say, it is most important that when production starts all communications are channeled through one person and that the producer is allowed to get on with the job. While committees may be valuable as part of the democratic process, their effect on creative production is disastrous.

Payment

The method of paying for productions varies, but is almost always based on stage payments. Payments are normally made at the various approval stages, but on a major contract may be made on monthly account based on certified work completed.

In principle there is no reason why a sponsor should pay money in advance, but sponsors should definitely pay for value received, so paying for the completed script and storyboard as a stage payment is quite reasonable. However, many – if not most – AV producers ask for a payment, often as much as one third of the contract price, at the time of signing. It is really up to the sponsor to decide if this is reasonable, but if the sponsor prepares the briefing document properly the problem can be avoided, as the document can specify timing of payments.

If an advance payment is insisted on, the sponsor should consider whether a reciprocal bank guarantee is required to ensure that any such payment is only used for purposes of the production.

Chapter 5

Making audio-visual shows

Compartmentalization of AV into film, video and slide/sound (or multi-image) has given the unfortunate impression that these are completely separate media with little in common. But, in the same way that rules for making visuals for visual aids or programs are exactly the same whether a slide, overhead transparency or video page is being made, rules for making an AV program, whether a film, videotape or multi-image show, are also the same. They all require discipline.

Film was the first medium to impose a discipline, because of the cost of film stock and the fact that results could not be seen instantly. On the face of it, video and slide/sound are easier because everything is so accessible. This very accessibility resulted in some truly dreadful productions, until users realized that these media also need a systematic approach. Not surprisingly techniques of video and slide/sound have moved towards methods of film making.

The procedure for making an AV program, regardless of medium, is summarized in Table 5.1. The first five points were reviewed in Chapter 4. The remainder are discussed in the next few sections.

Table 5.1 Making an audio-visual show

1	Define objectives
2	Choose medium
3	Agree budget
4	Prepare first treatment
5	Confirm costs
6	Write script and prepare storyboard
7	Carry out location photography and prepare special artwork
8	Rostrum photography
9	Prepare sound track
10	Edit or program
11	Preview
12	Correct if necessary
13	Make show copies
14	Show
15	Evaluate

Write script and prepare storyboard

This is the vital creative stage of production. Much of the succeeding work can be done by individual experts, all of whom can make a contribution to the quality of the show. What they cannot do is change the overall approach or basic script.

Scriptwriting is an underestimated task and a most valuable skill required for any AV production. A good script gives the other creative participants the best opportunity to shine, and, of course, is the key to the show meeting its objectives.

But the script cannot be written in isolation. The show storyboard must be prepared at the same time. A storyboard is a series of sketches showing visual development of the program. The principle is the same for a feature film or a five-minute training program. A feature film has an elaborate storyboard that looks something like a strip cartoon, and even includes precise directions for camera angles and likely final editing. The storyboard for a simple production need only consist of rough sketches or a series of written descriptions of the visuals that must be obtained.

There are no set rules as to how a storyboard is presented. Some producers use a looseleaf book with pre-printed pages, each page having, say, four frame outlines which can be sketched in and space next to them for the corresponding script together with a description of sound and visual effects required. Others use a system of pre-printed cards, with each card representing a key visual. This arrangement is more flexible because it allows extra cards to be inserted as the storyboard develops.

Writing AV scripts is *not* the same as writing a radio script or a piece of text. A radio script, for example, is written in such a way that listeners are forced to use their imagination. It must include enough clues to allow listeners to visualize a place, person or object. In an AV script visuals are used to convey a *precise* visual impression. The script must not describe in detail something which should, in any case, be clear to the audience. Thus the process of writing an AV script tends to be a two-part one: first a draft script, followed by heavy revision and cutting in the light of the storyboard development.

Although there are some valid uses of AV technique that rely on long verbal explanations of a single still picture (e.g. in maintenance instructions for complex equipment, or in the use of AV in giving assembly instructions in manufacturing) most AV shows must flow. In general, they have a new principal image every six seconds or so. This applies whether it is a moving picture (the new image being a shot from a different angle or a cutaway to a different image) or a series of stills. Some motivational shows use a much higher picture rate; multi-image sequences using one slide every two seconds are quite common.

This reveals another truism: the complexity of a message is in inverse proportion to the complexity of the means of its delivery. An educational slide/sound sequence, intended for an audience of one person at a time, may be used to explain a complex piece of organic chemistry, but needs only a straightforward factual script supported by clear visuals. On the other hand, a show designed to motivate a large audience and make them feel they belong to one happy family may require considerable resources, combining movie with multi-image, and calls for the ultimate in the scriptwriter's art. In the words of one AV scriptwriter, 'emotion costs money'.

The completion of the script and storyboard is another approval point. In fact, it is the last before the show is completed. From now on the sponsor should keep out of the way and let the producer get on with the job. If the sponsor has not got confidence that the producer can make the show, the producer should not have been appointed in the first place!

Location photography and special artwork

Now the actual process of production can start. Some shows are made entirely from existing material, others consist entirely of original photography. Most are in between. At this stage the producer must:

☐ Assemble existing material. This could include existing slides or artwork, and existing film or video material.

☐ Prepare special artwork. This ranges from the complexity of animated cartoon art, to the more usual requirement of individual pieces of art or typesetting. The choice of using conventional or computer methods is determined by the style of the production.

☐ Obtain original photography. Depending on the medium chosen this is still photography, film or video movie photography. A shooting schedule must be prepared showing exactly what is required, and where and when it will be shot. Much AV work is based on short studio sessions, but outside location photography may also be required.

☐ Obtain location sound. Any 'lip sync' sound forming part of a drama-style production will normally be recorded at the time of filming. Likewise any unique sound effects that cannot be created in the sound studio later, must be

Figure 5.1 The AV producer works from a storyboard. The show is planned in great detail. (Photo Tony Gidley Productions Ltd)

Figure 5.2 The production of a mixed-media show, like this one 'Energy for All?' at Technorama, Winterthur, Switzerland, requires the tracking of many strands of material – original and library slides, movie, model making, special effects and soundtrack. The discipline needed is just an extension to that needed for a single screen show. (Photo Rolf Frei and Ganz-AV)

Figure 5.3 One method of AV show creation is the use of the computer-controlled video rostrum camera. This system can transfer single images or programmed sequences direct to tape. (Photo Neilson Hordell Ltd)

recorded at the time the corresponding photography is done.

There may be an important administrative role for the show sponsor at this stage. If the production requires location photography or filming on the sponsor organization's premises, then the sponsor must arrange that this is possible at the time determined by the shooting schedule. It is most important that any staff who may appear in the production are fully briefed as to why the program is being made. If the program is likely to be shown publicly, it is advisable to obtain a written release from the staff concerned, consenting to their appearance free of charge. In practice money is not an issue, but some people prefer not to appear in programs and asking them first ensures that their wishes can be respected.

Rostrum photography

The use of a rostrum camera to prepare visuals has already been described in Chapter 2. Rostrum photography is an essential part of the production of professionally produced slide and multi-image shows. It is also often a major component of film and video sequences.

There are rostrum cameras that are fitted with movie or video camera heads. Besides allowing the usual single-image visual effects, they allow construction of image sequences where an original image is panned across, or zoomed into. Many documentary films are based on use of still images where movement is imparted by camera and compound table movements. An extreme example of the rostrum camera is the animation stand used for making animated cartoons. Although this type of work can in theory be done on a simple rostrum camera, in practice the demands of animation are more complex and require, for example, the ability to shoot backgrounds separated from foreground figures.

Rostrum cameras used for slide-making can benefit from computer control, but it is by no means essential. However, rostrum cameras used for direct production of video sequences require some level of computer control because of the technical difficulty of laying down video images frame-by-frame. When rostrum cameras are used for movie film production there is also a choice, with the bias in favor of computer control.

If direct computer-generated animation is required

for a production, the origination also takes place at this stage.

Prepare sound track

Some elements of the sound track may already have been prepared as part of the original photography, but further original material can be assembled in the studio. At this stage, procedures for slide/sound or multi-image programs may diverge from those of a moving picture program.

Figure 5.4 A properly equipped and staffed audio-visual sound studio is the best way to ensure cost-effective AV sound tracks

A slide, filmstrip or multi-image program usually involves making a sound track that is the final track, including narration, music and effects. The images are then matched to the sound track during the programming stage of production. With movie and video, the procedure tends to be the other way round; sounds are edited to fit the edited visual sequence. Thus, while live sound is simply edited together with pictures, music and narration may not be added until the main visual sequence is complete.

An exception may occur when a specially composed music sequence is used, or where a tight narration (as in a commercial) must fit a precise time slot. In this case, sound recording is done first, and the visual editor is left to solve the problem of making images exactly match words or music.

The problems of synchronizing many different source materials have been greatly simplified by the introduction of timecodes, referred to in more detail below and in Chapter 16.

Edit or program

This is where the show is finally put together. In a slide or multi-image show the sound tape is now complete and all slides are ready for loading into slide trays. Programming of a simple slide/sound sequence may merely involve loading slides in the required order, listening to the tape, and pressing a button at the moments when a slide change is required. Changes are made by redoing the whole or a part of the programming.

A multi-image show is more complex. Although the principle is the same, there is now too much material to do the programming in 'real time', so the process of loading slides tends to be combined with programming and previewing sections of the show (see Chapter 16). This allows changes to the timing of visual effects to be tried out and very precise synchronization to be achieved.

Film editing is an extremely disciplined process. Normally the film editor does all his initial work on a work print of the film. The process is time consuming, but is not capital-intensive, as the editor works at an editing table that allows the print to be viewed on a small screen. As part of the editing process, decisions are made on how the transition between one shot and another is effected, whether as a 'cut' or 'dissolve'. The actual effects can only be achieved at the time the final print is made. The film laboratory cuts the negative exactly as the editor cuts the work print, but into two reels – the A and B rolls – which consist of alternate picture and black. The optical printing machine is programmed to dissolve as required between the two overlapping pieces of film.

Traditionally, sound was edited by hand as well. Each sound element is recorded on to magnetic film, which is sprocketed just like the picture film. The film editor cuts each sound effect to match the film, and the resulting sound elements are dubbed on to a master magnetic film which has exactly the same number of frames as the corresponding picture film.

Video picture editing is heavily dependent on electronics. It is not practical to cut and splice videotape, although this was actually done in the early days of video recording. The reason is that videotape editing is one of linking electrical signals and, unless this is done at precisely the right moment, the result is a bad picture roll. In principle, each piece of video material is copied across onto a master tape, and the effect of switching in or fading in a new image is under precise electronic control.

Early video editing systems suffered from a lack of precision. Video editors soon realized that they had nothing equivalent to the discipline imposed by the sprocket on movie film. With movie film everything can be calculated in advance in terms of the number of picture frames. Great precision can be achieved in synchronization of picture and sound effects because each element can be precisely locked together.

The solution to the problem came with the invention of timecode. Here a separate track on the videotape is used to uniquely identify each frame of video material. The same code can be used on audio tapes. This allows for rapid location of material and for the ability to run different media in frame sync. It also keeps costs down.

High-quality video equipment is extremely expensive. A broadcast-standard editing suite can cost several hundred dollars an hour to use. The editor no longer has the luxury of a lot of time to make editing decisions, so editing may be rushed. The answer to the problem is to timecode all original material then copy it onto a low-cost video format. The editor then makes all basic editing decisions 'off-line', by noting all timecode readings of edit points – working, in fact, in much the same way as a traditional film editor. Often the 'product' of the off-line edit is, in fact, a complete edited tape, for example, on VHS. It shows all the picture timings, but does not include transition effects, final audio or special effects which can only be added at the final edit stage. In the same way that the traditional editor got the film laboratory to do the final negative cutting and printing, the off-line video editor goes back to the big edit suite to have the final tape made up from the original material, but does not waste any time taking edit decisions.

Many commercial video programs are now assisted by the technique of 'non-linear' editing. Here all the material to be used in a production is transferred into the hard disc memory of either a dedicated non-linear editing system, or a standard computer. Show editing is then done entirely in the digital domain. The method is particularly useful for interactive programs, and for applications where the same material may be used in several different ways, or where programs are required in different language versions.

Any video special effects are added at the on-line edit stage, and the sound track may also be completed. But maybe not; because the flexibility of the timecode system allows the original sound track to be made in a separate audio studio if appropriate.

Not surprisingly, the use of timecode has infiltrated the traditional movie world, allowing high-quality sound studios to prepare sound tracks on conventional (non-perforated) tape which can be electronically locked to the movie. Traditionally, final film dubbing is done in a dubbing theater where a work print of the final film is shown and the sound track is assembled on interlocked magnetic film transports. A new trend is to make a video copy of the film with the same timecode, and to make the sound track in a conventional audio studio, fitted with a video player that can be timecode locked to the master audio recorder. There is thus a great coming together of audio-visual media, with a wide choice of methods to achieve the finished result.

Preview

Now for the moment of truth. While someone previewing their own work may be happy to see the final show under any circumstances, those producing a show for a customer must ensure circumstances of the preview allow proper presentation of the finished product. It is essential to see it in a proper viewing theater, or, in the case of a complex multi-image production, in an environment that approximates to that in which the final show will be run. On no account should the client be allowed to feel that the production is anything other than complete.

Correct if necessary

If the production has been properly planned the preview should result in congratulations all round. Changes should not be needed, because any changes required should have been spotted and made at the editing/programming stage. Changes made at this stage are expensive in all but the simplest shows, but, if unavoidable, mean an extension to the editing process.

Make show copies

The show seen at the preview stage may represent the one and only needed 'show copy'. This might be the case with a multi-image show that has been made for a special occasion, or a simple in-house video or slide show. More usually, show copies must be generated in the quantity and format needed to reach the audience.

In the case of slide shows, slide copies must be made from the originals, and copies of the tape casettes prepared. Each show set should be provided with a spare tape, and show sets should each be run through on the same equipment which will be used in the field. Every item must be properly labelled so that there is no doubt how the show is intended to run. Slides should be numbered, preferably with an indelible marker pen (sticky labels come unstuck and can jam the projector) showing both the tray number and the position in the tray. If it was known at the outset that a number of show copies were going to be needed, they should have all been made at the same time that the original rostrum camerawork was done. This minimizes overall costs and maintains quality.

Film copies are made in different ways depending

Figure 5.5 If a lot of video copies are needed, it is best to go to one of the specialist duplicating houses. At Fraser Peacock Associates in London, banks of JVC triple-deck recorders are used for duplicating commercial videotapes

on the number of copies needed. The preview may be seen using SEPMAG sound, i.e. the soundtrack is run from a separate magnetic film running in lock with the picture. For distribution a married print with optical sound is usually more appropriate, so an optical master is made of the sound and a first 'answerprint' obtained. Final show copies are made from the negatives. The printing machine can make any necessary color corrections automatically.

It is unwise to allow master negatives to be used too often for making show copies, because if they are damaged they cannot be replaced. When a lot of copies are required it is usual to make either a direct copy negative (CRI or color reversal internegative) or a copy negative via an intermediate positive print. The film copy printing process is then straightforward.

By comparison, production of video copies is a simple tape-to-tape copying process. Obviously the aim is to copy from as high a quality original as possible, but it is usual not to use the actual master because, in theory, every playing degrades it slightly. Therefore, a copy master is made and this is used whenever copies are needed. If large numbers of videocassettes are required it is best to go to one of the specialist tape-duplicating companies.

Video material for exhibition or point-of-sale use is sometimes required in the form of standard videodiscs. These are 'pressed', and require mastering. The mastering process is quite expensive, but the per-copy cost is reasonable. D-1 or Betacam-SP master tapes are the preferred original media.

For prestige installation, component video (CRV) discs are used, and these are recorded individually. Discs are the preferred medium for video whenever shows must run continuously, or where random access to different program segments is required.

Increasingly video programmes are required on CDs, whether as linear CD-video programs, or, more usually, as part of a multi-media disc in CD-I or other CD-ROM interactive format. It is possible to make individual CDs on 'write-once' CD recorders, but if more than a few discs are required, it is again best to have the discs pressed. Material required for use in the various CD formats is most conveniently prepared using non-linear editing systems.

Now that CD-ROM mechanisms are an inexpensive accessory to a personal computer, it can be expected that many commercial and industrial video programs will be shown *via* the personal computer instead of from tape.

Show

On with the show. Success of an audio-visual production can depend as much on the circumstances in which it is presented, as on its innate quality. Show sponsors and producers must follow through the production to ensure that it is being used correctly. If the production is for a special occasion or fixed installation this process happens automatically, but if it is issued in the form of a videocassette in hundreds of copies it may be more difficult to ensure that it is being presented properly.

Evaluation

Showing the show is *not* the end of the matter. Both producer and sponsor need to evaluate the response to the show.

☐ Did it meet the objectives?

☐ What was the audience reaction?

☐ If the intention was that there be a quantifiable result (fewer complaints from the public, for example) what is the quantifiable result? Is it what was expected?

☐ Was the medium chosen the right one?

☐ Did the show get the intended exposure?

Only by objectively evaluating success of each production can users of AV gain the necessary experience to target its use more effectively.

The small-scale user of AV may regard this description of the production process as over-complicated. It is not. The same kind of procedure *must* be followed whether the production is a major sponsored film with a nine-month production cycle, or a simple slide/sound show made with in-house resources in a matter of days. It is only a matter of scale; part of the process that may take weeks for the big production may only take a few hours for the small, but no production steps can be eliminated just because it is a small production.

Do-it-yourself?

A person charged with commissioning an AV show has three choices:

☐ Do all the work him- or herself.

☐ Get the whole show made by an outside producer.

☐ Do part in-house and sub-contract the rest.

The general rule is that, however simple the show, unless the organization concerned has its own AV production department it is unwise to attemp a do-it-yourself job.

Most users get independent producers to make specialist training programs (which need a lot of resources) and all motivational, sales and public relations shows. However, many users do make training and product-orientated shows in-house, and many more have in-house graphics facilities for the production of visual aids. Nevertheless, when making shows in-house the experienced user avoids doing everything.

A professional sound studio completes production of an AV soundtrack in a fraction of the time of an under-equipped, occasionally used, in-house facility. It can be a good idea to make a rough version of the soundtrack in-house to check overall timing and effect, but the final soundtrack should be made in a studio. In this way advantage can be taken of the experience and facilities of the studio, including such things as a full music and sound effects library and professional narration, which not only gives a better image to the show, but saves on production time. Many sound studios catering for the needs of AV are also able to provide a programming service.

Location photography requires variable resources. If the organization has its own photographic department, it may well meet some AV needs. However AV photography requires a different approach to print photography, both in respect of format and the number of photographs that must be taken.

Preparation of an image sequence (whether for slide/sound, multivision, or some video) requires use of a rostrum camera. This is especially true where successive images must register and where any graphic treatment is involved. It is generally best to get this work done by a facilities house.

However, there is no doubt that the introduction of low-cost multi-media systems does make the production of simple video programs a practical proposition. Provided each element of the material is well prepared, for example, as a good photograph which can be 'scanned' into the computer, or as a competently prepared short piece of video which can be recorded as a file which can be accessed individually, then the multi-media computer can make better and more frequent use of material. Users need to judge the dividing line between the use of existing material in new ways, and the need for a fully fledged 'production' requiring outside help.

Facilities houses

A facilities house does not normally do a full production, but does one aspect of the production in a highly

Figure 5.6 The transfer of single screen multi-image programs to video is a well-established process. Professional studios use an optical multiplexer. This device directs the light from several slide projectors to create a single aerial image that is seen by the video camera. (Photo courtesy of The Creative Studio of London)

efficient manner. The service is available on a hourly rate basis or against quotation for a particular job. Some of the services now available are:

☐ Audio-visual sound studios. Not all sound studios are suitable: those that mainly make commercials or pop records are not always equipped to deal with the needs of AV.

☐ Slide production companies. These are at least equipped with rostrum cameras and film processing. They usually also have an art department, computer slide creation equipment, and an efficient means of typesetting. They produce or copy slides to order.

☐ Video editing suites.

☐ Film editing and dubbing suites.

☐ Transfer suites, to convert from one medium to another.

☐ Computer graphics studios. Many graphic slides can now be made on computer then transferred to film. High-quality slides made this way need expensive equipment and experienced operators. Similar techniques can be used to produce video images directly.

☐ Picture libraries. Conventional print-orientated picture libraries can be an expensive source. There are AV-orientated libraries where the pricing structure is more reasonable, and the pictures are of a more useful format.

☐ Film and video studios. Gone are the days when a studio was concerned with making films for only one production company. Most studios are now rented out by the day or even by the hour, with or without technical staff.

Transfer between media

There can be a requirement to transfer between one AV medium and another. This may happen because a user realizes he can make a better return on his show investment by targeting the wider audience the alternative medium gives, or because a deliberate decision is taken to make a show in one medium and show it in another. For example, a training film may be shown to groups on 16 mm, but to individuals on video. Multi-image production techniques have been found to be a highly economical method of making video programs.

Some possible transfers are:

☐ Movie to video. This is widely used. Movie film remains a highly satisfactory way of making films of the highest quality. Nonetheless, a larger audience can be achieved by also distributing on video.

☐ Video to movie. This is not so common, and requires technically sophisticated equipment to get good results. It is used either when a trick video sequence must be integrated into a film; or when a program must be available for world distribution, where 16 mm is the one sure standard.

☐ Slides to video. This is now very common, and is often an efficient method of original video production, as well as a means of converting slide shows intended for groups to an individual presentation format.

Slide transfers are carried out on an 'optical bench'. In principle they can be done by simple front or back projection, with the camera pointing at the screen. Back projection is better, provided a special screen with no grain is used. This type of screen material is known as CAT glass (copy and title).

Greater precision is claimed for the aerial image method of transfer. Here projector(s) create a virtual image seen by the camera. The method can be advantageous because it is easier to set up a number of projectors on the same optical axis, using beam-splitting mirrors. It also does away with any imperfections introduced by a screen.

An important point to remember when doing slide-to-video transfers is that the format is different. Original slides must be designed with this in mind. Not only does less of the slide get shown because of the format difference, but not all TV sets can be relied upon to show the whole picture. There is therefore a 'TV safe area' within which it is certain that all information is displayed. This is only about half the area of a full slide.

Low-cost movie-to-video transfers are done by the back projection method, but any professional work should be done on a telecine machine. There are two types: one uses a solid-state CCD device that transfers the image line-by-line, and the other uses a special cathode ray tube that scans the whole image with a flying spot. The flying spot scanner is more flexible because it allows scanning of only parts of the image when required. This is essential when transferring a wide-screen image or zooming in on part of an image.

Transfer of video to film requires highly specialized equipment for professional results. The service is only available at a few specialist laboratories.

Copyright

Those making AV programs, even if only for in-house use, must remember that any existing photographs, drawings, designs, music and recordings are likely to be someone else's copyright. There are established methods and fees that allow the use of copyright material, but producers must *never* use existing material without having cleared copyright.

There are special music libraries for AV shows and, if these are used, the procedure is very simple. There is a set scale based on the amount of music being used and on the intended audience. For example, the fee is much lower for a show that is only to be used in-house in one country, than for one that is to be used on exhibition stands throughout the world. It is possible to use music from normal records, but this use must be negotiated with the record company concerned and is usually very expensive. If music is specially composed for a production there is no problem, but the composer may only sell the right to use the music for the one show, and further separate exploitation of the music may require extra payment.

The situation is similar with existing photographs or artwork. Unless the material is already the property of the program producer or sponsor, or is specially made for the show, a fee must usually be paid. Picture libraries have a scale fee for the use of images in AV productions. There are also some so-called copyright-free art books available; purchase of the book entitles the user to use the designs in the book for AV and graphics work without further payment.

Chapter 6

Audio-visual in business presentations

The greatest use of audio-visual outside public broadcasting and film or video entertainment is in business presentations. This chapter examines why this is, and how to take advantage of AV for this purpose. A distinguishing characteristic of this application is that although the users are professional, in the sense that they are being paid to do a job, they are *not* AV professionals or professional actors, yet often they are called upon to be both.

The *Concise Oxford Dictionary* defines a presentation as 'a formal introduction'. The essence of a presentation is *formality*, and audio-visual methods can help meet this requirement.

A possible summary of the objectives of business presentations includes the communication of *facts*, *knowledge* and *emotion* in order to:

□ **Train**, e.g. sales training, use training, job training, service training.

□ **Inform**, e.g. press and public relations activities, staff and public information.

□ **Motivate**, e.g. staff, independent representatives, customers' staff, the public.

Figure 6.1 A proper environment for business presentations will make them more effective. The meeting/negotiation room at Mailer Oy in Helsinki is permanently equipped with visual aid and AV equipment. (Photo Audio Visual Systems of Finland)

□ **Educate**, e.g. staff, the public.

□ **Sell**, e.g. products, services, ideas.

Very few people have the ability to give a presentation just by standing up in front of an audience and giving an extempore talk. Equally there need be no-one so bad at presenting that their presentations are consistent failures. The trick is to understand how presentations work, and how use of visual aids and audio-visual programs can ensure that presentation objectives are met.

The discipline of presentations

Formality in turn means discipline. The idea that use of audio-visual in a presentation means less work is, unfortunately, not true. Preparation of AV means more work at first but, correctly used, AV ensures:

□ Better communication of the basic message.

□ When the same presentation is given several times, it is given consistently, with no accidental omissions. In this case AV is a time saver.

□ Better presentation discipline.

□ A greater likelihood that presentation objectives are achieved.

What is it that instinctive presenters know by intuition, and the majority find out by observation and practice? The discipline factors that affect the success of a presentation include:

□ The environment. The proper environment greatly helps the presentation, a poor environment can seriously damage it.

□ The structure of the presentation, which must be arranged to ensure that the audience retains the content. This means planning the order in which points are made or items revealed.

□ The realization that most individuals have a short attention span. The presentation must, therefore, be broken into correspondingly short sections of about fifteen minutes.

□ The realization that pace of the presentation must be varied.

☐ Rehearsal. All presentations should be rehearsed. For a major show this means a full rehearsal in the literal sense. For the small in-house event it may simply mean checking that everything is in place.

How audio-visual helps

Audio-visual programs and visual aids can make an enormous contribution to the discipline, and hence the success, of a business presentation. This applies whether the event is something as small as a training session for half-a-dozen people, or as large as a new consumer product launch with an audience of hundreds or even thousands.

☐ If the environment is correct for AV, then it will also be of a kind that ensures maximum attention from the audience.

☐ The use of visual aids and AV programs demands planning. This, in turn, ensures that the correct order of presentation is worked out.

☐ AV can make a big contribution to the presentation structure by acting as a 'divider' and as a means of varying the pace of the presentation.

☐ Provided the presenter sticks to material that is visually supported, there should be no danger of straying from the point.

☐ Presentations with AV and visual aids must still be rehearsed, but the experienced presenter finds the need for full-length rehearsal reduced.

Defining objectives

Discipline relates not only to the presentation itself, but also to the purpose of the presentation. Some presentations fail, even though professionally given, because the presenter does not properly define:

☐ *Why* the presentation is being given and what the presenter is going to get out of it.

☐ *Why* the audience should want to attend and what *they* are going to get out of it.

☐ *How* the audience is expected to behave or react as a result of the presentation.

The kind of thing that often happens is that a presentation which should give a lot of facts and figures to the audience is based on a 'mood' AV show; or, conversely, an audience which needs to be motivated and excited about a new product is presented with a dreary speaker with inadequate visuals.

In the first case there could be an over-investment in the wrong kind of AV support; in the second, a failure to realize that, if the presentation objective is not going to be achieved, it might be better not to give a presentation at all.

Choice of support

Asking some questions helps the presenter determine the support needed:

☐ Based on his knowledge of the audience, how can the message be tailored to meet both the presenter's and the audience's objectives?

☐ How long should the presentation run? This determines whether, for example, AV shows should be used to break the presentation into sections, or whether a second speaker is needed.

☐ What examples and illustrations are *available* and *relevant*?

The very act of serious advance preparation gives ideas to the support that will help. One visual can eliminate a page of confusing script. Choice of which visual aid or AV system to use is then made in the light of:

☐ Nature and size of the audience and environment.

☐ Considerations described in Chapter 1.

☐ Available material, and available time to prepare more material.

☐ Cost.

Chapter 2 made the important point that rules for graphic presentation are much the same for producing a flipchart as they are for making a slide. Thus, the use of AV in business presentations should be of the highest quality consistent with the occasion.

If the occasion warrants use of a presentation room, it probably also warrants use of slides or computer graphics instead of overhead transparencies. Conversely, if the presentation has to be given in an unsuitable office, an OHP may be more practical than a slide projector.

Figure 6.2 For round table presentations a device like the Wolf Visualiser facilitates the use of visual aids. The video monitor can then be used to show slides, OHP transparencies, objects and videotapes

Similarly, if an expensively produced AV show is being given to an important audience, an equipment operator or technician should be on hand. If the AV show is to be shown only informally to a group of three or four people, then it is easier to transfer it to video and show it in the office.

Later chapters give more specific advice about which medium to use for the different applications. This chapter is more concerned with *attitude*. If the attitude to the presentation is right, choice of AV and visual aids media are not difficult; they will almost make themselves. There is usually room for personal preference, so while rules of good presentation must always be obeyed, there is no need for rigidity in the actual execution.

Developing script and visuals

The question of whether to develop a script first and then the visuals, or vice versa, depends on the nature of the presentation. If the whole object of the presentation is visual, e.g. the introduction of a fashion or decorative product, then it is best to start from the visual.

However, most business presentations start from the script because the intention is to impart detailed facts. The script can then be used as a catalyst for defining needed visuals which, in turn, should reduce the script.

Organization of a presentation script is subject to the same discipline whether it is a complete script or a set of prompt notes. First define the objective, then develop a theme that meets the objective. When writing the script or script notes the presenter should observe the following order:

☐ Start by getting the audience's attention, by presenting an arresting piece of information.

☐ Tell the audience *why* they should be interested. Will what they are told save them money? Improve their job prospects? Prevent illness? Save time?

☐ State the main theme, and back it up with proof. Such proof needs visual or audio-visual support.

☐ Restate the theme, and at the same time answer the main questions expected from the particular audience. Again visual support is required.

☐ State the action expected from the audience.

Figure 6.3 The board room of the Australian Stock Exchange in Sydney uses slides, video and direct computer output to support presentations

Some audio-visual rules

The following rules apply to the use of AV programs used as the support to business presentations. The rules apply whichever AV medium is being used:

☐ **One show, one message**. One of the main failings of business AV shows is the attempt to do too much in one show. It is more efficient to make two short shows rather than one long one. This also helps the presentation structure.

☐ **Keep it short**. Eight to ten minutes is the ideal length, fifteen minutes is really a maximum.

☐ **Use the visual element**. Another failing is to record a written text as if there were no visuals. The visuals must be made to work. Often an original script can and must be cut in half.

☐ **Keep it moving**. With very few exceptions an AV show needs a new image every six seconds or so.

The presenter

However well planned the use of AV, the presenter can be a problem. There are two extreme examples of a bad business presentation:

☐ It is a big presentation. The presenter is nervous, not a good speaker, and it is almost more embarrassing for the audience to listen to him than it is for him to speak.

☐ It is a small in-house, almost routine presentation. The presenter feels he does not need to rehearse. He gives a muddled presentation, omits some items, runs out of time and leaves the audience confused, unmotivated and uninformed.

There can be some sympathy for the first, but none at all for the second. How can the presenter avoid being responsible for a bad presentation?

No-one *need* be a bad presenter. Anyone who can write a good letter and who can conduct a sensible conversation *can* give a good presentation.

First presentations are always difficult, because a lot of work is needed. As a presenter does more presentations confidence builds up and, in many cases, parts of presentations can be reused. This, and experience, mean that preparation work for any one presentation is reduced. For most people the rule remains: *if a presentation cannot be prepared, it may be better not to give it.*

Presenters can be nervous about speaking in public; they should be worried if they are *not* nervous. All the best actors and speakers suffer from butterflies before they go on, but, equally, they know what it is that they are going to do and say. They are *prepared.*

The audience

There is no presentation without an audience. In business an audience of six people may be as, or even more,

Figure 6.4 In business presentations an audience of six can be at least as important as an audience of 600. (Photo courtesy Missenden Abbey Management Training Centre)

Table 6.1 Do's and don'ts for presenting yourself

Do	*Do not*
Keep it simple	Fidget or wave your arms
Speak clearly	Mumble or drop your voice
Look at the audience	Look at the ceiling
Keep to time	Over-run
Keep to plan	Ramble
Prepare properly	Apologize
	Talk to the screen
	Wave the pointer around
	Jingle your loose change

important than an audience of 600. Successful presenters always put themselves in the position of their audience when planning their presentation. The attitude 'What's in it for me?' may sound selfish, but if the audience are listening to something that they want to hear they will happily tolerate the less-than-perfect speech.

An audience always starts by being on the side of the presenter, they *want* him to succeed. The presenter can benefit from this, but must not take the goodwill for granted.

Script or notes?

The best presentations are those where the speaker seems to be giving an impromptu speech, yet covers the ground and finishes exactly on time. These are the ones that take most time to prepare!

When a presentation is given for the first time a suggested procedure is first to write a complete script. This ensures that everything is in the right order and eliminates unnecessary repetition. Then, when the presenter is completely familiar with the script, he should make a set of key notes and give the presentation from these notes. The important thing when actually giving the presentation is to stick exactly to the original script concept. If a carefully planned sequence of visual aids is used the aids themselves can act as notes.

If a section of a presentation demands tight scripting and strong visual support, it is probable that this part should be given as an AV show.

Why rehearse?

The single most important element of success in presentations is *rehearsal.* There is nothing worse for a presenter than finding himself in front of a visual he cannot recognize, or running out of time.

The first-timer giving an important presentation should have a private rehearsal for self-evaluation. He should then ask one or two colleagues to sit in on a second rehearsal and invite their comments and suggestions.

'On site', the presenter must be completely familiar with his visual aids and AV shows. He should insist on having a technical rehearsal, and always remember that the audience are collectively spending a lot of time to attend the presentation. As a proportion of their man-hours devoted to him, his rehearsal time is small.

Chapter 7

Audio-visual in conferences

In the context of this book, conferences include two types of event. First, the general conference of a professional association or, one set up by a commercial organizer on a specialist subject using invited speakers. Second, the sales conference or product launch, where a company or organization calls together a group, usually of employees, sales representatives, dealers or customers, who are in some way financially interested in the subject matter.

It is no surprise that it is the second type of conference that is most suited to AV methods. Company sales conference organizers have come to realize that these conferences benefit from AV because:

- ☐ It helps provide a sense of occasion.
- ☐ It helps provide a structure to the meeting.
- ☐ It is cost effective, because it can save time and impart correct information efficiently.
- ☐ It can be a good motivator of people.

Use of AV methods is not a new idea. The USA led the way because, early on, they had a unique problem. Any US company trying to operate nationally was forced to bring the message to many outposts which might be not only many miles, but even days away. The sales meeting was developed as a means of motivating distant markets, and lantern slides and show business were pressed into service. A set of lantern slides from the early 1930s, used by the Pacific Gas and Electric Company to get the salesmen moving, can still be seen today and their quality is immediately recognized by any modern conference producer.

The Screen Works in Chicago, now a well-known manufacturer of collapsible fold-up screens ideal for conference work, was originally Wilcox Lange Inc. The splendid partners, Herb Lange and Kay Kallman, started the company in the late 1950s after nearly a lifetime doing road shows for automobile companies and other consumer product companies. They did not have the benefit of the automatic Carousel™ projector, so had to work like mad to feed hundreds of slides manually through 'push-pull' projectors, while operating hand-operated mechanical iris dissolve mechanisms. Ironically, they were able to project brighter images than are seen nowadays, using highly efficient optical systems with *f*/1.5 lenses.

The fact is that sales conferences are, and always have been, a branch of show business. Organizers should recognize this and plan the whole event accordingly, but should not lose sight of the objective(s) of a particular event. The well-known theater critic's line,

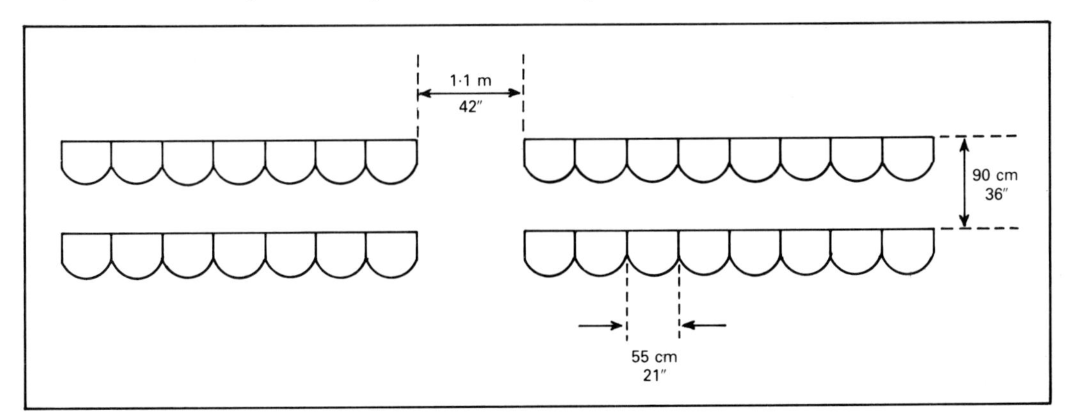

Figure 7.1 Typical spacing for theater-style seating in an auditorium or conference room

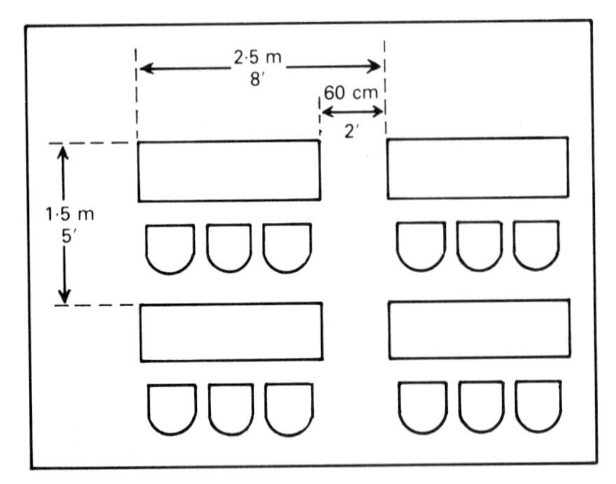

Figure 7.2 If conference or seminar delegates must be seated at tables, this is a typical arrangement

spoken on emerging from a well-staged but tuneless musical, that he 'came out whistling the scenery', can apply to sales conferences that have gone over the top. As in all things, organizers should be less concerned about using the latest gimmick – be it laser, high-speed computer graphics, videowall or multi-image show – than with what they want the audience to do as a result of having attended the event. If this is clear, the choice of medium is almost automatic.

Choice of media

The problem with sales conferences is that they are one-of-a-kind events, and often the information to be given is not available until a few days before. In principle, all AV media and visual aids systems can be used, but in practice slides are often the main basis of conference AV, for the following reasons.

First and foremost, slides can give very high quality visuals that can be seen by large audiences. A single slide image can be projected using simple inexpensive equipment. To get the same image brightness from, for example, a graphics computer via a video projector needs equipment costing over £20,000 ($30,000) and may be as much as £80,000 ($120,000).

Most conference work consists of speaker support, which requires good quality *still* images. So, it might be argued, why not use the overhead projector? It may possibly be suitable for the less formal sessions, but it is an untidy method for the main event, and is unable to give the smoothness of slides. The very discipline of having a slide sequence designed and made contributes to the overall polish and success of the event.

Slides are the most flexible medium because:

- ☐ The same system can be used for speaker support, visual aids and automatic audio-visual programs.
- ☐ The same system can show text, graphics and high-quality photographs.
- ☐ The projection system can be easily adapted to match the environment. Big screen images can be achieved at low cost using multiple projectors.
- ☐ The wide choice of lenses available make slide projection easy to design to match a set or a venue.
- ☐ Changes and additions are easily made.

Needs of the smaller sales conference are well met by a two or three projector single-screen system. This would be manually-operated for the speaker support sequences, and run from tape when showing pre-recorded AV programs or modules.

The larger conference or product launch usually uses multi-image with nine, twelve, fifteen or even more projectors. This is partly to give greater impact to the pre-recorded sections, but mainly to allow use of a screen area big enough to match the audience. For example, many hotel banqueting suites have relatively low ceilings, which may allow only a 2 m (6 ft) high image. A single 3 m by 2m (9 ft by 6 ft) screen is inadequate for a large audience. The wide-screen look of, say, a 6 m by 2 m (18 ft by 6 ft) screen is more appropriate to the room and allows new possibilities, such as comparisons and panoramas, on the screen.

Modern computer-controlled multi-image systems are very reliable, and allow mixed live and recorded sections. Speaker support can be as animated or complex as required. They also allow for those last minute changes which, in a perfect world, would not be needed, but in practice are. Their operation is described in more detail in Chapter 16.

Electronic presentation

The recent advances in video projection methods and the availability of cost-effective computer programs for presentations have now made all-electronic methods of support a practical proposition. The higher equipment cost is justified by the reduction in other production costs, and by greater flexibility. The new generation of auditorium video projectors can give images of 4–5 m wide on lighted sets, and 7–8 m wide in the dark. Videowall technique can be used to provide large images when very high ambient light is present.

This means that existing video program material can be run, and that speaker support is easily provided using standard presentation graphics programs. When very large audiences are present, the screen can also be sourced from a video camera to provide 'visual reinforcement' of the presenter.

Figure 7.3 The opening of Terminal 2 at Manchester Airport used videowall as the main presentation support to give sufficient image brightness in the high ambient light. (Photo Proquip Video Rentals)

However, there are some *caveats*. The seamless linking of mixed video and graphics sources does require both expertise and special equipment. Professional conference producers often use specialized programs or equipment, such as those from TVL (USA), Cadsoft (UK) and VideoShow™ to get round the practical problems.

These programs simplify the linking of text 'slides', scanned-in photographs and full motion video. It is not surprising to find that they work in a manner similar to the programs used for multi-image slide programming. The task of organization required is the same. The big advantages that the electronic method brings are:

- ☐ a wider range of transition effects;
- ☐ unlimited flexibity in re-programming (which should not be abused!);
- ☐ the ability for the operator to preview the next slide or sequence.

As described in more detail in Chapters 19 and 20, large screen electronic presentation usually needs better resolution than that obtained from a standard TV signal, where the line structure of the image becomes obvious on the big screen. It is for this reason that many conference producers still use a mixture of traditional slides for the still images and video projection for the moving images. However, devices like 'up-converters' (which increase the number of displayed lines in a TV image) and higher resolution graphics presentation systems are now making it subjectively difficult to distinguish between the optical and electronic methods for normal image sizes.

Industrial theater

Modern sales conferences have been referred to as 'industrial theater', and the largest of them may well use greater resources than a Broadway musical. They can be great occasions, so it is vital that they are well planned and executed. Because the cost of the event itself is often small compared with the cost of bringing the audience together (not to mention cost of their time) it is usually best to call in an outside conference producer to coordinate the event.

Some companies are big enough to have departments the sole job of which it is to organize conferences, and which may even have their own staging facilities for the smaller conference. Even they, however, would not attempt the big event without outside assistance. There is a range of firms to choose from. At one end there are theater-oriented companies who have the resources to mount a multi-million dollar launch of a new car. At the other are specialists who understand the needs of small industrial sales conferences and who will organize all aspects of staging, preparing visuals, making AV program modules, and so on.

The sense of occasion is important. The smallest conference should use a properly dressed screen, discreetly placed equipment with no trailing wires, correct lighting and sound – all items that can only be ensured if the venue has been thoroughly researched beforehand. Most conferences benefit from the use of a conference set.

Simple sets consisting of a screen, side flats and a matching lectern can be hired, and are easily customized to include the user's company logo or conference theme. The more elaborate occasion demands use of specially designed sets, and it is here that the talents of theater people are required.

The set should be seen as the item that brings cohesion to the whole event. However simple it is, as soon as they see it the audience know that the presenters mean business. It is all part of getting the most out of the presenters, the audience and the event itself.

Figure 7.4 A conference set makes a good impression. Here a standard hired set has been customized. (Photo Presentation Systems Ltd)

Figure 7.5 Sometimes the set is designed to reflect the theme of the conference, as in this set for the sale of shares in electricity generation companies. (Photo Imagination Ltd)

Some conference rules

Several suggestions for the smooth running of conferences can be made. The same rules apply, however big or small the event.

The first rule is to avoid a situation where one person speaks for more than 15–20 minutes at a stretch. If one person must speak for a long time the presentation should be broken up, either by alternating with another speaker or speakers, or by introduction of AV modules. Such modules need only be a few minutes long.

This highlights a common problem. Often the in-company conference organizer, or even the outside producer, is afraid to tell the presenter what to do, because the presenter is the chairman of the board or some other exalted personage. They simply let the poor fellow go ahead and give his rambling 90-minute speech because he asked for, or took, a 90-minute slot. Presenters should listen to outside advice given by people whose business it is to script and run sales conferences, because they have the experience to know what motivates audiences, and how to get the right mix of events within a presentation. Very often what the presenter perceives as important (and which may take up 70 percent of his proposed slot) may not actually be helping the conference objective.

Introducing too much detail is a sure recipe for failure, because the audience cannot absorb it and will quickly lose interest. This type of information should be given in a different way, perhaps as part of the written support material. However, it may be better discussed in the more appropriate environment of a parallel seminar.

A speaker should always have visual support but there will be sections of his presentation where visual aids are redundant. In these cases it is important not to leave up an irrelevant slide, but equally important not to have a blank screen. The company logo, conference theme slide or presentation title can be used to fill in.

Some organizers insist that their presenters never have a slide on the screen for more than 20 seconds. This is impractical for most users, but does make the point that the amount of information on any one visual should be limited.

All conference events should be properly staged. This means that even in the simplest event using, say, just one slide projector, there must be someone whose job it is to look after the AV. With the possible exception of single-slide advance, the presenter should not operate *any* equipment.

The conference venue should be thoroughly researched. The conference producer does this on behalf of a client, but if a venue has to be selected *before* a producer is appointed, the following minimum requirements must be met:

☐ There must be proper blackout.

☐ The lighting must be under dimmer control. This should be the push-button automatic type which allows control from different places within the room.

☐ There must be easy access.

☐ The room must be big enough to accommodate the intended audience, allowing them within the optimum viewing area for AV. (See Chapter 12 for these standards.)

☐ There must be sufficient height to get in a big enough screen. If necessary it must be possible to move chandeliers or other obstructions to projection beams or sight lines.

☐ Unless the conference company is providing a separate system, there must be a good speech reinforcement (PA) system. Microphones and cables must be in good condition, and back-up service available on-site.

Figure 7.6 The UK Army Presentation Team uses all-electronic support for its touring presentations so they can be kept completely up-to-date. This backstage view shows a VideoShow HQ device in use, which can seamlessly link custom graphics with disc- and tape-based video

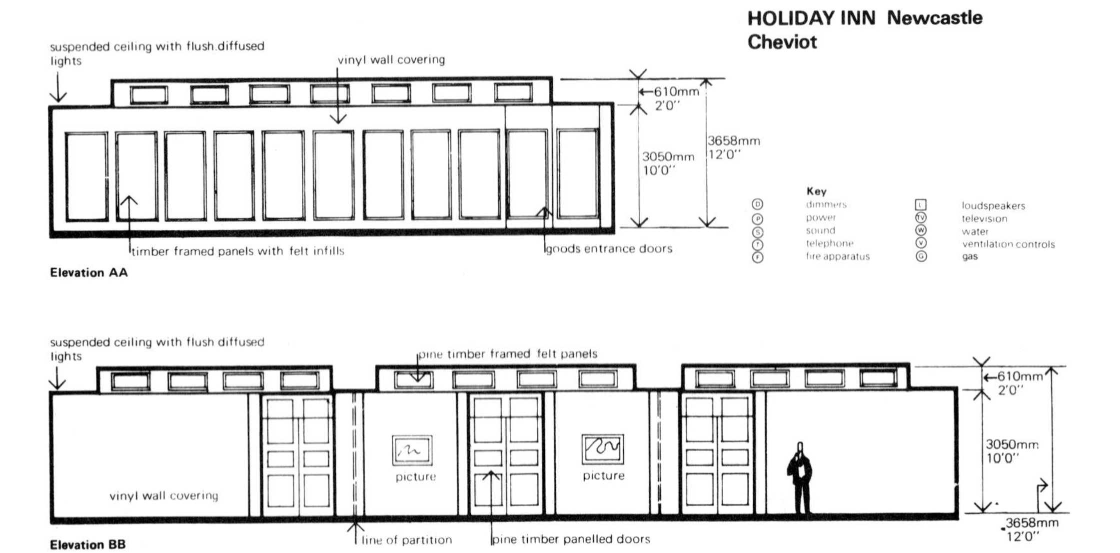

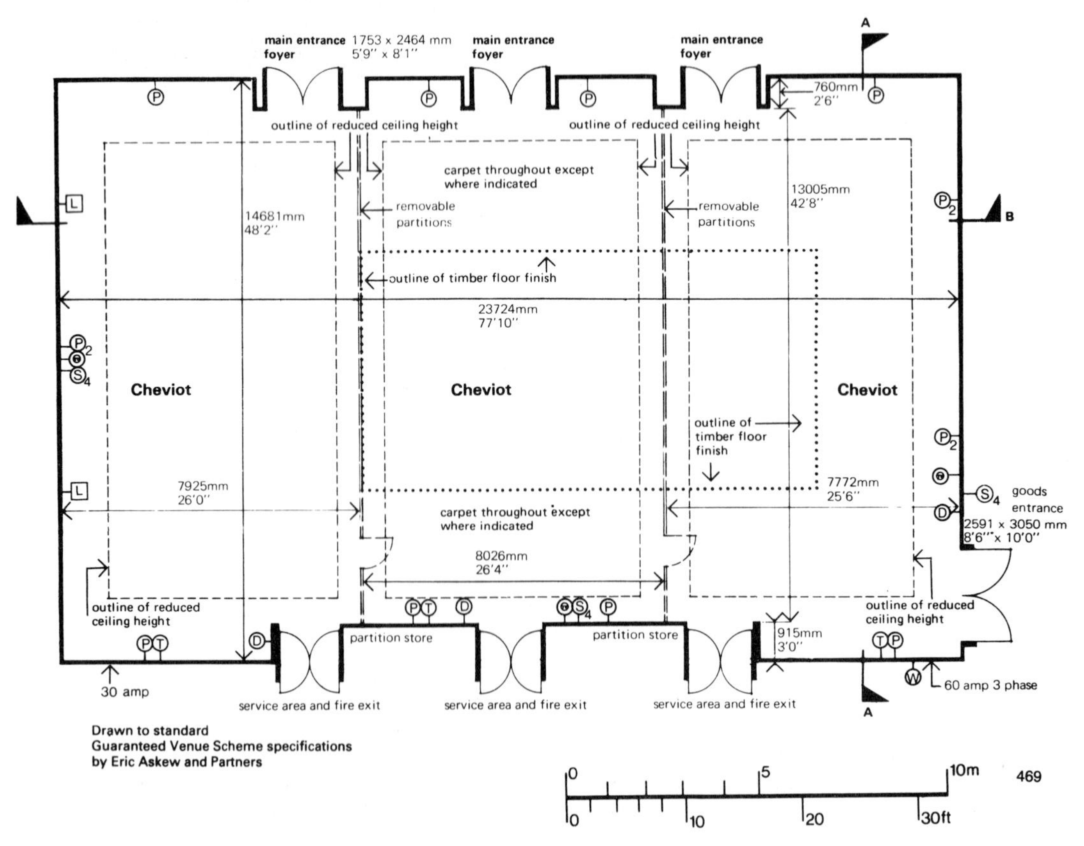

Figure 7.7 The conference venue should be able to provide an architect's drawing, so that the organizer can ensure that everyone and everything fits in. (Drawing reproduced from *The Conference Blue Book*)

□ If there will be no special conference set, there must be a good lectern, with separate lighting to light the presenter and not the screen.

A good rule to observe is that if the intended conference center or hotel is unable to provide an architect's drawing of the room to be used, showing all power points, controls, obstructions, access etc, it can be safely assumed that their venue should *not* be used.

Finally, time must be allowed to have proper rehearsals. The sales conference is, or should be, a highly structured event. Only by full rehearsal can everyone participating be sure that the event will work as a whole, in respect of content, timing, logistics and staging.

AV and the general conference

Unfortunately the general conference rarely has the organization to permit more than the simplest visual aids (slides or overhead projector) to be used, although many would benefit from better integration of AV methods into their proceedings.

Speakers at these conferences generally find that they have to fend for themselves. They are given a time slot to fill, and are expected to perform at the stated time without rehearsal. They must follow the rules on presenting themselves outlined in Chapter 6, in particular:

□ They must have their own private rehearsal to ensure that their presentation fits the time slot.

□ They should make maximum use of slides as visual support, taking special care that slides are legible at long viewing distances.

There are a number of commercial conference organizers who run special-interest conferences where they get together expert speakers. Some of these organizers are exemplary in ensuring that the event makes the best use of the audience's and presenter's time. Others could be accused of negligence, if not daylight robbery. They should run such conferences to the same discipline as outlined for the company sales conference, but usually do not.

A conference set to take place over several days may well be a means of justifying a bigger fee. In fact, most busy people would prefer a shorter, more effective conference – even if the fee was still the same.

Organizers of this type of conference *should* (but often do not) ensure that speakers are chosen for their known ability to present well to an audience. The use of visual aids, prepared to a uniform high standard, should be mandatory, and organizers should work with individual presenters to assist with their production or else should employ a conference AV producer to help.

The whole point of these conferences is that participants should gain more by attending than they could by simply getting a copy of the written papers. There is no point in a presenter simply reading from a written paper that is then handed out to the audience. The written paper should give all the detailed support, and should be written to be read. The presented paper is part of a performance, and should be written to be performed. The performance cannot be just the spoken word, it needs visual support. The aim should be to give the audience an overview of the information needed to allow participants to decide how they should proceed.

Figure 7.8 A laser pointer is a convenient accessory for the presenter at a general conference. This one is the Imatronic LP35, and is small enough to fit in the pocket

Prompting systems

Many stage-managed and fully scripted conferences use television-style automatic prompters such as Autocue™. These allow the speaker at a lectern to appear to know his scripted speech off by heart. The audience cannot see that he is, in fact, reading it from a virtual image seen as a reflection in one or more clear glass plates, which only he can see. It is best to rent these systems, because to be effective they need a skilled operator.

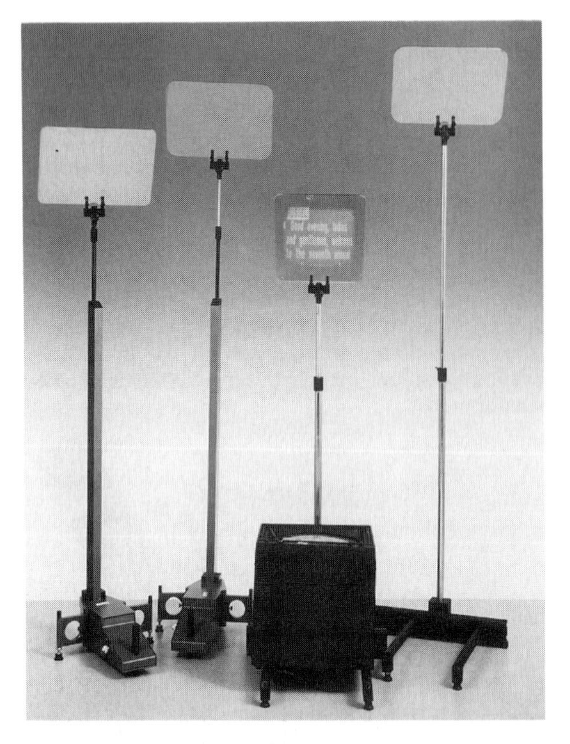

Figure 7.9 Prompting system for use in a conference. The speaker sees his script in the plain glass in front of him which can be programmed to automatically raise and lower to adjust to the needs of different speakers. (Photo Broadcast Developments Ltd)

Modern prompting systems are based on the use of a personal computer running a word-processing program for text entry and editing. A special computer graphics program then ensures that the text is scrolled and displayed in a large format with a font legible at a distance. The operator ensures scrolling speed is matched to the presenter's delivery.

Figure 7.10 Prompting systems work from personal computers. The viewing screen is a separate display from that of the operator (unless self-operation is in use). The operator's screen is multi-tasking so scripts can be edited even while the system is in playback. (Photo Broadcast Developments Ltd)

Whether these devices are desirable throughout a conference is debateable. Many of the best presenters work best from notes, and spontaneity can be lost by automatic prompting. However, in a major product launch using complex staging, they may be essential if a whole stage crew is depending on the presenter to read the script accurately and consistently.

Incidentally, the same secret viewing technique can be used to give the presenter a view of what is on screen behind him. One of the problems that conference producers have is stopping presenters continually turning round to see what is on the screen; they need reassurance that the right visual is there! Incurable cases can be helped by the head-up video reassurer.

Chapter 8

Audio-visual in training

Training in business and commerce, further education and deployment of labor has one or more of the following objectives:

- ☐ Briefing of an audience or individual on a situation.
- ☐ Teaching of a specific skill.
- ☐ Imparting of specialist knowledge.
- ☐ Imparting of knowledge about a company's product.
- ☐ Imparting of knowledge about an organization's services.
- ☐ Training in procedures.
- ☐ Teaching of preferred behavior.

Audio-visual methods can make a contribution to all of these. This chapter covers some topics relating to AV in training, although because training is one of the main uses of AV, most of the other chapters are also relevant to the subject.

Training is a structured process, so AV is a help in providing that structure. However, there is merit in simplicity, so trainers should use the simplest AV method that meets their objective, modified by considerations of distribution of the training material.

A specialist engineering company might consider use of short video programs to train staff how their products work. On further examination of the training problem, they might conclude that their needs would actually be better met by the use of the slide/sound technique. Apart from cost savings, the slide/sound method allows easy updating and program modification, and realistically allows short programs to be made in-house. On the other hand, a retail store group might have a simple training program on cash handling procedures that was easily made on slide/sound. However, they then have the problem of delivering the program where it is needed so conclude, rightly, that the simplest device to have installed at every branch is a TV set with video recorder. Thus they might create the training sequence on slide but distribute it on video.

Most training needs are highly specific. This means that there is rarely a big budget for preparation of training materials, although there should be a *regular* budget which, if wisely used, will continually develop the training resource.

The training room

The success of audio-visual depends as much on a commitment to its proper use, as to making the appropriate kind of program material. Anyone intending to use AV methods for training should ensure that a suitable space is available for them to be effective.

Figure 8.1 The training room at the Lutheran Brotherhood Insurance Company, Minneapolis

For some organizations the training room is the same as the presentation room described in detail in Chapter 12. Many users need only a training room, some will need both. In principle, the training room is a simpler concept than the presentation room, and its prime requirements are:

- ☐ It is big enough for the expected training groups.
- ☐ Basic visual aids such as writing boards and screens are permanently installed.
- ☐ There is proper provision for the expected AV equipment so that it does not distract. For example, projectors should be on proper stands, and use lenses of the right focal length.

□ There is simple lighting control to ensure that projected material can be seen properly, and that notes can be taken.

In other respects the training room is less formal than the presentation room. The important thing is that the use of AV material should be easy and require no setting up.

The overhead projector

An almost essential element of the training room is the overhead projector (OHP). Correctly used, this simple device makes a major contribution to training presentations by providing easy-to-use visual aids.

Figure 8.2 The overhead projector combines the facilities of a writing board and a projector. (Photo 3M)

The preparation of material for the OHP can be done in many different ways, most of which are inexpensive and quick. The OHP can be used in a fully lighted room. If there is no training room or if a training session has to be given in a less than ideal environment, the OHP is probably the best visual aid system for overcoming environmental shortcomings.

The OHP is a very personal method of presentation support, because the presenter can face his audience while operating the machine, and can do so either sitting down or standing up. He can point to an item on the transparency with a pencil without having to turn round to the screen. A correct sequence of OHP transparencies gives structure to his presentation, and lecture notes can be carried on the transparency frames to give further fluency to the presentation.

The OHP can be used as a writing board because the presenter can write directly on to a transparent foil with marker pens. Some people use two OHPs at a time, because this allows them to use one to display formal graphics, and the other as a writing board.

Animation kits are available to produce simple animation on the OHP transparency. For example, they can demonstrate mechanical movements and fluid flow. Items like mechanical linkages can be demonstrated directly by simple two-dimensional moving graphics. Fluid flow is best demonstrated using the polarizing technique. Here the light beam is interrupted by a rotating polarizing disk, which continually changes the plane of polarization of the projected light. The OHP transparency itself is conventional, except where flow is to be indicated. A layer of special film is superimposed over these places.

The special film consists of strips of polarizing material with alternating or varying planes of polarization. The effect of viewing this via the rotating disk is that the treated areas appear as travelling dark bands. The direction of motion depends on which way the material has been laid down.

The skilled OHP presenter quickly learns tricks that can be used to improve the presentation style. 'Successive reveal' is done simply by covering the transparency with an opaque card, and sliding it away as each new point is to be revealed. 'Build-ups' are achieved by starting with a clear transparency that carries only the basic information, for example, the coordinates of a bar chart. Successive transparent overlays are then added, each entirely transparent except for its add-on information, such as one element of a bar chart.

The principle of the overhead projector is shown in Figure 8.4. Most projectors use a tungsten halogen lamp, although very expensive compact arc-lamp models are available for those requiring big images – up to 8 m (26 ft) square – in large lecture theaters or briefing rooms. Behind the lamp is a concave mirror, and in the optical path ahead of it is a large fresnel lens. A fresnel lens is a light-gathering device made of glass or plastic molded with a carefully calculated linear or concentric prismatic pattern. In the OHP it serves as a condenser lens. A conventional condenser lens of the size required would be very heavy and expensive.

The transparency is placed on a transparent platform just ahead of the fresnel lens. Next in the optical path is a mirror used to divert the light beam, and finally there is an objective projection lens, usually

Figure 8.3 The OHP allows the presenter good contact with the audience, especially when the subject is complex. Here a group of engineers at Electrosonic are receiving an intensive course in the use of the C computer language. Note the use of twin OHPs

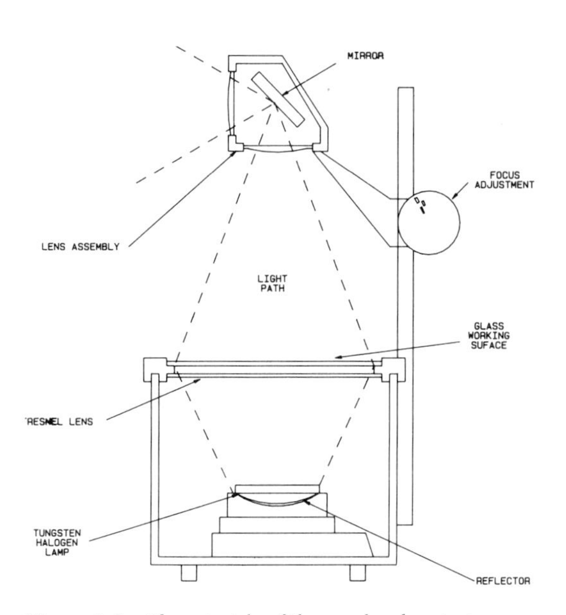

Figure 8.4 The principle of the overhead projector

with a focal length around 300 mm. In the best models the mirror is in a sealed unit. Some models reverse the order of mirror and lens. Care must be taken to avoid damage to the mirror, and to keep the lens clean.

Originally the stage on an OHP was 250 mm by 250 mm (10 in by 10 in), but in recent years projectors have moved over to A4 format (210 mm by 297 mm) which requires a stage of 297 mm (11.5 in) square if both landscape and portrait format transparencies are to be used. Transparencies in 250 mm by 250 mm (10 in by 10 in) card mounts project on either type of machine, but A4 transparencies are cut off on the long picture side when shown on an older machine.

The overhead projector is a relatively simple device, but there is a wide range of machines to choose from with a correspondingly wide range in price. It is not worth paying more than is necessary to do a particular job, so the benefits of special features offered should be evaluated accordingly:

□ The most economical units use lamps like the 24 V, 250 W one used in the Kodak SAV projector. This is an obvious advantage, but if very big images are needed (more than 2.5 m or 8 ft square) larger lamps are necessary.

□ When an overhead projector is to be used as a source of illumination for LCD panels to show computer output, it is best to choose one with a compact source arc lamp. This gives a much brighter image, needed because the LCD panel absorbs a lot of light, and it runs reasonably cool; this is necessary because LCD panels fail if they are cooked.

□ The more expensive projectors have more even illumination and better resolution. Projectors should definitely be evaluated for evenness of illumination, but high resolution is not needed if most of the material to be shown is bold graphics.

□ There is a choice of lenses within a narrow range, typically 280 to 360 mm, but special lenses of extra long focal length are available. These allow the projector to be placed at a conference table, or to be used in back projection systems.

□ Features such as twin lamps with a fast changeover switch, lamp economy switch, anti-glare filters and ease of stowage should be evaluated only in so far as they are of benefit to the particular user.

One important aspect of the use of the OHP is that, because the projector is normally sited so that its lens is below the bottom edge of the screen, the projection beam may not be at 90 degrees to the screen surface. This introduces keystone distortion to the projected image. The problem is eliminated by tilting the screen forwards as shown in Figure 8.5.

Training room furniture

There are now a number of furniture items that are simple in concept and can make a useful contribution to the success of a training room. Their usefulness depends on the user's exact training aims and training methods.

The 'teaching wall' is available either as a custom-made piece of furniture, or as a standard item. These walls contain writing surfaces and different types of screen. If back projection is used they may also house the equipment.

The 'video trolley' is a secure trolley which houses a video recorder with accessories and a large monitor. It has large castors to allow easy movement between different rooms.

Many of the top office furniture suppliers have

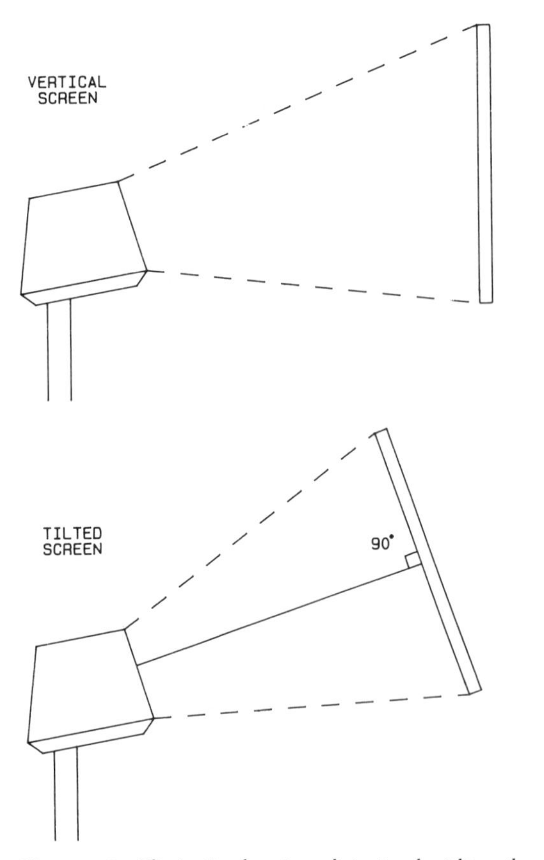

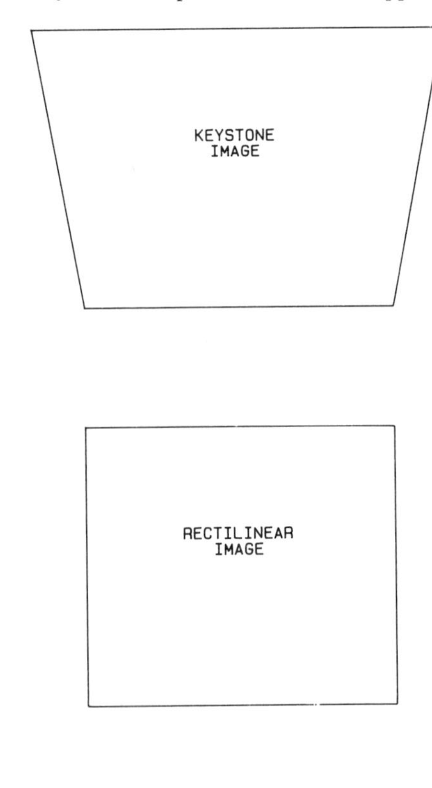

Figure 8.5 Eliminating keystone distortion by tilting the screen

Figure 8.6 The tilted screen in practice. A neat wall track mounting system from Teaching Wall Systems

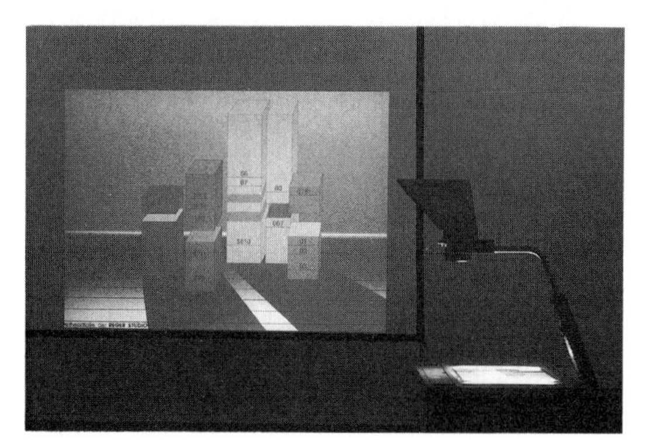

Figure 8.7 Audio Visual AG of Switzerland make a high specification OHP that not only corrects the keystone distortion optically, but also is able to be placed off-axis to give better sightlines to the screen

Figure 8.8 A teaching wall that combines various writing and projection surfaces makes a training room easier to use. Notice in the foreground furniture specially designed for the training room environment. (Photo Teaching Wall Systems Ltd)

Figure 8.9 A video trolley provides both security and convenience. (Photo Bretford UK Ltd)

chairs and tables intended for the training room. The tables can be arranged either as individual desks, classroom style, or linked together for meetings and discussions.

Where a full teaching wall is not justified, the installation of a 'picture rail' probably will be. There are rail systems available that allow the easy hanging of screens, flipcharts, writing boards, display panels and even equipment. The items are supported from the rail, and they can be easily slid to different positions as required.

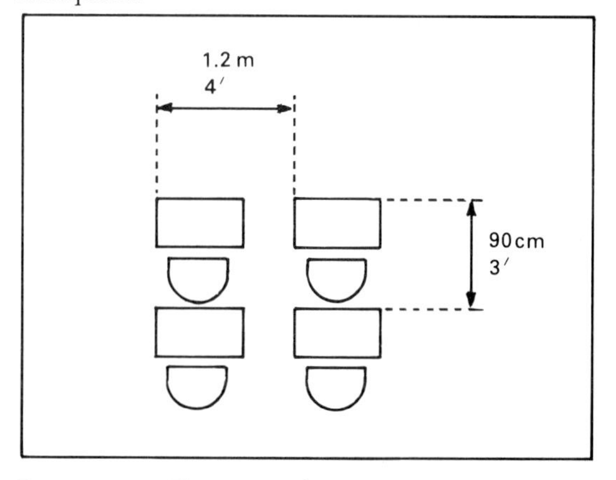

Figure 8.10 Classroom-style seating in a training room cannot be closer than this

Figure 8.11 The combined meeting and training room sometimes uses a rail system for AV aids (as opposed to a fixed teaching wall). This is the Bretford Walltrak™ System

The study carrel

Sometimes when self-instruction is a part of training, the 'study carrel' (a desk at which one person works and which has one or more means of AV communication built-in) may be of use.

These systems are highly effective, but *only* when there is proper software backup and commitment to the continuing development of programs. In practice, this means that either the organization is large enough to have a department devoted full-time to production of program material, or it is necessary to develop a strong connection with an independent specialist producer of this software.

Success of these kinds of programs depends on the continuity of the relationship between the organization which needs the training and those who create the programs. The nature of training programs is quite different from those made for sales promotion!

There is a choice of medium, and a real danger of choosing a medium for its apparent virtuosity and technical sophistication, when a simpler method suffices. The standard medium for the study carrel has been slide/sound, and for many users it will remain so for a long time yet. This is because it is the most practical medium for making and updating programs in-house, especially where the number of trainees who see any one program is small. The equipment need only consist of a simple AV cassette recorder, a slide projector and (usually) a rear projection screen arrangement. It can even be a self-contained unit used for individual presentation. If trainee numbers increase there is an argument for moving to videotape. At its simplest, this means transferring the slide program onto video, with, of course, the added possibility that movement can now be included.

It is now quite possible to make simple "slide/sound" programs on a computer. Those whose training departments are already equipped with high-specification personal computers with good graphics and audio facilities may well find this a cost-effective method. The actual discipline required to make the program is no different than making it by traditional methods. The main extra item of equipment needed is a means of scanning in images. But while the computer-sourced AV show is a quite practical proposition, it does not actually do anything more than real slide/sound show; so those who do not have the right kind of computers, or would prefer not to use them, need not feel they are missing anything by working conventionally.

Most program makers are best at making linear programs which run from the beginning through the middle and to the end. For many users it can be easier,

Figure 8.12 A typical study carrel. The furniture is made in a way that is not equipment-specific, so audio, slide, computer and video equipment can be installed as appropriate. (Photo Synsor Corporation)

and a better investment, to make a lot of short linear programs which each make a single point, than it is to make complex programs which include the possibility of branching, multiple choice and student assessment.

For the brave, however, all these things are possible at quite reasonable cost in equipment terms. The subject of interactive audio-visual programs is sufficiently important to warrant chapters of its own, so further discussion is reserved for Chapters 13 and 14.

The training film

Most company training is specific to the company, but training in behavior can be universal to many companies. For example, training in aspects of safety, hygiene, selling, general management, finance, service, customer relations and security could be the same for a company making electronic equipment as for one making tents. Many excellent training films are now available. They can be rented or purchased, and are usually available in both 16 mm and videocassette.

Figure 8.13 Training films are available on a wide range of subjects, from telephone selling to general management. (Photo Video Arts Ltd)

If these films are rented for a short duration they must *never* be copied. They are copyright material, and apart from the immorality of copying, penalties are high. Not only is there a heavy fine, but film companies make a point of advertising names of offenders. If a particular film is used frequently, it should be bought.

Films should not be used in isolation. They should be used to provide structure and to vary pace of an overall presentation on the subject. Often, suppliers of films provide written support material to help the user make best use of them.

If there is a large audience it is important to show the film in the right environment with the correct lighting conditions. An audience of 20 or more is more impressed by a big, bright projected image than a 22-inch TV set. As a group seeing the film together, they also respond to and absorb the training message better.

AV programs

Besides readily-available material such as training films, specially-made AV programs for group showing (as opposed to visual aids and programs made for individual training) may be needed. These programs should be short, and deal with only one topic at a time. They should be designed as an integral part of a training session, and be of a form that is easily updated. A point which needs repeating is that production of many short programs is usually more effective than producing a few 'Ben Hur' length programs.

Role playing

Another aspect of behavior training is to get trainees to act out a part, for example a salesman, stewardess

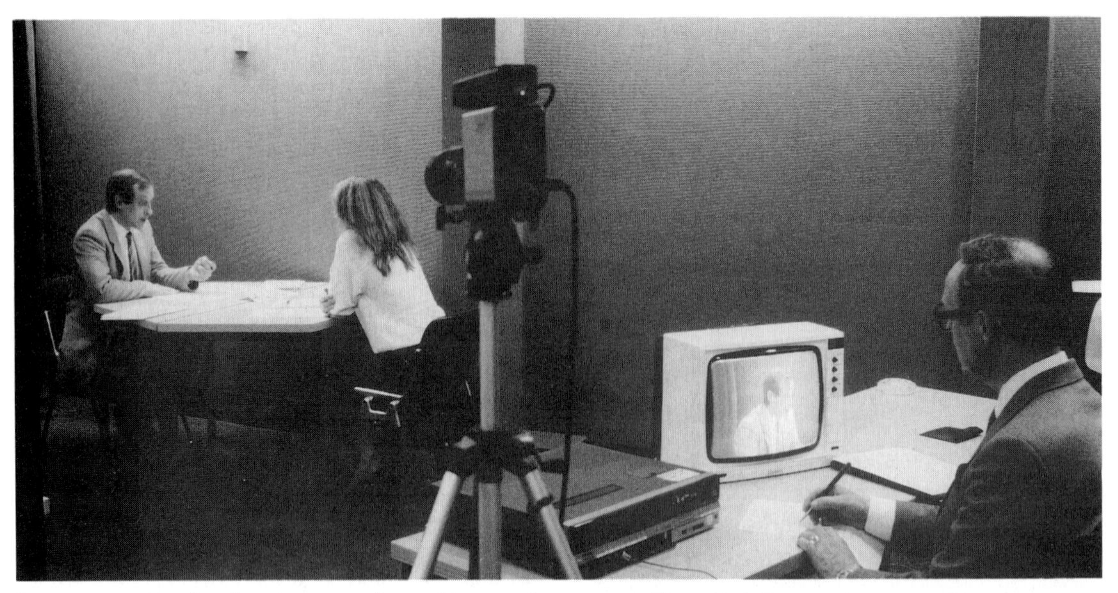

Figure 8.14 Simple video systems can be used for recording trainees. (Photo Tack Training International)

or bank clerk. This type of training is helped enormously if the trainees can see themselves. This can be achieved by simple video systems that, in principle, need consist of no more than a low-cost video camera and recorder. It is more important that the circumstances of the recording match those of the intended area of use (office, shop etc) than that the lighting and camera facilities are of a high order.

It is important to understand that these simple video systems are *not* suitable for making actual video programs. There may well be some users who can justify a full video production set-up, including a small studio, for in-house use. Of course they can then use this for the role-playing training, but care should be taken to ensure that trainees are not intimidated by the artificial environment of the studio, and training procedures should stay firmly in the hands of the training director and not get into the control of the AV producer!

Organizations that do not have a training manager with experience of these methods, but which occasionally need training based on role playing, may find it better to employ the services of a specialist training company to ensure effectiveness.

Networks

Large organizations such as banks and retail chains have the problem of training large numbers of staff each year. Some of the higher-level and specialist training should still be done at training centers because attendance at the center, and meeting with participants from other parts of the organization, increases effectiveness of the training process. However, most induction training, low-level training, and training in new procedures and products is best done at the place of work. In the case of a bank this can amount to several thousand sites.

Such large-scale trainers have to take several things into account before deciding on the precise method of dealing with the problem. The logical solution is to equip each branch with a video monitor and to provide program material either in the form of videotape or videodisc. The main problem is to ensure that each site has a suitable room for the training system and that local management is committed to using the training materials provided – in particular, that they allow the staff the correct time to use the system.

The question of tape or disc is an open one. If a large number of branches are involved, disc could be both more economical and easier to use because random access of program segments is extremely fast. It is also much better for those applications which need still frames. Another consideration is the possibility that the video system used for staff training is also used by the public as part of a point-of-sale display.

The possibility of running reasonable quality video from CD now means that CD-I and other dedicated players represent an economic dedicated platform for training video; less expensive than laserdisc and potentially more flexible than tape. Alternatively if the branches are equipped with high-specification personal computers, video-based AV material could be played through them via a CD-ROM player. However, the dedicated systems are considerably less expensive, and computers should only be considered if they are needed for other purposes.

Another possibility is that video and AV material is distributed by cable. Here the branch has a computer device which is on a network. This possibility is discussed in more detail in Chapter 9 and in Chapter 20.

Modern video equipment is reliable, but may need service like anything else. Those contemplating video networks should ensure that local suppliers have an interest in the sale to ensure after-sales back-up. This means either using multiple suppliers, or one main contractor who undertakes to organize local service on a subcontract basis.

Figure 8.15 Some training programs are better presented using slide/sound technique. The Telex 'Caramate'™ is a convenient unit for showing these programs

In spite of the advantages of video as a multi-branch training medium, there are other options, especially for the smaller user. Portable slide/sound units or standard projectors with an AV cassette recorder may well meet the needs of the smaller chain better than video, especially where it may be necessary to modify programs locally.

Student response

Some organizations must train groups of people on courses. Examples include airline and catering industries. In these cases it is useful for the instructor to know how much has been understood by the group.

Student response systems can be used by an instructor both to modify his own presentation in the light of the response received; and as a means of individual student assessment. These systems vary in complexity, but all essentially consist of an automated multiple-choice question system.

Figure 8.16 An audience response system in use. (Photo Group Dynamics Ltd)

Each student has a set of push-buttons, typically three or five, corresponding to the maximum number of choices likely to be required. When the instructor wishes to assess the group he presents a question, usually by slide or video projection, together with the numbered answer options. The students respond by pressing the appropriate button.

Depending on the type of system installed the instructor can see the totals for each answer, and/or can see individual responses. A record can be kept of each individual's response if necessary. The possibilities are extensive, but again the system is a waste of money unless there is a commitment to using it properly, and to generating the course material that makes it possible to use it at all. In those countries where there is written, as well as a practical, driving test this style of audio-visual group examination may eventually replace the written paper, especially because it allows motoring situations to be displayed to the examinees along with relevant questions.

Some advanced training facilities now have a personal computer at each trainee's position. At present this is usually only the case for training related to computers and computer methods, but some trainers

Figure 8.17 Temporary installation of response systems is greatly facilitated by the use of wireless keypads. The Reactive Systems Inc infra-red keypad allows the use of response systems in locations where wired equipment would be impractical

consider this a valid approach for other types of training. The communication then becomes two-way. The instructor can download programs and course material to all students, but can also elicit response or examine individual progress as required.

Simulators

The highest level of audio-visual training is probably encountered in the simulator. The best-known type of simulator is the flight simulator, where the most sophisticated versions may cost many millions of dollars. The flight simulator consists of a cockpit exactly like that used on the aircraft being simulated. It is equipped with a video projection system that presents the view

Figure 8.18 The flight simulator represents the highest level of audio-visually assisted training. This is the outside of a Thomson Training & Simulation Concept 90 Simulator, in this case configured as a McDonnell Douglas MD11 for Alitalia

Figure 8.19 A flight simulator from the inside. Notice the wide-screen computer-generated display of the runway. (Photo Thomson Training & Simulation Ltd)

out of the cockpit window, and the whole assembly is mounted on a set of hydraulic rams that can simulate take-off and landing accelerations, runway rumble, stalling, etc.

The whole system is designed to react like the real aircraft, and is used by pilots both for initial training

Figure 8.20 This mobile 14-seat entertainment simulator unit runs a fixed, but exchangeable, simulation program for public entertainment. The Venturer is from the Entertainment Systems Division of Thomson Training & Simulation

and for re-training. Full-length flights are flown, and the simulation is so realistic that participants often feel that they really *have* landed at the end of their 'flight'. Apart from routine training, the simulator allows pilots to react to simulated situations that would be too dangerous to try on a real aircraft.

These types of simulator bring together the latest in high-speed computer graphics (for constructing the moving video image), video projection (for presenting it), and computer control and mechanical engineering. They are used for all types of civil and military aircraft and spacecraft.

The same idea can be applied to other skills. Ship simulators are used for training tanker and freight crews, and for training ships' pilots, especially when a very crowded harbor or difficult approach is involved. Truck simulators are already being used for training truck drivers, and soon car simulators will be widely available. Relatively simple simulation methods are used in fields such as firearms training – it is obviously better for live ammunition training to be done on a simulation basis!

The same ideas have entered the entertainment field. Many of the applications are no more than extended linear programs, where groups of people experience a pre-programmed 'ride', with the visual 'software' carried on film or videodisc; but as image generation systems become more financially accessible, the entertainment systems will become as sophisticated as their industrial predecessors.

Chapter 9

Audio-visual in selling and public relations

There are only two reasons for using audio-visual techniques in selling:

- Reducing time taken to make a sale.
- Increasing the probability that a sale will be made.

If the AV show is not able to meet either, or both, of these objectives, then it is the wrong show for selling, even if it is the right one for staff training.

Person-to-person sales presentations

Audio-visual programs are being used successfully in individual sales presentations in many diverse applications. In all cases it should be possible to achieve *both* major objectives, and it must be remembered that it is often just as important to reduce the overall selling time from the buyer's point of view as it is from the seller's.

AV programs for person-to-person selling should be designed to reinforce what the sales person has to say and to give information in a way that he or she cannot. They should also give a structure to the salesperson's presentation. Therefore, it is no use making such programs unless the exact circumstances in which the show is to be given are taken into account. It is very important *not* to take over the salesperson's job, and the show must not attempt those elements of selling, such as 'closing', which can only be properly done by the salesperson.

The examples below illustrate those kinds of selling situations that can be helped by AV methods.

A manufacturer of drinks-dispensing machines wanted to enlarge his share of the hospital market. His competition was not the other manufacturers of vending systems, but the existing method of drinks dispensing in the hospitals. This was based on using traditional methods, requiring crockery which had to be washed, and therefore the service of hot drinks was limited to set times of day.

The manufacturer realized that he had to appeal to the operators of the traditional system. He had to overcome prejudice against machine-made coffee and tea, and the idea that dispensing systems using disposable cups were more expensive per cup than the traditional system. He therefore financed a trial at one hospital to allow the hospital management to get direct experience of the alternative. The trial was successful and a sale resulted.

It would have been uneconomic and a waste of time to repeat the process for all potential customers. So a video program was made which compared the two methods and, most important, recorded comments from the users including hospital management, catering staff, nurses and patients. This video program was then used successfully to sell many more installations. The video was particularly useful not only in convincing the managements of the benefits of the product, but also in defusing any criticism from the staff. Because they saw their counterparts enthusing about the system on the video (and the important thing here was that the people shown were real nurses and staff of named hospitals, *not* actors) the decision to change was not seen by the staff as a decision handed down from above.

It is not surprising that drug companies have been one of the biggest users of AV in selling. They have used, and continue to use, all media ranging from filmstrip to computer-controlled videodisc. Main use is for presentation of case histories and clinical data, the kind of information difficult for the salesperson to remember precisely when there is a continually changing and wide range of products. Similar thinking can be applied to any technical product that cannot be demonstrated on the spot.

Thus, at the other extreme, vendors of large machinery can use film or video to show their product in action. Even if the customer is able to see the product itself, for example a large piece of earth-moving machinery, it may not be practical to show it in operation in the way that the customer might use it. AV can help complete the picture for the customer.

Selling of services, such as consulting engineering, is most efficiently and convincingly done with AV help. This is not only because use of slides or video allows the consultant to show tangible results of his work, which may be several thousand miles away from that required by his new potential customer, but also because it permits use of a well-written and well-presented summary that consistently gives the same quality message.

Selling benefits is what all sales people are taught to do. Encyclopaedia Britannica frequently use AV support at the individual home-sale level to put over the benefits case for their product. This is an example of a case where the features of the product could be demonstrated directly, but the benefits argument is better made with AV support. Sometimes AV helps the other way around; it can underpin both types of presentation.

Choice of equipment

The need to ensure minimum overall sales time also dictates choice of medium for the person-to-person presentation. Usually the salesperson must visit the customer, so the medium chosen must use robust lightweight equipment, and must be able to give a show in an office or desktop environment.

Figure 9.1 Self-contained slide/sound units can be used in person-to-person presentations. (Photo Gordon Audio Visual Ltd)

Figure 9.2 A compact video presentation unit in use in an automobile showroom. (Photo JVC)

Fortunately there are a number of self-contained portable devices available based on videotape, slides and computers. Until comparatively recently devices based on filmstrip and Super 8 movie were also used for this purpose, but these no longer have a cost or convenience advantage, so are virtually extinct.

Portable video units combine a videocassette player with a small monitor. VHS is the most popular, but units based on the Video 8 cassette are also available.

Slide-based units are by far the most flexible, and allow the smallest of businesses to use AV as an economical selling tool. Most such units are available in record/replay versions, so it is quite practical for users to make up their own simple programs. While such an approach is wholly justified if the business concerned is a small one which wants a lot of short programs for use on only one or two machines, any fleet operator should get the sound commentary, at least, professionally prepared.

Another possibility is to use a portable computer. A top-specification laptop computer with full color screen and built-in audio facilities can be used for giving one-to-one AV presentations, and in theory gives the user great flexibility in tailoring the presentations to individual customers. However, this method is only suitable if the salesperson is using a computer anyway, otherwise it is an expensive gimmick.

CD technology can also form the basis of individual presentations. The most cost-effective platform is a dedicated player, for example, a CD-I or 3DO unit, but for those already using computers it is easier to fit the computer with a CD-ROM drive. The possibilities are discussed in more detail in Chapter 14.

Managers of large sales forces often use AV to support an individual sales campaign. For example, they may find it most effective if they use it two or three times a year, and their sales force ensures that

Figure 9.3 Bell and Howell are an example of a company using videotapes for selling. This one explains the benefits of microfilm systems

all regular customers get to see the show, which might promote the launch of a new product or introduction of a new service. Such occasional fleet users may decide to rent the equipment for the campaign period only.

The direct mail video

So far this chapter has been mainly concerned with the use of AV as a support to an individual sales presentation. The huge population of videocassette recorders which now exists opens up another possibility.

Several companies have successfully used video as part of a direct mail campaign. This is now practical in many countries because over 90 per cent of video recorders are VHS format and there is very wide home ownership of video machines within the target audience, as well as an increasing presence of video equipment in offices.

An example from the UK gives an idea of the method in action. The Thames Water Authority, which is responsible for water and drainage in the south-east of England, offered all its customers a videocassette at well below retail price. By clipping a coupon they could receive a standard three-hour cassette; however, at the beginning there was a film describing the activities of the company and how they affected the consumer. Those who received the cassette could reasonably be expected to watch the film at least once, before they used the cassette for other purposes. A neat piece of public relations, where the actual cost of distribution was self-liquidating.

In the commercial, as opposed to consumer, arena it is usual to mail out videocassettes in response to an enquiry generated by trade advertising; or to send them to named persons within a regular customer organization. Some important points about this approach are:

□ Actual cost of the cassette is unlikely to be more than double that of an expensively printed brochure (when applied to the relatively short print runs normal to industrial sales).

□ This approach *only* works if the video program itself is made to the highest professional standards. However, with careful planning this need not be excessive.

□ While those who receive the cassette at their own request can be counted on to see it, there may be some doubt as to whether those sent it unsolicited would view it. In practice it seems they do. Expensive literature often goes straight into the waste basket but items of a perceived intrinsic value, like a videocassette, do not. They are often viewed out of curiosity (even if there is a later intention of using them for something else!).

Point-of-sale audio-visual

There are two kinds of point-of-sale AV. Either it is intended to augment the efforts of the sales assistant, or it is intended to be seen by passers-by who, it is hoped, are moved to buy. In the former case the rules are the same as for a person-to-person presentation but, because there is no need to move the equipment, bigger pictures can be used, for example via 28-inch video monitors.

The second case consists of passive and active displays. Either passers-by see whatever is put in front of them, or they have a means of expressing a choice. In the case of the conventional linear program it is important that:

□ Shows are short. More reference to this in walk-past shows is made in Chapter 10:

□ Straight TV commercial shows are avoided. Shows should inform and, if relevant, demonstrate.

□ Any equipment used is designed for continuous operation.

The aim of point-of-sale AV is to increase sales by showing sales material in a way that cannot otherwise be achieved. This is best illustrated by the following examples:

□ Demonstration of do-it-yourself tools and materials in action, probably by video.

□ Explanation of how microwave cooking differs from conventional cooking, and how this can affect both lifestyle

Figure 9.4 This compact display uses slide projection to achieve a bright 180 cm × 60 cm (6 ft × 2 ft) image, while occupying only 0.6 sq m (6 sq ft) of department store space. (Photo Associated Images Inc)

and the variety of food available. This would be a promotion for the whole idea of microwave cooking, rather than a promotion for a specific product, and is best done on video.

☐ Illustration of furnishing and decorative fabrics, such as curtains and wall coverings, in use. Here the sample book is augmented by brightly-colored pictures of room settings using the fabrics. This is best achieved by slides on a screen, say, 90 cm by 60 cm (3 ft by 2 ft).

☐ Selling of houses. Visitors to a real-estate agent can see a selection of houses by random access to color slides, or by viewing short video tapes, or by random access of still video images 'frame grabbed' into a computer.

Figure 9.5 The Philips Super Projector is a means of achieving a big video image in relatively little space

☐ Selling of vacations. A video presentation system shows a wide choice of programs on different vacation venues. Price and availability information held on a computer can be superimposed.

☐ Promotion of software products. For example, the use of high brightness videowalls to promote video films, computer games and other computer software.

Random access is particularly suitable for point-of-sale work, either to allow customers to make their own choice, or to allow the sales assistant to show customers several alternatives on a quick-selection basis.

Good-looking display cabinets are available which give random access to 80 or 160 slides. They make a reasonable-cost, high-quality presentation catalogue with individual images available within two or three seconds of selection. Slides give excellent quality and allow inexpensive updating. Technically more than 160 slides could be used, but for this type of application, 160 is a realistic practical limit because an additional projector head is needed for every 80 slides.

Random access on video tape is possible, but is very slow and takes many seconds or even minutes. If it has to be done it is best to have a caption generator that can keep an image on the screen while the videotape recorder is searching. In general, tape is best reserved for conventional linear programs.

The best system for random access of both moving-image sequences and for large numbers of still images, though, is videodisc. The videodisc medium is particularly economical for multi-outlet displays, but the initial production cost of the disc may deter some smaller users. Random access on the videodisc is very fast; in a well-designed presentation

Figure 9.6 A random-access slide projection system, suitable for the point-of-sale display of services, property, furnishing fabrics and other items requiring a high-quality still image. (Photo Diekhoff GmbH)

Figure 9.7 The 'Electronic Cookbook', sponsored by the Meat and Livestock Commission, is an interactive display that provides details of more than 120 meat recipies. It is installed in Safeway and Co-op stores in the UK. (Photo McMillan UK Ltd)

it is often instantaneous, and never more than two or three seconds.

Systems which are both videodisc and computer-controlled can use 'touch screen' technology to simplify both the customer's choice and the technical installation. Displays using these technologies can be more than simple random access. If there is a situation where one choice leads to another – or, better, where the program segment shown depends on information provided by the customer – the display can properly be called an interactive display. The subject of interactive audio-visual is discussed further in Chapters 13 and 14.

Networks

Chain retailers selling media products have become major users of big video presentations in retail stores. Companies such as Warner Bros, Disney Stores, Best Buy and Music Land are using either big single screen video or, more usually, videowalls in order to get a bigger, brighter picture in minimum floorspace. The videowalls are used to promote software products. A typical chain may well have more than 100 stores equipped with big screen video replay.

The programs are conventionally distributed to the stores as videodiscs. However, the arrival of high-quality compressed digital video, ISDN digital phone lines and low-cost satellite distribution has opened up a number of new possibilities. All the following, either separately or in combination are possible:

☐ Program material is stored locally in the form of long-term material on videodisc and short-term material in digital form in a computer. Programs are automatically scheduled on a calendar basis.

☐ Scheduling information is sent by conventional phone line and modem.

☐ The same data link is used to monitor performance of the show system, with automatic fault diagnosis.

☐ Both scheduling information and new video program material is downloaded by ISDN line. This process either takes place at night when the display is not in use, or can take place while the display is running.

☐ Scheduling is linked to specific product promotions, and resulting sales are monitored.

☐ Where there is local control over what is shown, actual programs shown are logged automatically.

☐ Programs are directly distributed by satellite. (In fact continuous, real-time distribution would normally be uneconomic. More likely is the use of satellite as a fast method of downloading digital programs. An obvious exception is where the program material is itself a real-time event, for example a sporting event.)

The advantage of all these methods is that the control of what is shown lies entirely with the center. The use of 'physical' media like discs and tapes declines, with less likelihood that the wrong program is played, and with much greater flexibility in programming. The success of the programming can be judged against actual sales.

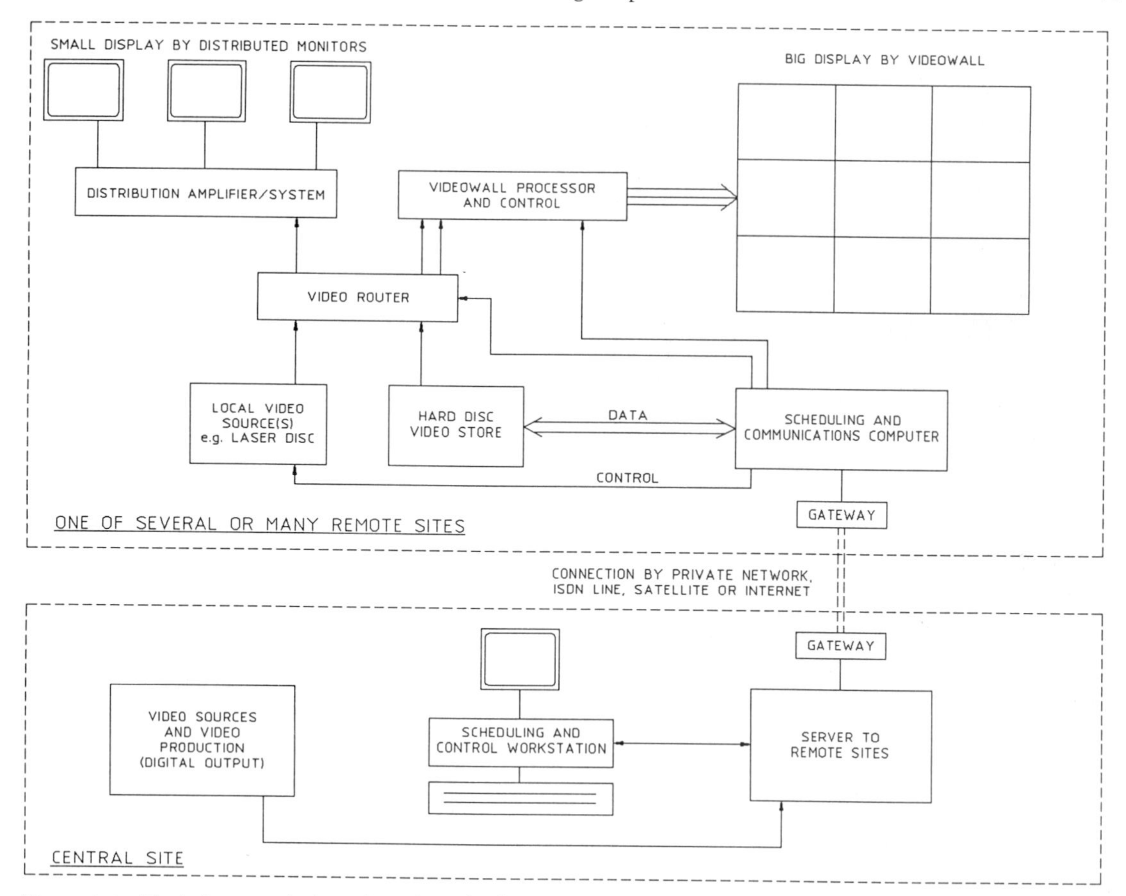

Figure 9.8 Block diagram of a hypothetical retail video system with central digital video sourcing. The 'gateway' will depend on the transmission method chosen

Mall TV

Another retail application where video is used in a highly structured way is the shopping mall. It may well be the case that within a shopping mall a number of retailers are using their own video displays within their stores. But in addition to this, the mall operators themselves may well have reason to run programming throughout the public areas of the mall.

An example is the company Western Developments which run a number of malls under the 'Mills' banner. Typical is the Gurnee Mills mall, a mile-long mall under cover in Illinois. There is distributed audio and video throughout the complex, and programming is both general and area-specific.

The sole object of the AV network is to keep people in the mall for as long as possible. A way to do this is to keep them informed. The programming consists of news of local and mall-based events and entertainments, store advertising and national brand advertising. The cost of participating for small stores is kept to an absolute minimum, the service being there to support tenants.

Until recently, the 'playback' media for such systems has consisted of a mixture of U-matic tape, videodisc, and still stores. New installations are tending to use all-digital systems without tape. These give greater flexibility in programming, lower running costs, and greater system reliability.

While all video monitor, self-contained video rear projection units and videowalls are suitable for use in shopping malls, their use in shop windows facing outside streets should be avoided at times of bright sunlight.

Sales meetings

Here a distinction is made between sales conferences and sales meetings. Events designed to motivate a

Figure 9.9 The mile-long Sawgrass Mills shopping mall in Florida has its own TV network, with an on-site facility for program preparation. Most presentation is by 35-inch video monitors. (Photo Electrosonic Systems Inc)

Figure 9.10 Projected videowalls are used in the larger concourse spaces at Sawgrass Mills

sales force are discussed in Chapter 7. A different type of event is the meeting where the intended outcome is a direct sale to individual members of the audience.

The first thing to remember is that it is very difficult to deal with a large number of people. Once the audience has seen the presentation, how can the presenters get all of them to buy?

There have been successful sales meetings to large audiences, especially in the travel business, but only when there has been adequate staff present to process the audience at the end of the meeting.

The sales meetings using AV that are usually unqualified successes are those where the audience consists of a group of people from a single customer. This method of selling is of increasing importance to all companies with a sophisticated product or service to offer. This type of meeting takes quite a lot of setting up. Precisely because of this, it is important that they get results. AV will help ensure this by:

- ☐ giving a proper structure to the meeting;
- ☐ giving a sense of occasion;
- ☐ minimizing overall executive time spent;
- ☐ improving communication of the basic message.

Examples of successful users of this technique are to be found in all industries, but especially those selling capital projects and equipment (e.g. builders of telephone systems and building contractors) those selling services (e.g. engineering and management consultants) and those selling property (e.g. renting of prestige buildings or selling of new developments).

There is a choice of media. Slide-based multi-image technique gives excellent quality, and is ideal for the bespoke presentation, especially when it is given in a permanent presentation room (see Chapter 12). If video and computer graphics are to be used, then it is important to ensure the highest presentation quality by using high resolution graphics and up-converted video (see Chapter 19).

In permanent presentation rooms, it is quite practical to make such presentations work fully automatically from the technical point of view. If shows are to go 'on the road', it is essential to have them properly staged either by competent in-house staff whose sole job it is to run presentations, or by an outside specialist company.

Public relations

Audio-visual is a powerful public relations tool. Some examples are described below and are illustrated in the accompanying figures.

A company already has a presentation room. It makes a point of having a good 'house show', not only for employees and customers but also for the general public, e.g. school parties, local residents and special-interest groups. Some users develop special programs for such audiences.

A company is invited to sponsor an exhibit in a museum or other public visitors' center. An AV presentation may not only meet requirements of the

Figure 9.11 The Swiss watch manufacturer Swatch has successfully used compact mobile videowalls to boost sales in retail promotions. (Photo Ganz AV, Zurich)

Figure 9.12 Kodak sponsored the multi-image show 'From Limelight to Satellite' shown as a visitor attraction at the Granada Studio Tour

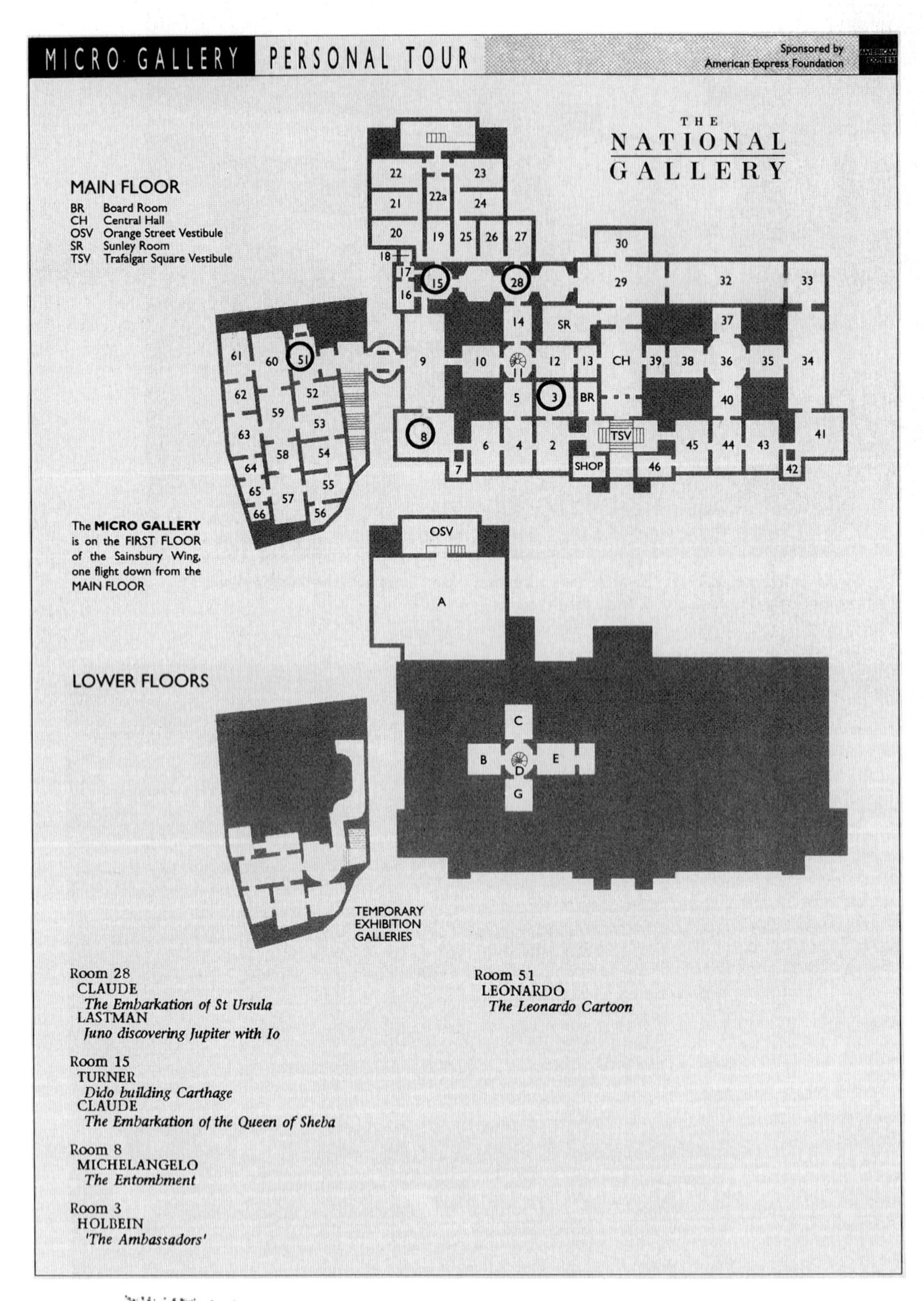

Figure 9.13 American Express sponsored the Micro Gallery at London's National Gallery. This is described in more detail in Chapter 11

Figure 9.14 Fuji film are a sponsor of Rock Circus, Madame Tussauds' history of pop music exhibition

museum, but allow the sponsor a method of putting its name in front of the audience in a way that is most acceptable. For example the Boeing Aircraft Company sponsored the exhibit and a specially produced multi-image show on the subject of flight, at the Singapore Science Center. At Britain's National Motor Museum several well-known companies such as Lucas, Alcan, Kennings and Ford sponsored exhibits that incorporated AV. An AV show can also help to give structure and a sense of occasion to press conferences and similar events.

The examples of AV use described here are open to all sizes of organization. Larger companies with household names can also consider sponsoring films and video programs, with a view to either having them shown on public broadcasting network or, more realistically, having them distributed to specialist audiences. In this case it is important to plan how the film is to be distributed at the time it is commissioned, because this may affect how a particular subject is tackled. Distribution of this type of film is best handled, at least in part, by specialist library firms who have both access to the audience and a streamlined system for issue, recovery and maintenance of film copies and videocassettes.

Chapter 10

Audio-visual in exhibitions

Trade exhibitions are a natural use of audio-visual. In fact, trade exhibitions in Europe were, for a number of years, the main application of some audio-visual media. There are two opposing attitudes to the use of AV on exhibition stands; one of delight at how effective the medium can be, the other of disillusionment because an expensively produced show has had little effect.

There are so many positive things that AV can contribute to a trade exhibition stand. Here are just a few examples.

Making more effective use of the visitor's time

Most exhibition visitors have limited time in which to absorb the content of a particular stand. An AV show may be used to summarize what is on offer. A different approach is needed depending on whether the show is aimed at existing customers paying their routine visit to a supplier's stand, or at potential new customers seeing the company's product range for the first time.

Attracting more visitors onto the stand

Although AV can be used for several different purposes in exhibitions, a particular show can seldom fulfil more than one objective. It is possible to use AV as an attraction to entice people on to the stand. However, it is unlikely that this kind of show would also be able to meet the effective use of time objective.

Solving a language problem

AV shows can have multi-lingual commentaries. This is only worth doing if the environment allows the visitor to see the show in comfort. It is more important that the visual sequence includes interpretation clues to aid foreign visitors.

Contributing to more effective selling on the stand

Most trade exhibitions are short. Anything that helps make more effective use of staff on the stand makes a better return on the exhibition investment. For example, at busy times visitors waiting to see a particular salesperson can watch an AV show. The right show ensures that when the salesperson is free the customer is well disposed to the exhibitor, and has already received some sales information.

Providing better public relations

Often an exhibition stand is one of the few places where an industrial company is on public display. It can be just as important for the company to look good in the eyes of third parties, as to make sales from the stand. An AV show can help the PR image of a company in the eyes of suppliers downstream customers, the general public, and direct customers. In some cases the AV can help entertain the public, who may have no need to take up the valuable time of sales staff, while serious visitors can be personally attended to.

There are many positive reasons for using AV techniques at trade exhibitions, but there are also some negative aspects. Negative results reported by some users are summarized in the following comments:

> 'The picture looks all washed out, far from being an AV spectacular, the exhibit looks weak.'

> 'We had lots of people looking at the show, and our stand seemed to be the hit of the exhibition, but we made no extra sales and got no more leads.'

> 'We had to turn the AV off as it was preventing our sales staff closing sales with customers.'

> 'We have this great AV that lasts 15 minutes and cost thousands of dollars. But on the stand nobody watched it for more than a minute or two and they usually missed the bit about the new product.'

> 'The AV was a pain. The sales manager is the only person who knows how to work it properly, and he was too busy. The fuse kept blowing, and the projector was in the way of the coffee machine.'

These complaints betray either a lack of commitment to making the AV element do its job or, more excusably, an ignorance of the factors that must be taken

Figure 10.1 Exhibition shows with a story must have a proper viewing environment. (Photo Electrosonic GmbH)

into account when using AV in the special environment of the trade exhibition.

Because any company or organization participating in a trade exhibition is hoping to make a reasonable return for the investment in stand space (whether the return is measured in money terms, or in indirect benefit terms), it is important that those using AV do not fall into unnecessary and wasteful traps. How, then, is disappointment to be avoided? The key to this is understanding the nature of the problems:

☐ Any stand is competing with many others. Does the potential audience have time to see a complete AV show?

☐ There is no way to control *when* a visitor will arrive on the stand. An AV show half-way through is more likely to make visitors move on than attract them.

☐ Can the exhibition stand be designed so that the AV show is seen and heard properly, yet does not interfere with other activities and displays on the stand?

If the objectives of the exhibition stand are fully understood, it is easier to see how problems can be solved. There are specific points that the exhibition manager must consider.

If portable audio-visual equipment is to be used, then it must be used in the same way as it is on the road. This means that sales staff must be in attendance to introduce the show and talk to visitors. Therefore the AV units must be sited to allow individual presentation. The exhibition stand must become an extension of the salesman's territory. If this condition cannot be met, it is much better not to use the AV program at exhibitions.

If a conventional AV show designed for group communication is to be shown, there must be a small AV theater or semi-enclosed viewing area. The exhibitor must make sure, in this case, that he can hold people until the show starts and that there are enough staff on hand to talk to each group after the show.

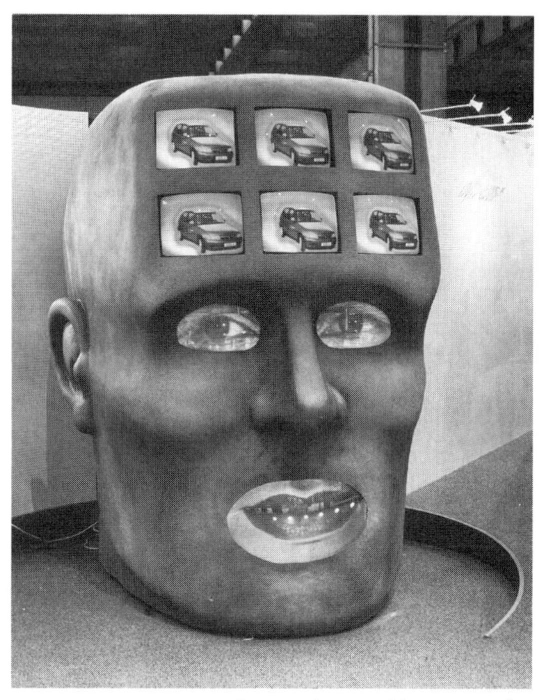

Figure 10.2 An unusual screen format adds interest. The 'Ecohead' was a feature on the Opel stand at the 1991 Frankfurt Motor Show. It uses nine video monitors of different sizes. (Photo Adam Opel AG, HP-ICM Ltd and Interactive Television Ltd)

A theater-type show is really only suitable if the exhibitor has a highly targeted audience, or the show is fulfilling a PR role. An example of the former would be a manufacturer of telephone exchange equipment selling in a country where the telephone system is a state monopoly. In that case there is only one customer, so the show can be targeted at the people who matter in the buying organization. An example of AV fulfilling a PR role is a public-image show put on by a consumer goods manufacturer. Here the show can be considered part of the overall PR effort, and its success measured by size of audience and audience appreciation.

If the controlled environment of a theater is not possible, and often it is not, an effort must be made to integrate the AV into the exhibit fabric and to ensure that viewing conditions are correct. This means paying attention to details of staging, such as avoiding direct light on the screen, by using back projection, providing a leaning rail or barrier for viewers, and ensuring the correct screen height.

A show on an open stand must be completely different from a 'house' or sales meeting show. It must be *short*. Ideally two minutes long, but no longer

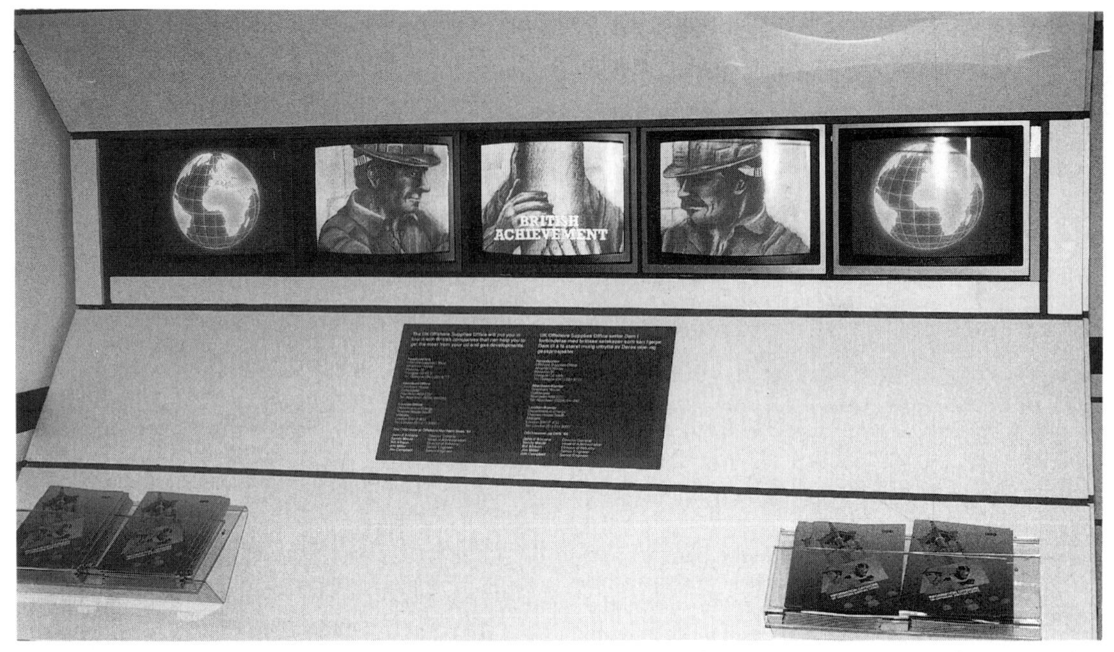

Figure 10.3 A single video monitor on an exhibition stand can look insignificant. The UK Department of Energy used a clever five-screen video production on its stand at offshore technology exhibitions

Figure 10.4 The adverse effect of ambient light can be minimized by setting the screen back. Here the designer has used the simple trick of side mirrors to make the compact AV display look bigger

than five minutes. It must be written in a circular fashion, so that it makes sense to visitors within a few seconds of joining in and ensures that they will see the whole sequence through, regardless of where they started.

Two important ideas emerge from these points. Firstly an existing program investment can be re-used. If a company has already commissioned a show from a producer, e.g. a house show, a sales meeting show, or any other show intended for a controlled audience, using the same producer and same basic material it should be possible to create a special exhibition version which is shorter and sharper. This kind of show can be very successful.

The second idea is to use AV as a 'come on'; an – invitation on to the exhibition stand. In this case the AV show must be integrated into the decor. In fact, it becomes much more a part of the stand design than a detailed communication method. This approach requires short, simple and bold program material and is one of the best applications of multi-screen technique. It is particularly suitable for general public exhibitions.

Choice of medium

In principle all AV media can be used on exhibition stands, but there are some guidelines which should be followed.

A TV set is just a TV set and its use as a display may, by its very familiarity, devalue a show on an exhibition stand. The exhibition designer must take care to 'design in' video. Some rear-projection TV sets may be more suitable than a conventional set. Better still, if space permits it, is use of video projection with high-gain rear projection screens. These screens are available in sizes equivalent to a 45-inch or 67-inch

Figure 10.5 GM Canada using a videowall as a 'come-on' on their stand at the Canada Autoshows. (Photo Bankten Management Services which designed the stand)

TV set. Bigger sizes are available, but they are intended for darkened viewing areas.

Individual communication units, such as portable single slide/sound units, or compact video units should not be used as display devices. Instead, an environment must be created in which they can be properly used by sales staff.

Multi-image technique using slides is an effective medium for exhibitions, and is easy to adapt to special needs. Back projection is the only satisfactory method of presenting optical projection on an open stand, and it is best to keep individual image areas small and bright. This simply emphasizes the point that the multi-image show made for a sales conference is unlikely to be of any use on an exhibition stand. On the other hand production costs of exhibition multi-image *should* be low in relation to effective use of the stand. Shows should be short and based heavily on existing high-quality visual material. If a show producer offers a 15-minute show using only specially created material for an exhibition stand, either he knows little about trade shows or imagines he has a gullible customer. Equally, the exhibitor should specify very simple objectives for these shows and not complicate the message by asking them to do too much.

However, it is the case that multi-screen video techniques have largely taken over from slides. The cost of these exciting and fast-moving techniques, once considerable, is now quite reasonable.

Multi-screen video comes in two forms. The best known is the 'videowall', where a large number of video monitors or back-projection modules together produce a very big, bright image. In addition to creating the big image, by splitting the incoming video signal across many monitors, modern videowall systems allow the programming of special multiple-image effects.

Monitor-based videowalls take up very little space, typically only 60 cm (2 ft) of depth. The gap between individual screen elements is not a problem with the right source material, and the monitor-based videowall has a dynamic of its own. Projected videowalls need 2 to 2.5 m (6 to 8ft) depth to allow for access, and care must be taken to ensure sufficient viewing distance. On big exhibition stands they are very effective.

The standard videowall is TV-format so always has the same number of monitors vertically as horizontally, for example 5 by 5. However, the exhibition stand is the ideal opportunity to use unexpected formats, such as wide-screen displays and unusual juxtapositions of monitors.

Figure 10.6 Sharp used a videowall display in the dealer area of their stand at the consumer electronics exhibitions in Vienna, Austria. (Photo Silberberger AV)

Figure 10.7 An interactive videowall for a specialist audience. At the Frankfurt Automechanika exhibition, Herberts Standox, a supplier of paint finishes, used the technique for answering visitors' questions

This leads on to the second type of multi-screen video display, where a number of video screens are synchronized together, each showing different, but related, material. More information on videowalls and multi-screen video is given in Chapter 19.

The prestige exhibition stand, especially one at a show open to the general public, can benefit from a mixed-media approach, where a synchronized sound and light technique is combined with projection. There is no limit to the possible size or complexity of such shows, so it is entirely up to the sponsor to decide what is justified based on his own criteria of effectiveness.

An attractive idea is to have displays that offer the visitor random access to different images or different programs. The problem here is that only a limited number of people can use such displays in a crowded exhibition. They therefore tend to be of greater value in permanent or long-running exhibitions, or in specialist trade shows where the exhibitor can justify a high cost per serious visitor. The possibilities are described in detail in Chapters 13 and 14.

Multi-language displays are of special benefit at international exhibitions. These can use either multilingual recordings or multilingual graphics on the screen, or both. They are, however, only a support for the selling effort in the customer's language, not a substitute for it.

Product demonstration

Some exhibitors use their exhibition stands for formal presentations of new products, given at regular intervals. Usually some kind of clock indication is given for the time of the 'next show'. This kind of demonstration has proved particularly effective for suppliers of computer and games software, but is equally effective for other technical and consumer products.

The main priority here is to ensure that the presenter can be heard. The best type of microphone to use is the headset, which ensures clear, intelligible speech in the noisiest of surroundings. Loudspeakers should be directed to cover the audience area only — not the neighboring stands!

The presentations should be thoroughly rehearsed, and in most cases it is best to use the services of a professional presenter, with company staff on-hand to answer detailed enquiries after the show. Normally visual support in the form of a large back-projected video screen or videowall is required, either because this is the way to demonstrate the product (for example, a software product) or because this allows a close-up of what the presenter is doing. If a video camera is to be used, it is essential to have a professional camera operator.

Staging

The main practical problem in using AV on an exhibition stand is the less than ideal circumstances of installation. A wise exhibition manager takes steps to insulate himself from the problem by appointing a competent staging house or rental company both to install the equipment, and to ensure its proper running throughout the course of the exhibition. Even if the exhibitor owns the equipment, it may be best to contract out its installation and maintenance unless there are staff whose sole job is AV support.

It is important to allow enough time for installation. This means that other trades, especially the electricians providing power, must finish early enough to ensure

Figure 10.8 Product demonstrations at trade exhibitions benefit from a proper environment, good sound reinforcement and visual display. 3M had all three at the COMDEX exhibition

that the AV contractor has time to complete the installation, which can only be started after everything else is finished.

One common problem is lack of adequate power. Simply having enough power sockets is not always enough. If only one or two video units are involved, there is unlikely to be a problem, but if a videowall or twenty slide projectors are to be installed, the story is different. In theory power may be sufficient to run all monitors or projectors, but in practice the switch-on surge may trip circuit breakers, or a high-impedance supply may result in mutual interference between projectors. These problems are entirely avoidable if correct power supplies are specified and installed.

It is a false economy to skimp on space provided for the AV. Either it forms an important part of the exhibition stand, in which case it should be properly staged, or it does not, in which case it is best omitted. Therefore, the rear projection area should *not* also serve as the coffee room or catalog store. Equipment *must* be accessible for service without inconveniencing visitors.

It is quite possible to arrange for AV shows to be staged in overseas countries. The question as to whether it is best to take equipment from the exhibitor's country or to hire locally depends entirely on circumstances, and should be reviewed with the staging company. In general, simple shows using standard equipment configurations are best done on a local-hire basis, but only if program material is sent to the rental company well ahead of the exhibition date. This allows the show to be tested on the actual equipment it will run on. Complex shows often need equipment and staging from the exhibitor's country of origin.

Chapter 11

Audio-visual in museums and visitors' centers

The old idea of a museum being a place where dusty archaeological remains are viewed in an equally dusty cabinet is, fortunately, now out of date. Today there are museums on subjects as diverse as geology, tennis, science, horse racing, natural history and beer. There is also an increasing number of sites which cannot be called museums, but which have many of the same requirements, both in terms of what the visitor expects, and in their day-to-day management. It is easiest to refer to these as visitors' centers. They may vary from an exhibition gallery attached to a tourist site such as a National Park, to a permanent exhibition at a company headquarters. A traditional museum can serve as:

- ☐ A place of study for scholars of the museum's subject.
- ☐ An archive, where storage of antiquities or artefacts is more important than their display.
- ☐ An educational institution.
- ☐ A place where items are on display and an effort is made to interpret their importance and relevance to visitors.

Major institutional museums serve all four objectives, but smaller museums and all visitors' centers only attempt to display or interpret exhibits. It is in this area that audio-visual can make a significant contribution, both to the visitor's enjoyment and appreciation of the subject, and to ways in which visitor sites are managed.

Audio-visual techniques can assist the interpretive function in many ways, for example:

- ☐ solving language difficulties;
- ☐ creating an atmosphere;
- ☐ changing the environment;
- ☐ allowing comparison with outside material;
- ☐ setting a context;
- ☐ providing historical background;
- ☐ developing an idea;
- ☐ introducing a subject;
- ☐ demonstrating a process.

They can also help in management of visitors. In some cases AV can make the difference between a particular site being a worthwhile place to visit and an unsatisfactory visitor experience.

Museums and visitors' centers represent one of the most interesting and potentially successful applications of AV techniques. The fact that they are permanent installations enables the best value to be extracted from a program investment, but this permanency also introduces some special problems and disciplines. In principle all AV media have a place. This chapter reviews some points that are unique to the permanent installation.

Show environments

AV can either be shown in special auditoria or can be integrated into an exhibit. In the first case special attention must be paid to ease of exit and entry to the auditorium. The question of whether the audience stands, sits or has leaning rails depends on the length of the show. Shows longer than twelve minutes should have seating.

An auditorium is only used where the show stands on its own. Many applications require that AV is built in to the exhibit fabric. If this is done, shows must be *short* (see also Chapter 10.) It is also important for the show to be properly 'designed in'. It is not good enough to install a piece of equipment like a TV set in the middle of a gallery – it must be designed in so that the screen forms a logical part of the display.

Separate auditoria have the advantage that lighting conditions can be controlled to ensure optimum viewing of the show, but the separate show is only one possibility; often an integrated show is more appropriate. In this case special care must be taken to ensure that viewing conditions are correct. Particular attention must be paid to prevent front lighting falling directly on screens, and to make sure that screen size is suitable for prevailing ambient light conditions.

Figure 11.1 An elegant example of AV integration is at the National Museum of Racing in Saratoga Springs. Historic photographs are projected as slides in the 'window' of a set

Choice of media

Choice of audio-visual medium is made on the usual bases. However, in permanent exhibition work:

□ 16 mm movie projection is no longer used. Video projection can now give equal or better quality, with lower running costs.

□ 35 mm and 70 mm movie are the best media for big moving-image presentation in a theatrical environment, with screen sizes up to and beyond 20 m (66 ft) wide. Fully automatic equipment is available to help minimize running costs.

□ Video using monitors is reliable, but familiarity with the TV set can devalue the show. Clever designing in can help solve the problem.

□ Front-projected video can only be used in small auditoria with controlled lighting conditions.

□ Small liquid crystal video projectors have some quality and light output restrictions, but do provide great flexibility in placement within exhibits. Front projected images of 60 cm (2 ft) and rear projected images 90 cm (3 ft) wide are practical.

□ Conventional three-tube video projectors can provide excellent rear-projected images up to 1.8 m (6 ft) wide using special high-gain screens.

Figure 11.2 For big impact the National Museum of Racing uses a 40 ft × 10 ft (12 m × 3 m) screen for a specially commissioned extra widescreen movie called *Race America*. Here 35 mm represents the most cost effective method of presentation

☐ Auditorium high brightness LCD and LC lightvalve projectors can support rear projected images 3 m (10 ft) wide in reasonably high ambient light conditions. The replacement lamp costs should be taken into account.

☐ Multi-screen video and unusual video formats can greatly enhance the effectiveness of video programs.

☐ Slides are an excellent visitor or single-site medium. They allow shows, both single-screen and multi-image, to be made that are unique to the venue. These shows can have a great impact on the audience, and are easily kept up-to-date.

☐ The permanent exhibition allows use of the synchronized sound and light or 'son et lumière' technique. This can bring a display to life. The system may be as simple as lighting up a few transparency boxes in sequence, or as complex as having fully animated figures with programmed theatrical lighting.

☐ Media can be mixed. A frequently successful example is where a slide program accompanies the description of an animated map or diagram. In principle, it is possible to automatically program the activities of any device that can be directly, or indirectly, controlled electrically.

The auto-present concept

Some first-time users of AV in permanent displays do not appreciate the difference between running an AV program as part of a person-to-person presentation, and running one in a fixed installation on a day-in, day-out basis. Equipment suitable for one may be quite inadequate for the other.

Ideally fixed shows should run fully automatically at the touch of a button. There can be no operator. This is where the idea of the auto-present show is introduced. Its attributes are best illustrated by the example of a permanent exhibition with a small auditorium show based on the multi-image technique. It has houselights in the auditorium and a set of motorized drapes across the screen. The auto-present method of operation works as follows:

☐ A *single button* is pressed. This causes the following:
○ the projector power to switch on;
○ the tape to start.

☐ The show automatically runs from the tape, which can also program auxiliary devices. Therefore it is used to:
○ fade down houselights;
○ open screen drapes;
○ run the main multi-image show.

☐ At the end of the show the sequence is again fully automatic. The tape is programmed to:
○ close screen drapes;
○ fade up houselights;
○ instruct all projectors to return to zero or 'home';
○ rewind tape to the beginning;
○ when all projectors are 'home', switch off projector power.

(Tape would actually only be used for temporary or infrequently run shows. Permanent shows would use

Figure 11.3 One of four audio-visual theaters at the Citadel Historic Visitors' Center, Halifax, Nova Scotia. Each show describes one century in the Citadel's history and uses a mixture of slides and son et lumière technique. Shows are in English and French

CD or digital soundstore, with a separate show controller; but the principle remains the same.)

These principles can be applied to any automatic AV show, whatever the medium used. Video media are easy to auto-present. Slide/sound and multi-image are easy only if the right equipment is installed. Movie film is also auto-presentable, but does require special projectors or projector attachments.

Where a mechanical device such as a slide projector is used in an auto-present system, it is essential that the homing procedure *independently* checks the device is in the zero position. For this reason the projectors used must have zero-position sensing microswitches, and the control equipment must be of a kind that uses them. The same principles apply to other devices, such as turntables, that can take up several positions. It is easy to see that continuous shows can be given by allowing the system to restart itself, but only when it has checked that all equipment is properly reset.

A simple addition is to add a clock system that runs the show automatically at preset intervals or at preset times of day. An additional feature is a countdown clock which shows the audience how long they must wait until the next show.

It is vital that proper running of these shows is monitored, but one of their benefits is supposed to be unattended operation. The best thing is to have a routine inspection. One helpful feature that makes this possible for shows based on projection is an automatic lamp changer. This device is available for current models of automatic slide projectors.

Enemies of reliable operation of AV equipment are *dirt, dust* and *excessive heat*. It can be a worthwhile investment to provide a special environment for the equipment. For example, projection rooms should have their own filtered air supply. Video players can be centrally sited in a clean, air conditioned, or at least cool, room, and slide projectors in individual exhibits can have their own protective box with positive pressure filtered air supply. Correct installation of such equipment can make the difference between negligible or appreciable maintenance costs, and can increase equipment life by years.

It is very important that equipment be *accessible* for maintenance. Designing an AV show without providing proper space and access for equipment is like designing a restaurant without a proper kitchen. Maintenance or repair of an item of AV equipment should never require use of ladders or access equipment, and should always be possible in a way that does not impede visitor flow. There is nothing more embarrassing for a maintenance engineer, or more likely to reflect badly on the museum, than to have to struggle with a recalcitrant projector while perched on the top of a ladder in the middle of a sea of visitors.

Figure 11.4 The video replay system at the Museum of the Moving Image, London, England. The disc players are properly rack mounted, well ventilated and controlled by standard disc controllers that give continuous run, demand start, synchronized multi-screen play or random access program selection as required

The role of the videodisc

When a video program is to be used in a museum or visitors' center, the rule now is 'no tape'. The reasons why tape is not recommended are as follows:

☐ Continuously running shows can only be achieved using two machines in tandem. Otherwise there is a significant rewind interval, which means a blank screen during rewind if an interval caption or program to cover the interval is not provided.

☐ To minimize wear on the cassette and player of a continuously running show, programs should be repeated as many times as possible down the cassette. For example, a five-minute video sequence should be recorded six times on to a thirty-minute cassette. This, however, increases the cost of dubbing the cassette.

☐ Tapes wear out after a few months and must be replaced.

☐ Picture quality is seriously degraded by dirty tape heads. Regular tape head cleaning maintenance is therefore required.

☐ Worst of all, tape heads actually wear out. They may need replacing every one or two years, which is expensive.

☐ Museum and exhibition designers like systems that allow visitor choice and give good still-frame performance. Videotape is a poor performer for still-frame use, and is very slow in any random-access application.

Advent of the industrial videodisc player (described in more detail in Chapter 19) has transformed the way in which museum and permanent exhibition video is done, as it overcomes many of the disadvantages of tape. Videotape is now only used for temporary exhibitions with single programs.

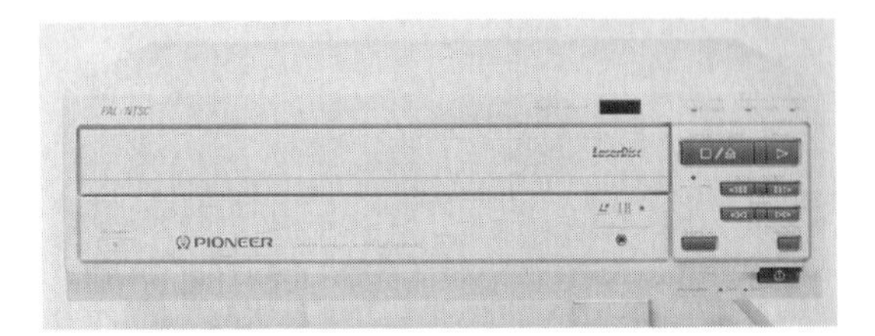

Figure 11.5 The Pioneer LD-V4500D is one of the latest generation of videodisc players. It can play both NTSC and PAL videodiscs

The first reaction of a museum curator to the use of videodisc might well be one of horror at the cost of creating the videodisc itself. However, if overall costs are examined, it can nearly always be shown that videodisc is actually less expensive than videotape, as well as introducing operational advantages.

Simple industrial videodisc players are less costly than industrial videotape players. Even if only one year's operation is taken into account, elimination of both routine maintenance and replacement cassette dubbing yields a worthwhile saving.

The videodisc approach makes even *more* sense when considered on a system basis. This can be illustrated by an example. An imaginary craft museum might wish to use audio and video techniques to achieve the following:

☐ Run a continous three-minute video program to describe the museum as a whole.

☐ Run three other continuous video programs, each of two minutes, and each accompanying a particular exhibit. They might show the machine/artefact/tool being used.

☐ Have an exhibit giving push-button selection of ten one-minute video sequences, each demonstrating a different facet of the use or manufacture of objects related to the museum subject, but which for some reason cannot be exhibited themselves.

☐ Have an exhibit giving push-button selection of 100 (or 1000!) still images.

☐ Have one or more simple interactive displays, testing visitors' knowledge about particular subjects or, at the end of the museum tour, testing what they have learnt in the museum.

☐ Run several continuous sound tracks to accompany other exhibits. For example, birdsong, people in a reconstructed dwelling or shop, or machines.

With careful planning all these items are easily fitted on a single disc side as one disc side can carry 54,000 frames, representing 30 minutes of full-motion video in NTSC, or 36 minutes in PAL.

If, on the other hand, the project uses videotape and sound cartridge machines, preparation work still has to be done so the only extra work is originating the disk. Cost of this is quickly saved in lower system capital cost and lower running costs. Benefits also include better picture quality and the ability to achieve fast-responding interactive displays unachievable with any other method.

Figure 11.6 shows equipment needed to achieve the hypothetical example. Clearly it is up to the exhibit designer to integrate video screens into the exhibit fabric and, although video monitors are shown, there is no reason why some sequences cannot be shown by projection.

The pre-eminence of the laserdisc as the video source for museum/exhibition applications is now being challenged by various forms of compact disc which in one form has the advantage that it is easier to make 'one-of-a-kind' discs. Many of the exhibit principles remain the same, and the technical differences and possibilities are reviewed in Chapters 13 and 14. There is also now the possibility of sourcing video from computers. This is discussed later in this chapter and in Chapter 14.

Interactive displays

The word *interactive* is somewhat overused, and for museums it is really better to talk in terms of visitor-operated displays. Sometimes, all the visitor is asked to do is express a choice, for example, a choice between a number of video programs or a number of still images. These displays are very useful where a lot of facts must be available, almost on a catalog basis. The possibilities are:

☐ Push-button random access of slides. Random-access slide projectors should be used where big bright images are needed, and where changes might be needed. One random-access projector can hold 80 slides, although it is possible to have several projectors working on the same screen to give a larger capacity. Four projectors, 320 slides is a practical maximum.

☐ Random access of still images on videodisc. The possibilities here are almost limitless as one disc alone accommodates up to 54,000 still images. In both this case and that of slides the simplest method of selection is to offer the visitor a decimal keypad, usually with a simple two- or three-digit display confirming the slide selected. It is also possible to have one button per selection, which can make for a better display.

☐ Random access of video sequences. This is now done from some kind of videodisc. The standard laserdisc holds

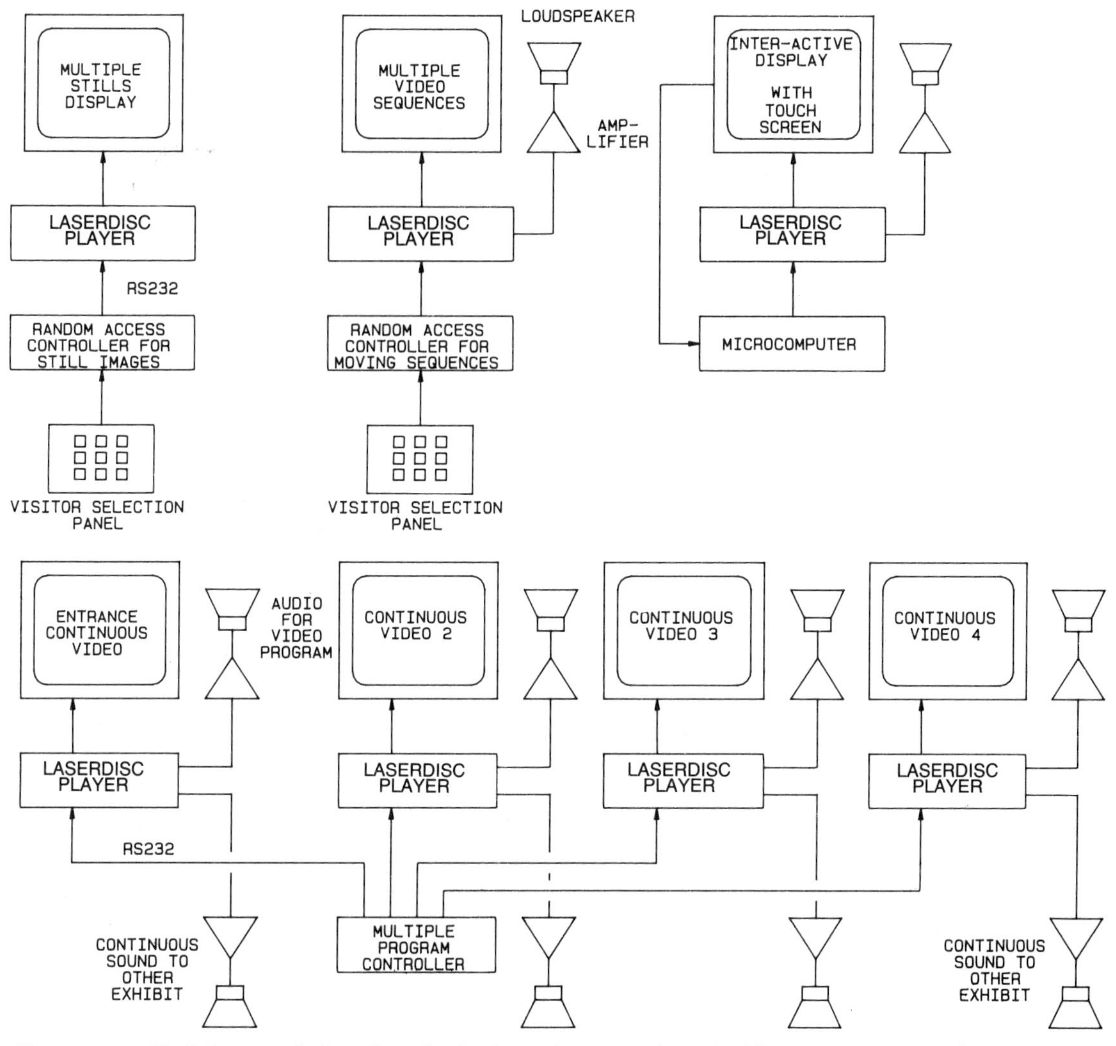

Figure 11.6 Block diagram of a hypothetical videodisc replay system for eight different exhibits in a small museum

up to 36 minutes of material (PAL format) or 30 minutes (NTSC format), so provided total run time of all segments does not exceed these maxima, all that is needed is a single disc player with a suitable selection panel. This is normally a special control which permanently stores sequence information. It could, for example, give visitors the choice of ten two-minute sequences and five three-minute sequences. Actual durations are not important, but the total must not exceed the disc capacity. If required, more players can be used.

☐ Another interesting possibility is to use a videodisc or CD-I disc to give a much larger capacity of program choice using still pictures with audio. Special disc production techniques allow many hours of audio to accompany still pictures; but actual capacity depends on the mix of audio and picture. Easily accommodated, for example, is a choice of 100 two-minute programs, each containing a sequence of 20 still images. Chapters 13 and 14 give more information of such techniques.

☐ Random access is not confined to slide or video images. Randomly accessed computer data can be in graphic or, more usually (because, in general, real images are best presented on slide or video), in text form. Such displays are ideal for facts and records (e.g. sporting records) as they can be kept up-to-date on a daily basis if necessary. It is important, however, that any such displays format data presented in an attractive form, and that the amount of information on any one page is limited.

A truly interactive display is something more than just a choice of images, and much more difficult to do well. Thus, simple random access should be

Figure 11.7 The 'Youth Culture' exhibit at the Museum of the Moving Image in London, England, allows visitors a push-button choice of twenty-four video sequences. Here group viewing is encouraged

thoroughly explored first, because it meets many of the display requirements of exhibitions. Then, when a display is required that demands the active *participation of* and *input from* the visitor, real interactive displays should be introduced. This complex subject is covered in Chapters 13 and 14.

Before leaving the subject of visitor-operated displays, a word about visitors' grubby fingers and propensity to destroy push-buttons. Although thought of as a 'small boy' problem, experience shows that boys and girls of ages 5 to 85 are equally likely to damage the input end of such displays. Therefore, the following points should be considered:

☐ Touch screens are popular with many designers, but need regular cleaning. They can also make all interactive displays seem the same, so some designers prefer a custom push-button panel that can be better related to the exhibit design. Buttons are less expensive too.

☐ Push-buttons should be chosen for toughness. Those used by gaming machine manufacturers are a good choice. For really difficult cases, buttons designed for use on heavy machinery can be considered.

☐ If a museum or exhibition has a number of displays with the same visitor panel; it can pay to have a membrane keyboard custom-made. These can have artwork of any complexity, with as many or as few buttons, indicators and displays as required, *and* are waterproof. Once the tooling is done they are inexpensive. They are easy to keep clean, tamper-proof and economic to replace.

Audio systems for permanent exhibitions

Rules for audio systems of museums and permanent exhibitions have changed rapidly in the last few years. Before defining technical solutions it is best to define the requirements of these installations. They vary

Figure 11.8 In the Numbers and Forms Gallery at the Taiwan Museum of Natural Science, the interactive displays are housed in polyhedral housings that declare that the exhibits are for individual use. (Photograph courtesy of James Gardner Studio)

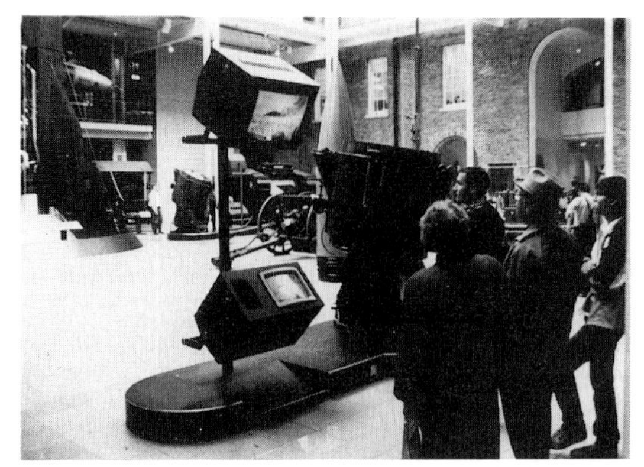

Figure 11.9 At London's Imperial War Museum, video exhibits use a two-screen arrangement to allow both individual and group viewing. Program selection is by the bottom screen which is a touch-screen

enormously in terms of the resources needed to give the required results, and include the need for:

☐ Continuous sound effects to enhance an exhibit, for example, sound of locomotives in a transport museum; sounds of birds and animals in natural history exhibits; sound of people in environmental or room reconstructions.

☐ Sound accompanying small-scale audio-visual exhibits. These include video, single-screen and multi-screen slide shows, and small-scale sound and light shows.

☐ Sound accompanying large-scale and theater-based audio-visual shows. For example, larger multi-image shows, sound-and-light sequences and animated figure shows.

☐ Exhibit commentaries delivered on a localized basis, sometimes in several languages.

☐ Acoustic guides where visitors are given a device to carry round with them giving an individual commentary on each exhibit area, often with a choice of languages.

There is now a wide choice of storage media for audio. Which to use is a choice made on cost-effectiveness grounds, with particular emphasis on minimizing running and maintenance costs. Ignoring, for the moment, the special requirements of the acoustic guide, the following choice of media exists:

☐ Reel-to-reel tape, in $\frac{1}{4}$-inch, $\frac{1}{2}$-inch and 1-inch widths, able to carry, typically, four, eight or sixteen parallel audio or control tracks.

☐ 16 mm movie film carrying either an optical or a mangetic sound track, or both.

☐ 35 mm movie film carrying either an optical sound track (mono or stereo) or up to four magnetic sound tracks; also 70 mm film able to carry up to six magnetic sound tracks.

☐ 35 mm magnetic film carried on a separate transport, able to carry up to eight sound tracks.

☐ $\frac{1}{4}$-inch lubricated magnetic tape, carried in an endless loop cartridge. Best known is the NAB cartridge, widely used in the broadcast industry.

☐ Compact cassette.

☐ Videocassette.

☐ Videodisc, both as a source for accompanying a video program, and just as a sound source.

☐ Audio compact disc, and computer variants such as CD-ROM.

☐ Hard magnetic disk, as used in computers.

☐ Recordable optical data disc.

☐ Solid-state audio store.

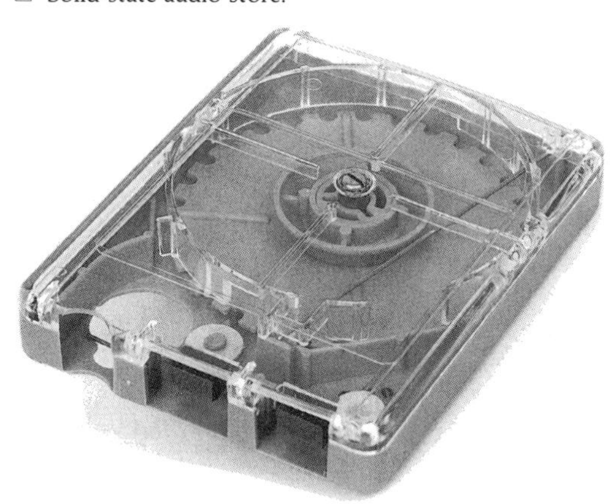

Figure 11.10 For many years the NAB tape cartridge was the standard method of sourcing continuous sound, but in all new installations it has been superseded by EPROM, videodisc or CD

Sound accompanying movie films in special venues like museums is usually carried in digital form on a medium separate from the film. This can be multi-track digital tape but is more usually CD-ROM or hard disc. The techniques are described in Chapter 17.

It is now the case that audiotape carried in standard tape cassettes and in endless cartridges should not be considered at all for exhibition and museum use. The rule is definitely 'no tape in permanent exhibitions'.

High-quality shows given in theaters or large exhibition areas may still use tape, especially multi-track digital tape, because of the ease of production and because modern multi-track recorders are reliable and represent excellent value. However, any system using tape requires regular maintenance, so it is preferable to use a more robust medium, for example, CD, which gives superb stereo sound, and it is now possible to have CDs made individually. They are especially suitable for exhibits requiring long sound tracks. It is possible to synchronize CDs to sound and light displays and mixed-media shows, using the techniques described in Chapter 16. Only CD players suitable for programmed remote control should be used in exhibition applications.

There can be practical problems in synchronizing a number of CDs to create a multi-track sound replay system. The way a CD works can result in an inconsistency in start-up, meaning that several CDs started together may run a fraction of a second apart. This does not matter for some background sound applications, but certainly does matter where precise synchronization is required. More expensive professional machines get round the problem.

This raises the point that 'consumer' CD mechanisms are only designed for a few thousand hours playing life. The choice for the museum designer looking for an exhibit life of ten years or more is:

- ☐ Choose cheap CD mechanisms, but warn his client that they will need replacing at intervals.
- ☐ Choose expensive CD mechanisms (but even these will need maintenance).
- ☐ use solid-state sound replay systems of CD quality (see later).

Another possible sound source is the standard videodisc. If a museum is already using a number of videodiscs for video, it can make sense to standardize on videodisc for both audio and video so only one type of equipment is used. Some professional videodisc players can be precisely synchronized, and some can also play CDs, again helping with equipment standardization.

The preferred sound source for museums and permanent exhibitions requiring sound tracks running continuously or on demand is the digital sound store. This is a maintenance-free device when correctly installed.

As explained in Chapter 23, sound can be stored in either analog or digital form. The fact that it can be stored digitally allows it to be stored in standard memory integrated circuits (chips). Until quite recently the memory capacity needed was considered enormous, and was certainly expensive. The demands of the computer industry have now ensured that memory cost is no longer an issue.

The memory used in sound stores is 'non-volatile', meaning that it does not get lost in the event of power failure. Until recently the most practical chip type was eraseable EPROM (electrically programmable read-only memory). Eraseable EPROMs hold their memory unless erased using ultraviolet light. For most museum applications, this arrangement is satisfactory, but it can be a little tedious to have to remove the chips in the event that a recording requires replacement.

EPROM has now been joined by 'flash-EPROM'. This type is non-volatile, but can be 'rewritten' (re-recorded in audio terms) *in situ*. It is now also available in a convenient 'credit-card' style packaging, the PCMCIA format now widely used in personal computers. This yields the convenience of a replaceable, re-recordable medium, with the reliability of no moving parts.

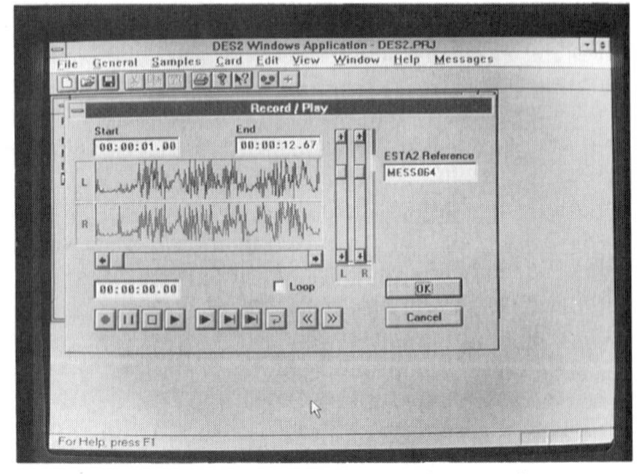

Figure 11.12 Transferring master audio recordings to EPROM or flash memory is easily done in a personal computer

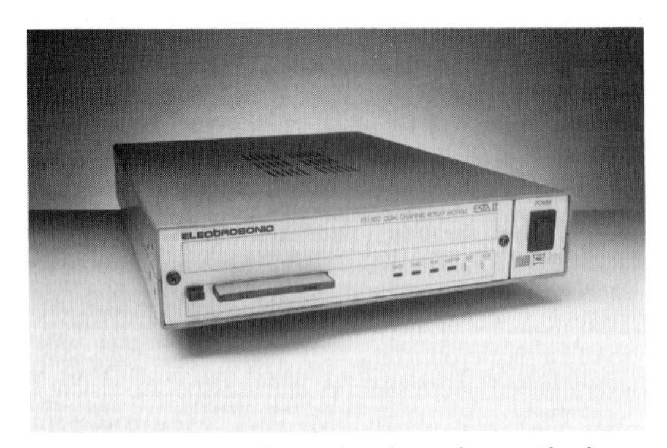

Figure 11.11 A solid-state digital sound store. The device illustrated uses PCMCIA 'flash' memory cards, and can play back up to 64 minutes of CD-quality stereo sound. It can also control simple lighting effects

A distinction must be made here between speech or sound synthesis and digital recording. Synthesis has a place in some special exhibits, but is of limited interest to most users. The sound store should use a true recording technique that does not color the sound in any way. Typically, master recordings are made conventionally, and then transferred to the EPROM or 'flash' chips using a personal computer fitted with digital audio facilities. If the soundstore uses PCMCIA cards, the process is particularly simple.

Digital sound stores are available with a wide range of specifications, and an equally wide range of price. For public exhibitions it is important to use equipment which gives good presentation quality, and this rules out the cheapest equipment which may be adequate for other applications, such as security message systems.

Digital sound stores are specified in the following way:

☐ By audio bandwidth. For example, 4 kHz, 6 kHz, 12 kHz, 16 kHz, 20 kHz.

☐ By dynamic range. For example, 60 dB.

☐ By duration, dependent on frequency response. For example, a unit may have a 60-second capacity at 20 kHz, or a 300-second capacity at 4 kHz.

☐ By the size of memory it uses. EPROM sound stores use multiple individual chips, typically eight or sixteen chips each of 2 or 4 megabit. Flash memory sound stores using PCMCIA cards can use cards in the range 1–64 megabyte (or more, when such cards become economically available).

☐ By the compression method used (if any).

☐ By control facilities.

Modern compression methods, discussed in Chapter 14, mean that it is now possible to store long programs of CD quality at reasonable cost. A typical sound store, using MPEG audio compression at 12:1, gives stereo audio indistinguishable to all but 'golden ears' from the original uncompressed audio at CD quality (20 kHz bandwidth). In this particular case it is easy to remember the memory capacity needed, each megabyte of memory holds one minute of stereo or two-channel sound.

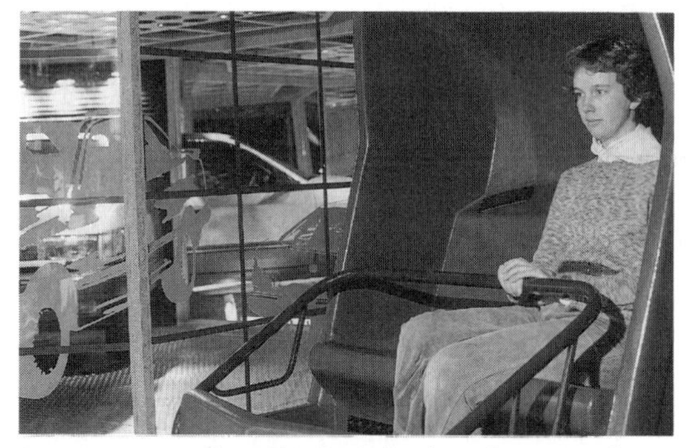

Figure 11.13 The 'Wheels' dark ride at the National Motor Museum, Beaulieu, England, uses a solid-state sound system. Each pod in the ride is fitted with its own commentary loudspeaker which keeps in exact synchronization with the exhibits being passed

If this proves too expensive, then the frequency response should be selected according to application. A 4 kHz response is sufficient for speech delivered through telephone handsets. Depending on the compression method used, the same megabyte of memory could hold between 5 and 20 minutes of mono audio.

Some digital sound stores are simple replay devices, only able to replay a single message or limited choice of messages. Those with microprocessor control may offer a number of additional facilities, for example:

☐ The ability to link a number of units together, to provide true multi-track replay.

☐ The ability to select a large number of individual messages or message segments, either by remote hard-wired control, or by remote serial control (RS232 or RS485, for example).

☐ The ability to follow an external source of timecode, for example, from a movie projector or show controller.

☐ The ability to operate relays at designated points within the message. For example, to light up displays. (More complex mixed media shows would normally use a separate show controller.)

☐ The inclusion of an override facility for emergency or service announcements. Especially useful in applications like dark rides.

Installation practice

One question that the museum or exhibition designer must answer at the beginning of a project is whether the AV source and control equipment is to be centrally sited, or distributed round the exhibition. In theory, installing each item (disc player, amplifier, etc) next to its exhibit results in a lower cost installation but, in practice, this arrangement usually ensures that the equipment is inaccessible and operates in an unsuitable environment.

It is strongly recommended that all video and audio sources and central control equipment is built-in to proper equipment racks, and that these racks are sited in cool, clean rooms. This greatly simplifies maintenance, and allows any equipment replacement to be quickly carried out without disturbing visitors. Any control or projection room area should be temperature-controlled so that, ideally, ambient temperature does not exceed 28°C (82°F). Videodisc players are not able to operate in temperatures of higher than 35°C (95°F).

An example of centralized equipment is seen in the Natural History and Life Sciences section of the Taiwan National Museum of Natural Sciences. This is a complex of fifteen exhibition galleries, and although it is a large-scale example, it does contain lessons applicable to museums of all sizes.

The designer was the late James Gardner, who had considerable experience of integrating AV into exhibit design. Out of a total of 2000 separate exhibits, 120 need AV support. Some technical and design decisions taken at the outset resulted in the following:

☐ no audio or videotape used anywhere in the museum;

Figure 11.14 'The Dinosaur's Breakfast' at the National Museum of Natural Science, Taiwan. The animated models discuss the merits of being carnivorous or vegetarian. All sound is carried in digital sound stores. The design was by the late James Gardner Studio

Figure 11.15 One of the galleries at the Taiwan National Museum of Natural Science. In this gallery it is good lighting and display that enhance the exhibits; AV has correctly been sited in separate study areas

- all video sourced by videodiscs;
- all show audio sourced by videodiscs;
- all small-scale sound effects sourced by EPROM-based digital sound stores;
- a central air-conditioned and dust-free control room to house the majority of the 90 videodisc players;
- use of slide projection to achieve a quite different 'look' to exhibits, to contrast with those using video;
- all slide projectors fitted with automatic lamp changers, and all installed in easily accessible housings;
- any interactive exhibits housed in separate 'pods' to prevent interference with main visitor flow;
- subjects requiring a full AV presentation having a proper small theater environment, big enough to take a visitor group.

The AV exhibits themselves range from a parrot that talks to visitors when approached (using a proximity detector to start the message), to a multi-image theater using nine slide projectors synchronized with two videodisc players. In the latter case the disc players are used as a high-quality, minimum-maintenance, multi-channel sound source.

This design philosophy keeps the number of different types of equipment to a minimum. For example, one standard videodisc controller is used that can be programmed to achieve any of the following:

- Independent control of four videodisc players, so that each player plays a specified disc segment on either a continuous run or demand start basis.
- Random access control of a videodisc player, to allow selection of any one of, say, twenty programs.
- Synchronized control of up to four videodisc players, allowing automatic presentation of multi-screen video; for example a three-screen video panorama.
- Computer control of up to three videodisc players in a multi-media or multi-image show.

James Gardner always paid great attention to the way in which visitors circulate within an exhibition space, by, for example, ensuring study exhibits do not impede flow. This thinking is applicable to all kinds of museum and visitors' center. When audio-visual shows are an integral part of an exhibit, it is essential to pay attention to their effect on visitor flow. Are *all* visitors expected to see a particular show? If not, there must be room for those who do not want to see the show to get round those who do. If yes, there must be sufficient viewing area to cope with the planned visitor flow.

For example, if shows are to be given in a small auditorium its size must be matched to visitor flow. In some cases this is related to the way in which visitors arrive. Thus, the busload may be the unit which determines auditorium size. How long do visitors wait to see a show? Anything longer than ten minutes demands that the waiting area has some exhibit features of its own.

Figure 11.16 'Sounds in Nature' is a multi-image program at the Taiwan National Museum of Natural Science. It is shown in a theater big enough for the circulation pattern. Sound is from laserdisc

These kinds of questions are important, not just for successful running of AV but also for successful operation of any kind of visitors' center.

The acoustic guide

Sound can enhance a museum or exhibition, but it can also be a nuisance if it leaks to where it is not wanted.

Where there is a major show, or where there is ambient sound, loudspeakers are used to deliver sound. Choice of which kind of loudspeaker is then made on normal cost/performance criteria. The problem arises when there are a number of exhibits close together where one may interfere with another.

Loudspeakers still are the preferred choice because they require no maintenance and no special action on the part of the visitor. Provided the exhibition is carpeted, it is surprising how close it is possible to have exhibits with their own loudspeaker commentary, especially if care is taken to point loudspeakers down on to the listeners, and to use strategically placed display screens to isolate sounds.

For especially difficult situations, parabolic loudspeakers can be used. These work rather like a car headlight, but direct sound instead of light into a beam. To be effective, the dimension of the parabola must be similar to the sound wavelength to be directed. This means that small loudspeakers of this type will only direct high frequencies, but this can be sufficient, since it is the higher speech frequencies that carry most of the information.

Big parabolic loudspeakers (up to 6 ft, 1.8 m

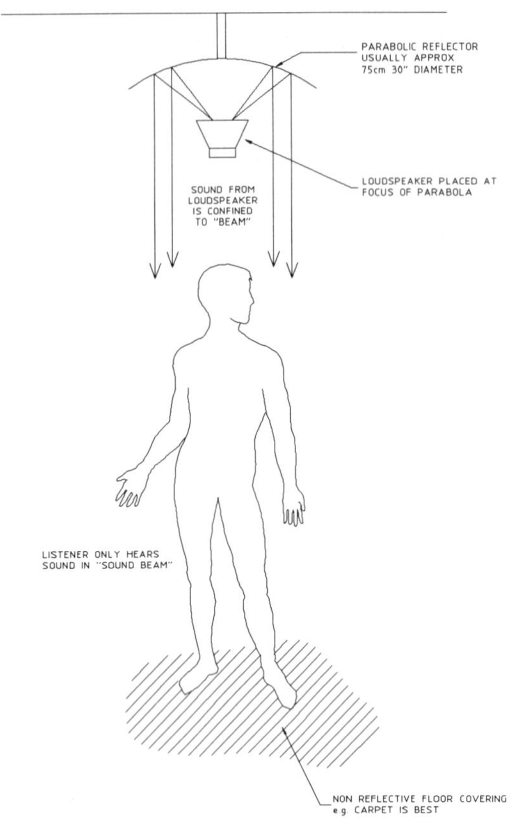

Figure 11.17 Parabolic directional loudspeakers can help confine sound to a small area next to an exhibit. The floor should still be carpeted to minimize sound spread by reflection

diameter or more) can solve sound direction problems in large open galleries, but their installation is something of an art, and their use is normally overseen by an acoustics consultant.

Ultimately the problem remains, either because the museum management does not want a lot of commentaries on loudspeakers (e.g. in a picture gallery) or because of the need to offer a choice of languages. The choice then becomes that of either having earphones attached to the exhibit, or of giving visitors their own personal means of hearing commentary.

One or more sets of earphones attached to the exhibit is technically the simplest, and also the cheapest, method. The 'phones are normally similar to a telephone handset, or of the lorgnette or stick variety. The obvious disadvantages are that: they require the exhibit designer to allocate exhibit space to accommodate them; the number of people listening to one exhibit commentary at one time is limited; and the 'phones, especially their cables, are subject to wear and tear.

Another possibility is to fit exhibits with an acoustic jack, and to issue each visitor with a pair of acoustic headphones of the type used on airliners. Such an approach is really only valid if a charge is made for the headphones, because it has no significant advantage over the attached 'phone method and requires extra work to issue and collect the headphones.

Figure 11.18 At the Science Museum in London, England, handsets are used for exhibit commentaries. These commentaries should never be longer than two minutes

Issue and collection is a problem of any system which requires visitors to take devices round with them. It is at its worst when visitors are issued with small cassette tape recorders, a system that is used in many museums. The major disadvantages are:

- □ Serious wear and tear on the equipment, especially on cables.
- □ Headphones are not convenient for all visitors.
- □ Visitors have to carefully follow operating instructions. If they do not, they get out of sync with the exhibits.
- □ It is not possible to run an audio commentary synchronized with an exhibit show (e.g. video).
- □ There may need to be a cash deposit system to ensure return of the recorder.

Nonetheless, tape-based systems are used with success at museum and visitor sites with specialist interest. They have the advantages of low capital cost and great flexibility in program and language variety. For some sites it may be best to franchise the provision of this service, as this relieves the museum of the need to fund it.

A system with fewer disadvantages is the wireless transmission system. Here each visitor is given a small receiver, and wherever they stand in the museum

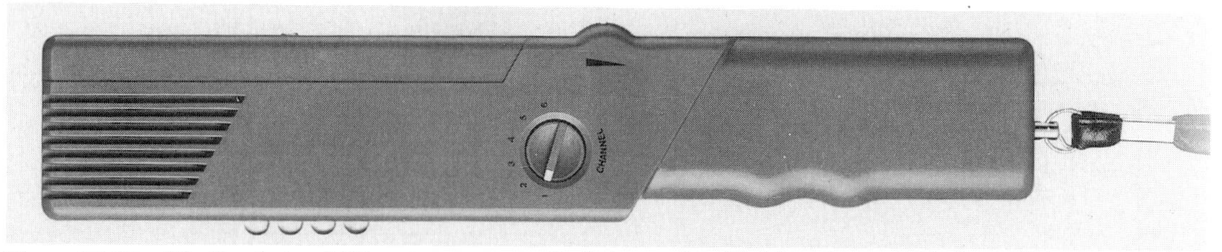

Figure 11.19 The Sennheiser MIA system uses a particularly neat hand-held receiver suitable for multi-language and multi-zone working

Figure 11.20 Complete infra-red headphones can give a sense of isolation, but if stereo music is required, as at Rock Circus in London, they are essential

Figure 11.21 The Sennheiser MIA receiver in use. The lack of cables makes it an easy item to issue and collect

they hear a commentary related to the exhibit nearest. The ideal attributes of this type of system are:

□ The receiver has minimum controls and no loose wires for earphones.

□ The receiver uses rechargeable batteries with sufficient power for all-day use and overnight recharge.

□ Sound quality can be made good enough for speech and even music, without background noise.

□ There is no crosstalk between adjacent zones.

□ Where there is no sound, the system cuts out cleanly on leaving a sound zone – without any whistles, crackles etc.

□ The system permits multi-language operation.

While some music-based installations require use of headphones to give stereo sound, these do give the visitor a sense of isolation. Therefore for most installations these requirements result in the acoustic stick being the best device to give the visitor. This can have clean lines, and is easy to use. The visitor simply holds it to his ear, and differing hairstyles should be no problem. Although it is possible to have volume and language selector controls on the stick, it may be better not to. Volume can be adjusted simply by how close the stick is held to the ear. Language selection can be achieved by having different colored receivers for each language or by having a secret, staff-operated, selection switch. Less there is for the visitor to play with, less the likelihood of user difficulty and complaint. When not in use sticks are stored in a combined storage and battery-charging rack.

Today the most practical method of wireless transmission for this application is infra-red. Combination of low-sensitivity receivers, and low-power infra-red transmitters allows very close spacing of sound zones. Modern receivers ensure a clean cut into silence between zones. These installations must be carefully planned, but they can be extremely effective. Transmitters are usually overhead, projecting a cone of infra-red radiation. Anyone holding a receiver with upward-facing infra-red detectors and standing within the cone hears the message. Multiple transmitters are installed to cover the required area.

Combination of a digital solid state sound store and an infra-red transmission system represents one ultimate in acoustic guidance systems for museums. There are, however, two technical limitations. First is the available infra-red spectrum; present technology limits each zone to a maximum of twelve speech quality channels or two high fidelity (20 kHz audio bandwidth) channels. Second is possible interference from other sources of infra-red. While an art gallery is

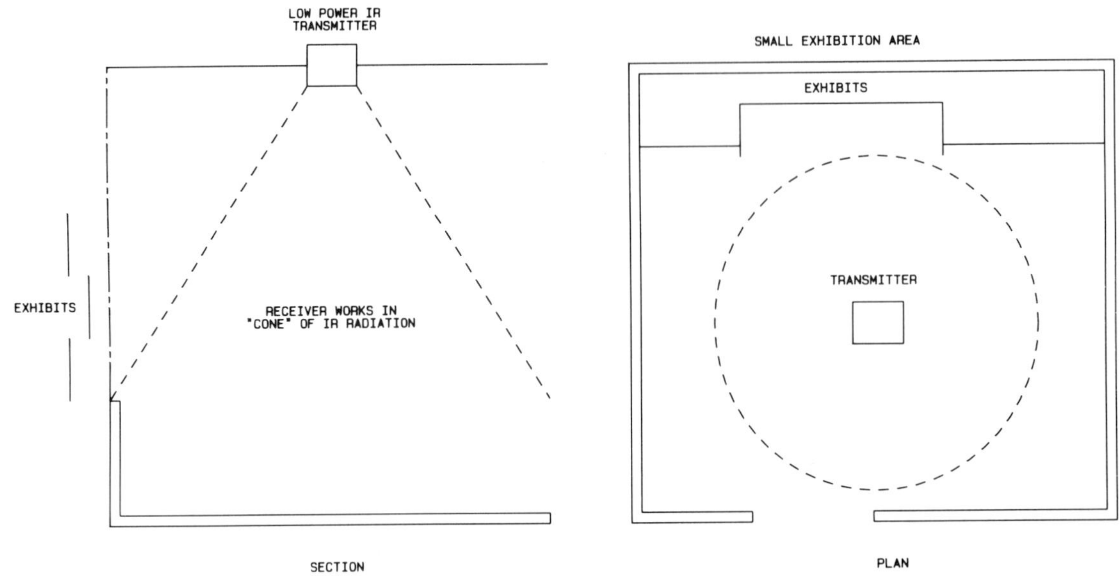

Figure 11.22 Infra-red wireless sound transmission has the advantage that it can be matched to small exhibit areas. This is because IR radiation behaves like light

Figure 11.23 The Sennheiser infra-red radiator. It is easy to install on its own or amongst lighting fittings

unlikely to suffer from these, a museum of pop music with flashing strobe lights might well. This results in a crackling noise heard in the receiver.

A new contender combines the virtues of the individual stick listening device with those of the digital sound store. It has the advantage of not needing any transmitter infrastructure. At least two specialist companies in the museum tour business, Sound Alive and Acoustiguide, have introduced listening devices based on RAM computer memory.

Listeners select messages by push button, like using

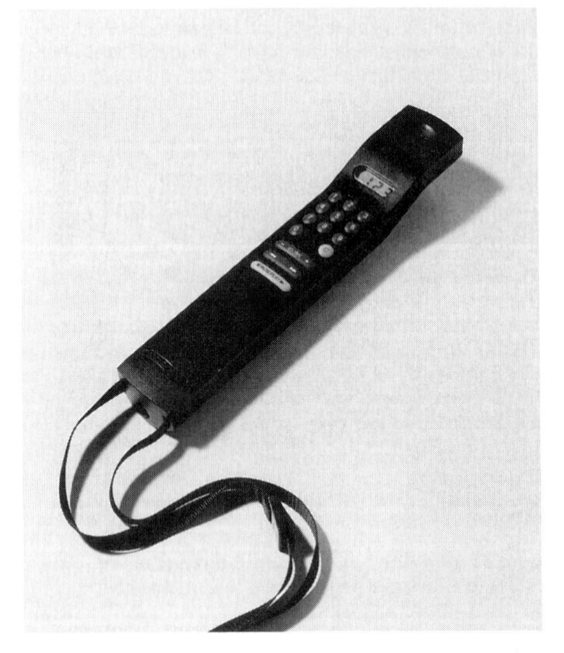

Figure 11.24 The Acoustiguide 'inform' carries sound commentaries in computer memory. Users can select any of 256 messages, and the unit can be 'reloaded' with alternative programs in a matter of seconds

a telephone. The required message numbers can be identified on the exhibits, pictures, etc., concerned.

Figure 11.25 The Micro Gallery at the National Gallery in London, England is one of the most successful applications of interactive displays to be seen anywhere. The $800,000 cost was covered by sponsorship from the American Express Foundation. Approximately $150,000 on equipment, and $500,000 on computer software and editorial input

Figure 11.26 The Micro Gallery uses high-resolution touch screens, and this off-screen photograph shows the 'classic' design of the screen layout. The color reproduction of the paintings is remarkable, it is based on 8-bit digitized images with a customized color look-up table (CLUT)

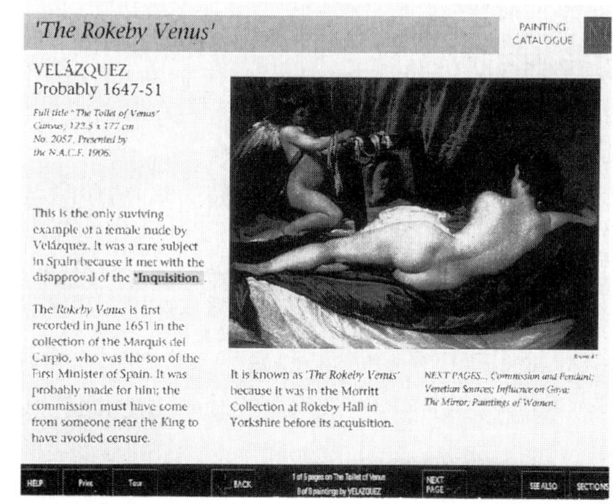

Figure 11.27 Another off-screen image from the Micro Gallery. The system uses Apple Macintosh computers and 1.6 gigabyte hard discs. This method of storage ensures that nearly all of the 4000 screen pages can be recovered in one second or less. Touching a word like '*Inquisition' in the above example can cause either a complete new screen or a 'pop-up' explanation to appear

The use of RAM memory means that messages can quickly be 'reloaded', covering the needs of multi-language operation and the use of a single inventory to cover more than one site. Memory capacity is of the order of one hour with 4 kHz speech quality bandwidth.

Archive and study systems

Some museums have identified a need for study areas, outside the main exhibit flow, where individuals or family groups can sit in a study carrel, and choose from a wide range of archival, program or interactive

Figure 11.28 The Wexler Learning Center at the National Holocaust Memorial Museum in Washington DC was one of the first museum study area systems to be based on a computer server and multiple touch-screen workstations, providing full motion video, audio, text and graphics. The top photo shows the study area with family-sized workstation booths. The bottom photo shows the remote video server and hard disc drives

material. The material is usually presented on a video monitor.

Anyone embarking on commissioning such a system should not underestimate the enormous amount of work needed to make it fully effective. They should first consider the following questions:

- ☐ What will the system do that a good library and librarian would not do as well or better?
- ☐ Who is going to run the system?
- ☐ Who is going to keep it up to date?
- ☐ What exactly are the benefits to visitors using it?

These questions are particularly pertinent as the age of 'multi-media publishing' dawns. There is a danger that a museum's offering may appear inadequate or amateur compared with already published material available, for example, on CD-ROM. It is certainly a mistake to try and emulate the latest computer game techniques, since this is both expensive to do, and will quickly look out of date. However, there are a number of positive points to be made:

- ☐ The material is more effective by being close to the museum displays. The environment is conducive to learning.
- ☐ For many people, not equipped with computers or inimical to them, the study area can be a revelation.
- ☐ There are archiving and presentation tasks which a museum can do, but which are inappropriate or commercially impossible on the open market.
- ☐ In an increasing number of cases, a museum or institution will be the originator of a database, which they can not only present in their own study areas, but which they can make available to other individuals and institutions by such means as published media (e.g CD-ROM) or on-line computer access.

The last point demonstrates how fast technology has moved. In the early 1990s, any museum study area system would have been based on laserdisc or computer hard disc on the grounds that this yielded affordable cost per station. At the time the integration of full motion video into the computer environment was an expensive, experimental and generally unhappy experience for all except those with an unhealthy amount of taxpayers' money.

Rapid developments in compressed video and in computer networking techniques, described in more detail in Chapters 14 and 20, now mean that the larger study area system is based on storing all material, whether it be photographs, text, audio, graphics or full motion video, in a digital form. The material is distributed by a computer 'server' to the necessary number of workstations, via a computer network.

With such systems it is possible, for example, to store hundreds or even thousands of hours of full motion video material, and allow almost instant access to any part of it to many users simultaneously. The tasks of organizing the material and providing easily intelligible access to it should not be underestimated.

There are now many different techniques available, and the choice may appear difficult; but provided a given technique works, is affordable, and provides a good visitor experience, then go for it. Museum curators should not feel it necessary that they pioneer technology, they should just take maximum advantage of whatever is available. So if 'set-top box' technology, paid for by the telephone and cable TV companies, yields a cost-effective method of presenting material which would otherwise be impossible to show, there is every reason to use it.

Chapter 12

The presentation room

A poor presentation environment creates frustration for audience and presenter alike. Previous chapters have reviewed use of AV in exhibitions and public venues; this chapter is devoted to use of AV in the commercial field, and how to eliminate problems of the presentation environment.

Frequently a commercial presentation must be given either in your own offices, or in those of a customer. If it is an individual presentation, it may literally have to be given in an office, but this should be avoided if possible. The very least required is a meeting room.

The serious presenter always:

☐ Has someone visit the room beforehand to check black-out, screen arrangements, seating, power points, etc.

☐ Arrives early to ensure a smooth set-up.

☐ Has a quick technical rehearsal before the audience arrives.

Any organization that takes presentations seriously soon recognizes the need for a *presentation room.* Ideally this is a room used solely for presentations but obviously, in a smaller company, it may also do duty as a boardroom or general meeting-room.

The presentation room creates the right atmosphere for effective communication. Correctly designed, it makes the presenter's job easier and the audience more receptive. The major benefit of the presentation room is *effective communication* with, for example:

☐ *customers,* resulting in more or sustained sales;

☐ *employees,* resulting in a more effective and better informed workforce;

☐ *suppliers,* resulting in better terms of supply and better service;

☐ *press,* resulting in better public relations and accurate reports of business;

☐ *public* (if applicable), resulting in a well-disposed public and hence customers.

Presentation rooms can also be used to greatly improve efficiency of management meetings. If meeting participants are introduced to the discipline of preparing all their material in a form that can be shown to a group in a presentation room, the other participants' time is saved, and they all work from the same data. Often the only reason that this approach does not work is reluctance of participants to make the effort to prepare material in the correct form. Nowadays there is really no excuse for this, because many visual aids, such as overhead transparencies, computer text displays and slides, can be prepared almost instantaneously – so provided the meeting is planned, the support material can be there too.

To achieve the intended benefits of the presentation room, there are several aspects of the room that need special attention. Usually the room is in a normal office space with restrictions on ceiling height etc. However, even with such restrictions, it is quite possible to create an acceptable presentation environment. The following topics must be considered.

Seating

This must be comfortable and have good sightlines to screens. Usually it is theater-style with movable chairs, but it may be necessary to allow for schoolroom-style as an alternative if the room is also used for training, or boardroom-style if the room is also used for management meetings. The room should be big enough for the largest regular audience expected, but not too big as it must feel comfortable. Some possible layouts are shown in Figure 12.1. Useful guidelines include:

☐ A room about 7.5 m by 9.6 m (25 ft by 32 ft) accommodates up to 45 people theater-style or 24 people schoolroom-style.

☐ A room about 6 m by 7.5 m (20 ft by 25 ft) accommodates up to 30 people theater-style or 15 people schoolroom-style.

☐ For audiences up to 25 people the room should be at least 6 m (20 ft) wide.

☐ For large audiences and in auditoria the practical minima for theater-style seating are 90 cm (3 ft) between rows, 54 cm (21 in) seat width and 106 cm (3 ft 6 in) aisle width.

☐ The audience should not be outside a 60-degree viewing

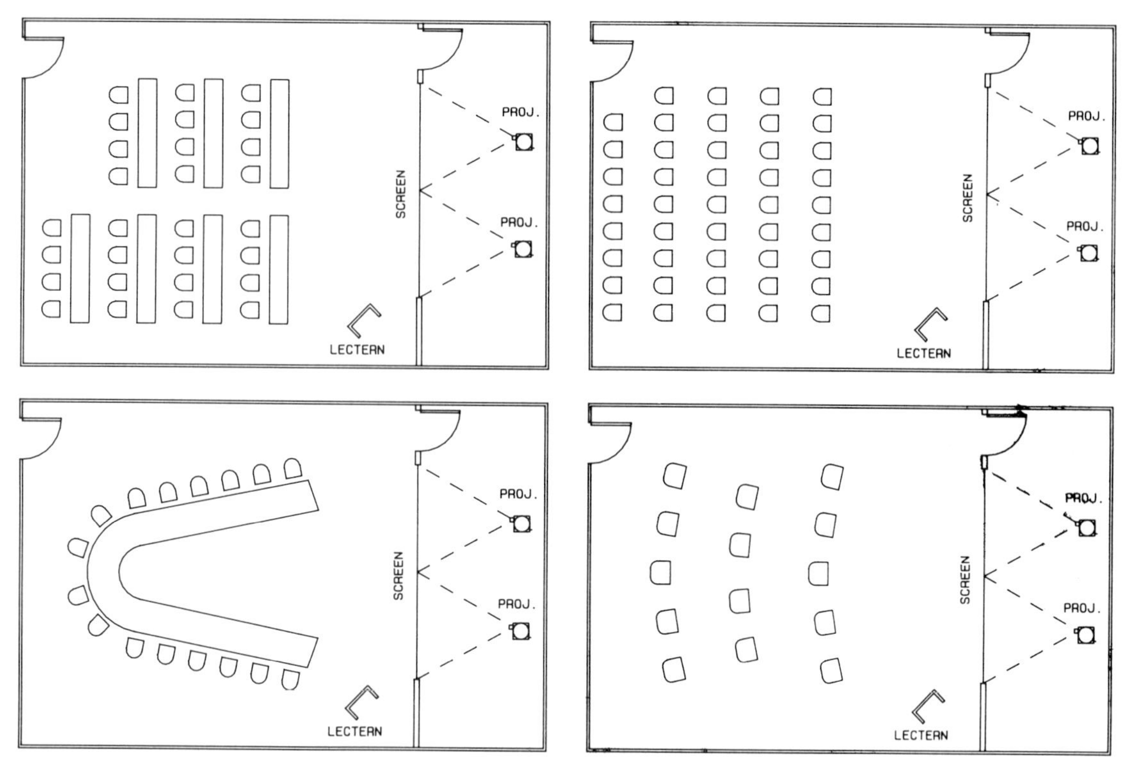

Figure 12.1 Four ways of using the same space. The nature of the intended meetings and presentations will determine whether schoolroom, theater, boardroom or island seating is appropriate

Figure 12.2 Organizations giving regular presentations to large groups need a theater-style layout. (Photo of their presentation theater courtesy Statoil of Norway. Photo from Audio Grafisk)

area (i.e. 30 degrees either side of center line). This may mean shorter seat rows at the front.

The room must have good air circulation. If it is being built into a new building, consulting engineers take into account the room's occupancy and ensure that air conditioning or air handling equipment is correct for the task. However, if an area in an existing building is converted – which previously was an office with only a few people in it – it is worth reviewing ventilation. Otherwise the presenter may find herself or himself talking to 25-sleepy people.

Lighting

The room must black-out properly. For this reason presentation rooms are ideal for using up space in the building core or basement! All lighting must be under dimmer control, and dimmers should be push-button automatic, *not* rotary-knob or slider, because they usually need to be operated from more than one place. The automatic dimmer ensures a smooth fade at all times.

In presentation rooms that are multi-purpose, or which use several different media, the lighting should be multi-scene. Usually this type of room has several different light sources, e.g. downlighters, fluorescent, separate lights for lighting the presenter, and display lights. It is confusing to have a separate set of controls for each, and much better to define the various ways in which the room can be used and allocate one 'scene' for each use.

For example, there might be a scene for meetings, when a boardroom table is in use and no visual aids are required. This would be full lighting. Another scene might be for a lecture. In this case main room lights would be dimmed, but still bright enough to take notes by, a lectern spotlight is on, and all other lighting adjusted to ensure that no unwanted light fell on the screen. Other scenes could relate to video, product display, presentations and so on. The important thing is that there is only *one button* for each scene.

Figure 12.3 Some presentation rooms must provide writing facilities for the participants. (Photo of their East Finchley training theater courtesy McDonald's)

Figure 12.4 When AV is used to support meetings, the boardroom layout is appropriate. Notice the twin screens for comparisons. (Photo Multivision Electrosonic Ltd, Toronto)

Figure 12.5 Prestige selling presentations to selected corporate customers demand island seating. (Photo of their Toronto presentation room courtesy Bell Canada)

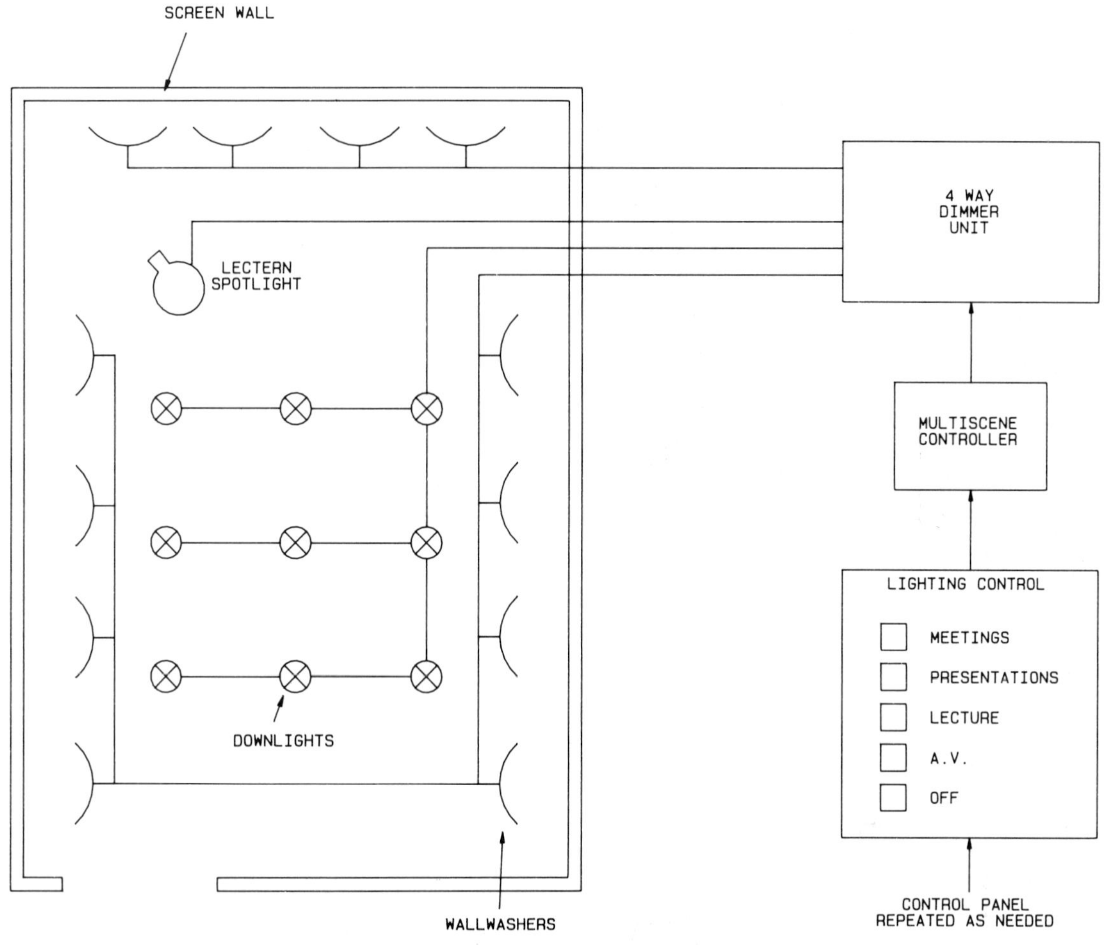

Figure 12.6 Block diagram of a multi-scene lighting control system for a small presentation room

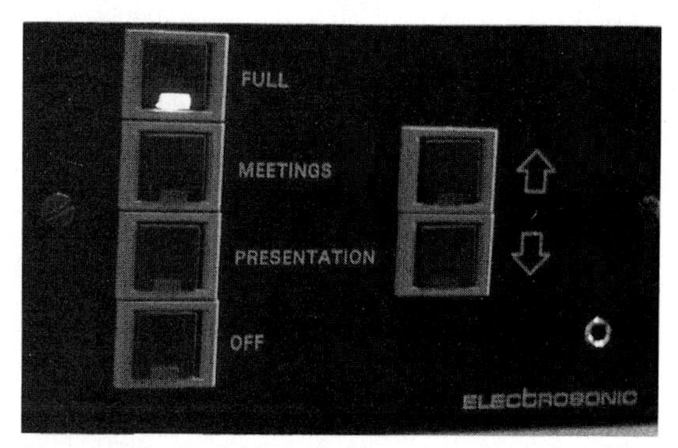

Figure 12.7 Push-button lighting control is essential for presentation rooms. It allows consistent results to be obtained from several different control points

Lighting control is normally undertaken from the lectern, from controls at entrance doors and maybe from some central control panel. If there is a separate projection room it must be possible to control lighting from there too.

In presentation rooms also used for discussion meetings there may well be daylight present. The control of the daylight, by blinds, etc., is just as much part of the lighting 'scene-setting' as any other part of the lighting control. Formal sales presentations of limited duration benefit from tightly controlled lighting, but long-running discussion meetings are better held in high ambient light. Modern high-gain rear projection screens permit the use of video and graphics projection in high ambient light.

Audio

AV equipment often has its own sound system. In small presentation rooms it can be best to keep each

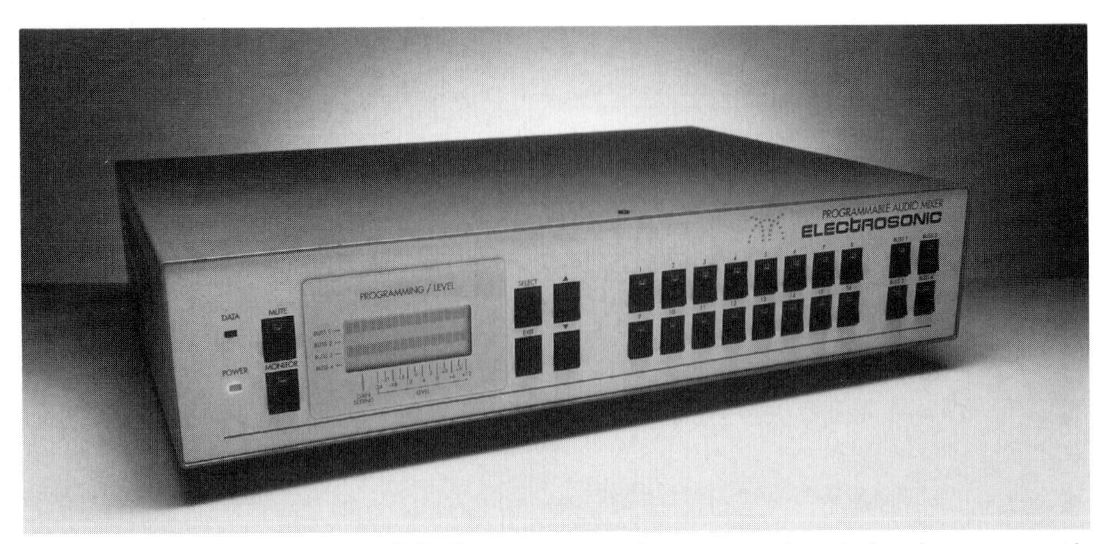

Figure 12.8 This microprocessor controlled audio mixer stores equalization settings for multiple audio sources, provides 'soft' switching between them and provides a remote volume control facility. It can also be configured as a multiple microphone controller with automatic mixing without the need for an operator

item separate with its own loudspeaker. If there are many AV sound sources or it is a big room, better results are obtained by a specialist installation to route the various sources through a common amplifier/loudspeaker system. Each source can then have pre-set equalization and level adjustments that optimize the program material to the room.

The program sound must be completely separate from any speech reinforcement. Because of the high absorption of sound by people and furnishings, even quite small presentation rooms need speech reinforcement to reduce listener fatigue. This can be done simply, either by using ceiling loudspeakers, or by a lectern amplifier.

One of the main user problems in presentation rooms can be complexity of the audio systems. It can be the case that a consultant or supplier, in good faith, recommends an audio system of highest quality which the end-user has difficulty operating because of its complexity. Very few users of presentation rooms can afford to have an operator present whenever the room is in use, so the ideal is clearly to have a system that operates itself.

This is now feasible. A typical prestige presentation room is now fitted with a speech reinforcement system with as many microphones as required, arranged so that microphones are automatically controlled, i.e. if no-one is speaking into them, they automatically switch off. The system also ensures optimum microphone gain, and never goes into feedback (howlround).

The program sound system is usually a high-quality stereo system playing through studio or auditorium-

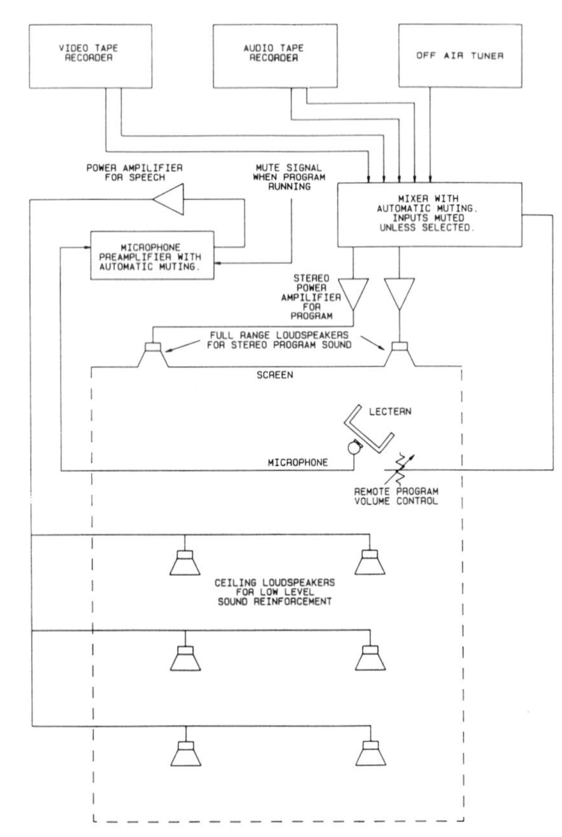

Figure 12.9 Simplified block diagram of an audio system that is suitable for presentation rooms

type loudspeakers placed correctly in relation to the screen. This system also has an automatic mixer which is muted except when playing show material. It might well have to deal with several sources, such as tape recorder, multi-image sound, video sound and so on. It also ensures that the speech reinforcement system is off while shows are run.

A feature usually required is remote-control of show sound volume. This is best done with an automatic push-button control, as this kind of system resets to a preset volume for the next show, thus avoiding the problem that a presenter might spoil a succeeding show by mistake.

The aim of good audio design in a presentation room is to make the user unaware of any technical complexity.

Lectern

All presentation rooms intended for formal presentations have to be equipped with a lectern. Many standard lecterns are now available although, often, a user prefers to have one designed by his architect or interior designer to match the room. The lectern should give the presenter confidence, and act as a focus for the audience.

The lectern should include the following items:

- ☐ A reading light. Ideally, this should be on the same dimmer circuit as the lectern spotlight, to ensure that it goes out whenever an AV show is run. Otherwise it may throw unwanted light on the screen.
- ☐ Basic room lighting controls (i.e. the main 'scene' buttons).
- ☐ Simple remote-control facilities for slides and for starting AV shows.
- ☐ A microphone.

The lectern can include many more items. The general rule is that if the lectern is going to be used by many different presenters, it should be kept as simple as possible. On the other hand if only one or two presenters are going to use it on a regular basis, they may appreciate more extensive facilities. Some of the many possibilities are:

- ☐ clock and/or digital timer;
- ☐ motorized height adjustment;
- ☐ control of motorized curtains and other display devices;
- ☐ video prompting system;
- ☐ video monitor;
- ☐ video 'writing' facility;
- ☐ random-access slide selection;
- ☐ random-access video selection;
- ☐ push-button control of audio volume.

Figure 12.10 The Lectrum™ is an example of a standard lectern. For most presentation rooms it is advisable to keep the number of controls on the lectern to a minimum

Many of these are now discussed in more detail below.

Screens

Good presentation room design results in the audience being unaware of the technical equipment. They should be interested in the message, not its means of communication. Therefore, if at all possible, there should be a separate projection room.

If many different formats are used, back projection may be best because:

- ☐ It allows projection to be used in relatively high ambient light conditions. This is essential for meetings.
- ☐ It is easier to mask, to ensure that the screen is the right size for the medium being used.
- ☐ It allows a much bigger image where there are low ceilings.
- ☐ The audience cannot see any projection equipment.

In principle, a back-projection room should take up no more space than a front-projection room. Some other points to consider include:

- ☐ Whether using back or front projection, some form of curtains or cover panels should cover the screen(s) when not in use. Preferably, but not necessarily, motorized. Blank screens must be avoided.
- ☐ If using back projection, special lenses are available to limit projection distance and eliminate keystone distortion.
- ☐ Small presentation rooms can use a projection wall with built-in back projection AV devices to further reduce space. Alternatively, front projection equipment can be mounted in a suitable wall cupboard with projection ports.

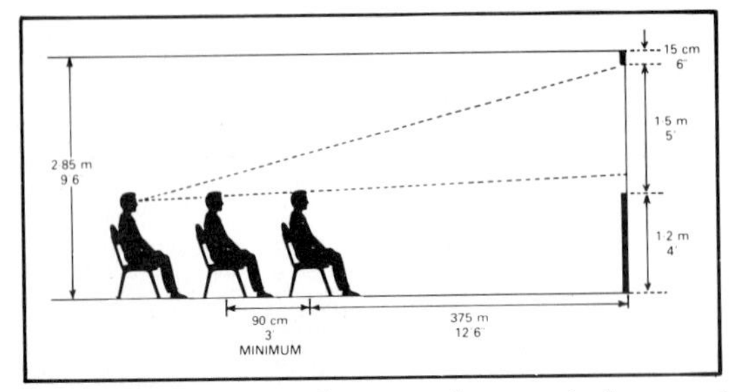

Figure 12.11 Even with staggered seating the bottom of the screen needs to be 1.2 m (4 ft) above floor level to avoid unacceptable head interference

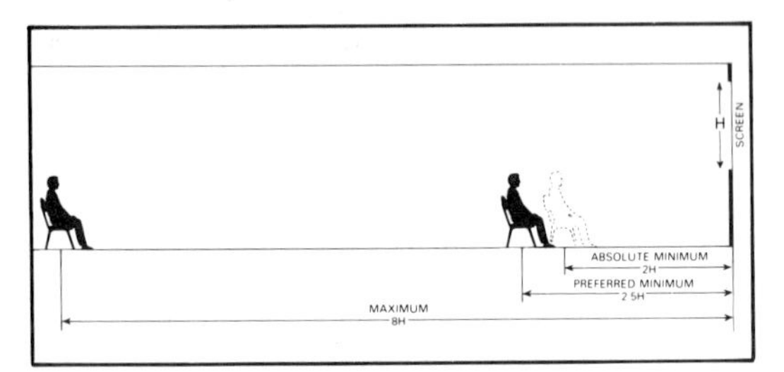

Figure 12.12 The audience should not be further away than eight times the picture height

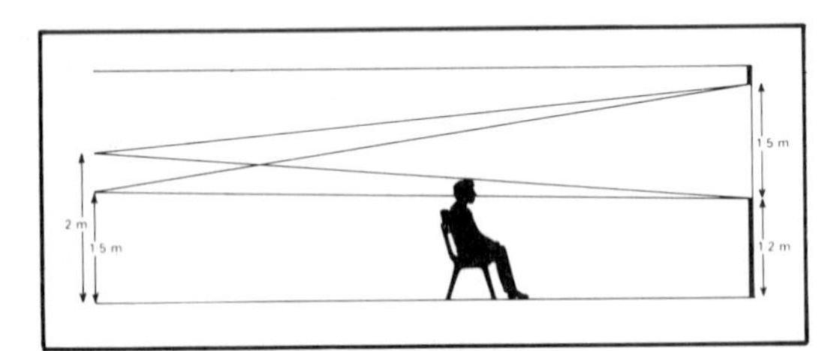

Figure 12.13 When front projecting, the projectors must be well up. The convenient operating height of 1.5 m (5 ft) is too low to avoid head interference. Back projection is often a preferred arrangement in presentation rooms to avoid this kind of problem

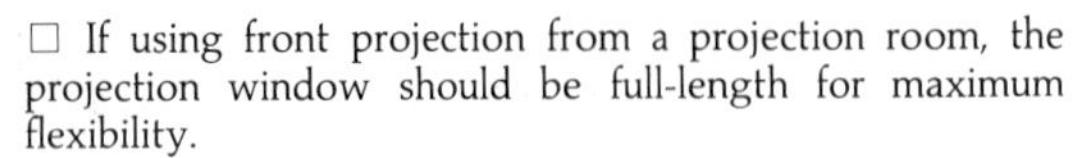

☐ If using front projection from a projection room, the projection window should be full-length for maximum flexibility.

☐ If only video/data front projection is being used, then there may be no need for a projection room because the projector can be ceiling mounted. However, this will mean that projection can only be used in low ambient light.

The question of screen format naturally depends on the way in which the room is used. In practice, two screen areas side-by-side are a popular arrangement because it allows comparison of images – for example, two slide images, or one slide image and one video-projected image.

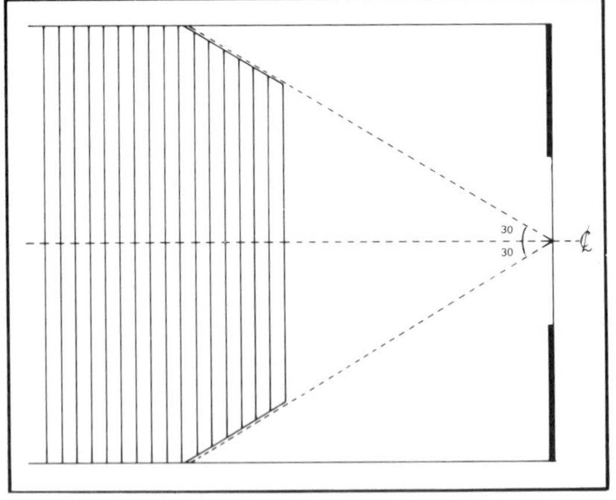

Figure 12.14 The audience should be within 30 degrees of the screen axis. In temporary installations with matt screens 45 degrees is possible, but not recommended

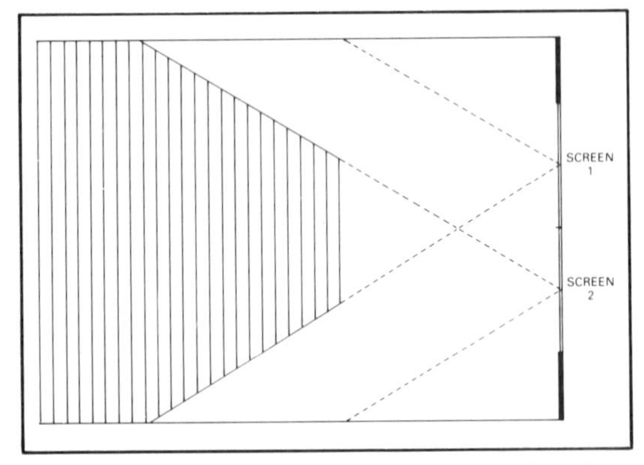

Figure 12.15 The 30-degree rule means that when there are two screens side by side the optimum viewing area is reduced

Audio-visual equipment

Besides lighting control, audio and screen equipment a presentation room normally includes at least:

☐ A single-slide projection system for visual aids work.

☐ A means of showing video programs and computer data preferably projected, otherwise via monitor(s).

The problem is where to go from here. Provided the room has been designed with flexibility in mind there is no need to over-equip it at the outset, as extra items can be added when a proven need develops. The following observations give some guidance as to the possibilities, and recommendations as to the place

Figure 12.16 The principle of the back-projection OHP. (Drawing courtesy Audio-Visual AG of Switzerland)

within a presentation room of specific devices. Properties of individual subsystems (e.g. video presentation systems) are dealt with in more detail in other chapters.

Overhead projector

One way of distinguishing a presentation room from a training room is that the training room is always equipped with an overhead projector, whereas a presentation room should never have one. The aim should be that all presentations are given to a uniformly high standard. This, in turn means that all presentation material is back projected from slides or video.

In practice it may not be feasible to enforce such a rule. Notwithstanding the fact that slides can be made instantly, there is always the board member who insists on bringing in his own hand-crafted OHP transparencies. However, introduction of the OHP projector can cause an unfortunate degree of clutter in the presentation room design. If the use of the room can be defined so the OHP is not needed, so much the better. If occasional use is unavoidable, it is important that a suitable screen arrangement is incorporated in the initial room design.

One very expensive possibility is to use a back-projection OHP. This device allows the presenter to face the audience and use the OHP in a conventional way. The projector itself is fitted with a very long focal length lens, directed at an opening to the side of the screen. Large mirrors are then used to get the image on to the back-projection screen. This

Figure 12.17 The AV Visumaster-Retro gives bright back-projected OHP images in high ambient light

arrangement is not only expensive in itself, but also makes installation of slide and video projection equipment more difficult. Nonetheless it does allow a very high quality of presentation.

A more realistic alternative for many users is the video overhead. In principle, this is nothing more than a small video camera aimed at the document concerned; the resultant image presented by a video projector. In practice, the results are variable. In particular, it is difficult to get good and consistent results with actual OHP transparencies which simply reinforces the point that it is better to prepare (or convert) the material into slides. However, if a properly designed top-lit system is used, quite acceptable results can be obtained from hard copy and even three-dimensional objects.

Figure 12.18 The Wolf Visualiser™ allows documents, objects and OHP transparencies to be presented on video systems

A sophisticated example of the video overhead is the Wolf Visualiser™. This uses an elegant system of illumination in which the video camera and object illumination share optical path. A rectangle of light is presented on the working platform, and the user only has to position text or object to be shown within the rectangle. Illumination can be zoomed over a wide range, so can be adjusted to match the object. The camera and illumination projection lenses are electrically coupled, so the object can always be made to exactly fill the screen. The illumination system can also be used for presentation of single slides.

The quality of camera used in video overheads is important. For example, the Wolf Visualiser is available with either a single-chip CCD camera, or with a 3-CCD camera. The 3-CCD camera gives a significantly better performance.

Slide projection

Slides form the basis of the best quality visual aids used in a presentation room. For many users, a single-slide projector is sufficient, because this arrangement is the easiest to use and to prepare for. However, it is really much better to use a 'dissolve pair' of projectors, because this looks more professional and properly exploits the build-up and sequence ability of slides. It should be no more difficult for the presenter to use, but does require use of two slide trays.

All slides should be glass-mounted in the same type of slide mount. Autofocus systems and remote-focus control should be avoided.

However, there are always some users unable to enforce this kind of discipline, and suffer from visiting presenters who arrive with a scruffy pile of card-mounted slides. If this is likely to be the case, remote-focus control should be designed in from the outset as it is difficult to add later.

A popular format is the use of two slide screens side-by-side because this allows comparisons. In this case care should be taken to keep the presenter's remote-control as simple as possible. It may also be necessary for the slide projection system be able to run two or three projector slide/sound shows, but unless these are permanently installed it is best to have an operator on hand to assist the presenter. This also applies if complex multi-image speaker support sequences are required, when the procedures become those of the sales conference.

An additional refinement is the use of random-access projectors. These allow push-button selection of any slide with an access time of about three seconds. In practice, use of these devices in presentation rooms is limited as most presentations are linear. However, they are useful either in the meeting situation where a presenter is asked to reprise a slide ('Let's see last year's figures again, Jim') or in the sales situation, where the course that a presentation takes is dependent on audience reaction.

The house show

Some presentation rooms accommodate the house show. If an organization receives visitors on a regular basis, the running of an audio-visual presentation both saves time and gives the best introduction to the organization. Unless there is a separate visitors' center, the obvious place to run the house show is in the presentation room.

Because this type of show is unique to the organization, it may well be made as a multi-image show because this medium is the most flexible for updating and gives the best presentation quality. Even if it has been

Figure 12.19 Side-by-side slide projection at the NCB Bank in Amsterdam. Notice that this room also uses a conference microphone system. (Photo courtesy NCB Bank)

transferred to video for portable use, the show should be run in its original form in the presentation room. In order to ensure that it is easy to run, it should be installed on an auto-present basis (see Chapter 11).

Similarly, if the house show is a video program, it is well worth ensuring the best presentation quality. This can be done by running it from CRV disc, giving both broadcast quality and very precise show starting (no waiting for the tape to plough through several seconds of mush). When a big screen is used, the line structure of the TV image can be eliminated by the use of an up-converter (see Chapter 19).

Computer graphics and computer display

Nearly all presentation and meeting facilities now require that it is easy to display computer data. The problem is that this can become very expensive if every possible format must be allowed for. The technical aspects of this subject are covered in more detail in Chapter 20.

If audiences are sitting at reasonable viewing distances, and if the graphic information has been prepared to be legible, there is no need for 'super high resolution' when only a single screen is being used. Video projectors with VGA resolution, suitable for both video programs and standard computer data, are reasonably priced, so for many users it is a sensible restriction to confine the projection system to this capability. However, it is essential to check with the prospective users because if, for example, map displays or product design graphics are required, there is no choice but to use the more expensive 'graphics' projector.

Because video projectors must be set up to match particular sources, it is best to have a permanently installed computer (or computer-like device, for example, VideoShow™) for displaying computer-generated material. Users of the facility can then be instructed to prepare their material, using a program with a replay version that is resident in the presentation room computer. When this is done, the showing of a sequence of computer 'slides' is no more difficult than showing conventional slides, indeed for many it is easier.

For meetings there may be a requirement for participants to be able to 'plug-in' their own lap-top computers. Provided this is done on a regular basis, so that the system can be configured for the particular machines

Figure 12.20 Presentation direct from computer is possible using a data projector. (Photo Sony UK)

to be used, there is no problem in principle provided the appropriate interface connection has been installed. However, participants must be advised on the limitations, so they only show material suitable for group viewing. Most spreadsheets are quite unsuitable.

Room control

What starts out as a simple concept, a room to give presentations in, can develop into something of a monster. The apparently simple requirement for the audience to be able to see and hear everything properly; for the presenter to be able to show a series of visuals; and for it to be possible, for example, to be able to run a movie, show a videotape, see a program on TV or display a teletext page can result in a control panel that looks like the cockpit of a 747.

The room itself should be designed so that the audience are not aware of specific technologies, only of the information that is being communicated. However, an audience can be all too well aware, especially if, for example, presenters have to juggle a lectern control panel, a slide projector control, and three different infra-red controls for different parts of the video system. To eliminate this problem, the correct procedure is to have a room control system that minimizes the number of controls presented to the user.

An example that was mentioned earlier is the idea of scenes for lighting. In a larger or more sophisticated room the principle can be extended to cover operation of the entire room. Modern presentation rooms use room controllers that are programmed to meet the needs of the individual user.

The principle is that each device in the room is able to operate independently (this is essential to allow for maintenance and the, happily remote, possibility of total system failure) but is also linked to the room controller. This is a microprocessor device that can be expanded to work with any foreseeable combination of equipment. The devices could be:

- ☐ lighting dimmers;
- ☐ motorized panels;
- ☐ blackout blinds.
- ☐ curtains;
- ☐ turntables;
- ☐ video projectors;
- ☐ video monitors;
- ☐ videotape and disc players;
- ☐ display computers;
- ☐ slide projectors;
- ☐ movie projectors;
- ☐ anything that can be electrically controlled.

The running of a videotape requires the following sequence:

- ☐ Switch on the video projector and videotape player (if not already on).
- ☐ Switch off any competing device.
- ☐ Close blackout blinds.
- ☐ Fade lighting to video scene.
- ☐ Open screen curtains.
- ☐ Start the videotape.

Figure 12.21 The Electrosonic MRC is a room controller for small presentation and training rooms. It combines the functions of audio and video switching, video source control, control of lighting etc. in a single compact unit

It is obviously much easier for the presenter if he only has to press one button labelled 'start videotape', than it is to remember all the other buttons to be pressed. This is the facility that the room controller can provide.

This kind of approach makes it more likely that the room is used properly. It greatly reduces the number of user controls, and makes use of the room less intimidating. It can include all the necessary logic to ensure that everything is in the correct state for a particular part of a presentation. Provided the system is of the programmable kind, it allows changes in the way the room works if, in practice, it is found that different timings or level settings are needed.

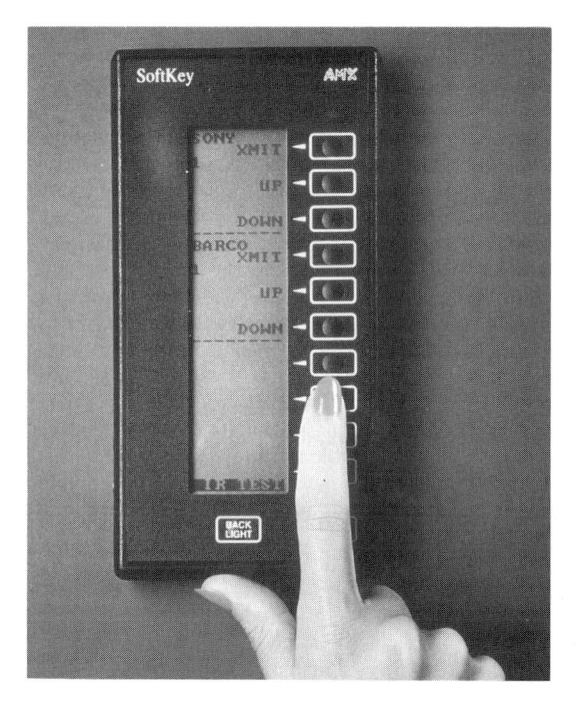

Figure 12.22 A programmable cordless remote control. Each button can be allocated to a different room function, and the liquid crystal display can show what it is

A typical room control system is delivered as a compact rack-mounting unit for permanent installation. Programming is done either on the unit itself or on a personal computer; in this way different logic rules, settings and timings can be tried out. If computer programming is used it may be possible to model the room on the computer, so it is often possible for the user to spot snags before the main system is built. Once programming is complete, the program is transferred to the room controller in the form of non-volatile EPROM memory.

When planning programming of a room controller it is important to include not only the main devices to be controlled, such as slide projector, VHS videotape and off-air TV, but also the idea of 'room reset'. There must be an easy way to restore the room to its normal condition at any time, and the easiest way to do this is to have a button marked 'reset'. In the case of rooms with blackout blinds it is best to have two different kinds of reset. A full reset that calls up a lighting scene that includes opening the blinds, and a presentation reset that calls up a scene where blinds remain closed. It can be very distracting if, in a presentation or meeting that is switching between different AV media, or is only partially supported by AV, blinds keep opening and closing.

Remote control

An incidental advantage of the room controller concept is that it can greatly simplify control wiring. Typically, a lectern control panel of any complexity needs only four thin wires. This allows the room to be more flexibly laid out, with possible control points spread round the room.

Control panels can take many forms, and users should decide which suits them best, both from design and practical use points of view. A simple panel with a few large, clearly marked, buttons may be all that is necessary. Alternatively, use of a 'soft' control panel may be preferred.

A typical soft or software-driven control panel uses a liquid crystal display with a touch screen overlay. The liquid crystal display can be formatted to present several different 'screens', each one of which may either represent control of a particular device, or may be tailored to the specific needs of an individual user. Touching the appropriate area of the display causes

Figure 12.23 This is the Texaco Executive Boardroom in Houston, Texas. Notice the 'soft' AMX liquid crystal touch screen control panel in the foreground. (Photo from AMX which makes control systems for rooms of all sizes)

Figure 12.24 An infra-red hand control (foreground) is used to control all the AV facilities in the meeting/negotiation room of the Outukumpu Company in Espoo, Finland. (Photo AV Systems of Finland)

Figure 12.25 When the AV facilities are not needed, it is important that they are neatly concealed, as here at the Outukumpu meeting room. (Photo AV Systems of Finland)

the specified device to operate, or selects another screen page.

Wireless controls

A further possibility, not confined to rooms with a room controller, is to use infra-red or radio wireless remote-controls. These are similar to TV remote-controls, but have a much longer range and are made in forms specially suited to presentation use. Typical characteristics are:

- A range of 30 m (100 ft).
- They are specially engraved to match the requirements of the particular installation.
- Hand control versions control about fifteen different functions.
- Desktop or lap controllers control 30 or 45 functions.
- When used in conjunction with a room controller the number of functions can often be kept down to 15 or less.

In some installations it can be a good idea to have more than one kind of wireless controller. One multi-function controller can be operated by regular users of the room to achieve any required condition. A second controller might be fitted with only the controls for a slide projector, and be used by a visiting presenter.

Control, decision and high-level presentation rooms

Computer graphics technology has led to the introduction of a new kind of 'control and decision room'. The techniques used in such spaces are becoming of direct relevance to high-level presentation rooms. Indeed in some cases the decision room and high-level presentation room will become the same thing.

The management of utilities, telecommunications networks and transport undertakings is now facilitated by networked computers. People operating or observing the system under control sit at workstations with high-resolution screens. In many cases, there can then be a requirement for a large high-resolution display which can show 'the big picture' to all present; the operators who need to have an idea of what is happening beyond their own screen, senior management and visitors.

A large-scale example is the control room of the Texas Department of Transport in San Antonio, Texas. This is ultimately intended to control a large part of the Texas Highway System – monitoring traffic flows, adjusting signalling, and directing police and emergency services.

Facing the ranks of operators is a big high-resolution display. This can be used to show multiple video images (or a small number of big images) from the hundreds of surveillance cameras which will cover the highway. In addition, it can show the output of a number of workstation screens, carrying mapping and signal status information, at high resolution.

The large room is complete with a glass-walled visitors gallery. The entire display system can be used to give visitors a clear idea of what is happening, both under normal and emergency operating conditions.

The significant point about such a system is NOT that the business presentation room may suddenly

Figure 12.26 Silicon Graphics' 'Reality Centre' in the UK is an 'ultimate' presentation room, ready to run standard presentations or try out new concepts. Resolution on the compound curved screen is 3840 × 1024 and images are generated entirely from computer. The concept is of direct relevance to the late twentieth-century business presentation room

require a huge number of video projectors (although some control rooms certainly will) but that it is now possible to integrate many sources on to one large display.

In 'small' rooms, the large display may need only one projector to give the required resolution. In a larger room, or where very detailed information must be displayed, multiple projectors are needed, but the whole display can be considered as one.

It is then possible to source the display from many inputs, for example, video conferencing, conventional video, high-resolution graphics, personal computers, etc. Using sophisticated processing equipment, the various sources can be synchronized together, and each 'windowed' into each other as required. For example, if the matter of the moment is a conference call to New York, the New York participant takes center screen, with other imaging moved out of the way. If, next, the item of importance is the CAD graphic of a new product, this moves to the center — while the New York participant moves, for example, to screen top left.

Another example of what is possible is to be seen at Silicon Graphics' 'Reality Centre' near Reading in the UK. This uses a huge compound curved screen to present the output of three graphics projectors served by Silicon Graphics 'Reality Engine' image generators. Besides being able to demonstrate their own simulation products, the facility points the way to how architects, product designers, engineers and others can present their concepts either for approval or to facilitate decisions. Buildings which have not yet been built can be 'walked round' for client approval. Emergency evacuation procedures can be reviewed on oil rigs which have not yet been built. The big image is an essential aid to understanding – a single desktop monitor simply cannot provide enough information.

Systems of these kinds are practical now, but at present they are both expensive and require some skill in operating. It can be expected both that prices will come down and that they will become easier to use, partly because the generation of people who will make most use of them will themselves be unafraid of computer technology. But, however complex the technology, the presentation principles will remain the same.

Conclusion

A presentation room is more than the sum of its parts. It is essential to have the technical aspects of a presentation room engineered as one system, so it is usually best to make one supplier responsible for its successful operation.

Chapter 13

Interactive audio-visual

Interactive audio-visual is fashionable. People who have not used it or who do not know what it is are worried that they may be missing out on something. It is true that some kinds of interactive AV are leading to new ideas in point-of-sale and other display applications. It is also true that in the training area, where interactive AV is likely to be of the greatest benefit, it is very important that any intending practitioner is fully conversant with linear AV methods first.

The essential differences between normal AV and interactive AV are:

□ A normal AV program is linear, it has a beginning, middle and end and is seen in its entirety. Interactive AV is non-linear, in that a given presentation is unlikely to use all the material available, and the order of presentation can be different for every viewer.

□ A normal AV program does not require any input from the viewer. The viewer is passive. Interactive AV requires the active participation of the viewer.

□ The participation itself can be on two levels. At its simplest, the viewer is simply making choices, using the random-access facility which is at the heart of any interactive system. More truly interactive systems require the viewer to give information about himself.

Generally, interactive AV is intended for an audience of one. The resulting presentation matches the one particular viewer's requirement. It is possible to run a large audience show, with branching points in the show determined by a majority vote of the audience, but such programs are necessarily simple and certainly do not represent mainstream use of interactive AV.

Interactive AV depends on the random access of program material. In response to the direct request of the viewer, or as a pre-programmed requirement based on an evaluation of the viewer, it must be possible to retrieve segments of material. These can be still images, moving images, or audio segments. Branching within a program arises when a program reaches a stop point. There may then be several different ways forward. The choice of which branch to take can depend on the choice of the viewer but, in training applications, it is more likely to depend on an assessment of the viewer's knowledge. A typical situation might be where the trainee is asked a question. If he gets the answer right, the program continues with the next linear segment; if wrong, the program loops, to repeat the previous segment. If he gets the answer wrong again, the program branches to a remedial program, only coming back to the main line when this has been successfully completed.

These programs are not easy to write, and it can be important to monitor how each presentation is actually run. If trainees have consistent difficulty with a particular question it may turn out that there is a deficiency in the program rather than in trainee!

Flowcharts

Any interactive AV system consists of a minimum of four components:

□ a visual display;

□ a store of images;

□ an input device;

□ a device to determine what is to happen as a result of the input stimuli.

Strictly speaking, if it is an AV system (as opposed to an interactive visual display) there is also a store of sound segments. Finally, there can be additional output devices. Thus, one particular configuration could have a video monitor as the display device; videotape with a random-access search system as sound and image store; a personal computer as the system organizer; a keyboard as the viewer input; and a printer as additional output. This is only one of many possibilities. Another system could be built-up using an entirely different set of devices. However, before letting the hardware impose any limitations, it is instructive to examine the nature of an interactive program in more detail.

As with all AV programs the intending user must be quite specific about what the viewer is going to gain by participating in the program, and how quickly it is reasonable for this objective to be met. For example, the following types of programs might be developed:

☐ A program designed to teach people how to complete computer records in inter-company transactions.

☐ A program to encourage shoppers to buy from a range of garden products.

☐ A program to allow museum visitors to find out what is in the reserve collection.

☐ A program to train an automobile mechanic in the servicing of a particular model of car.

☐ A program to teach elements of microbiology.

These all represent perfectly valid applications of interactive AV, but the nature of the programs would vary greatly. The microbiology program would assume that the student has a reasonable amount of time to work through the course (hours or probably days) and would include many tests of the student's understanding. The public programs, like the point-of-sale program or the museum program, have to work very quickly to be of value. Their success is partly measured in how many people per hour can use the system, as well as in the increase in sales or visitor satisfaction.

It is convenient to describe a particular interactive program by a flowchart. This is in addition to the story-boarding of the AV segments. The flowchart shows how the program is intended to work: how the user gets in to the program and, just as important, how he gets out of it. It has been known for producers to inadvertently introduce dead ends into a program, resulting in a complete system lock-up and a very confused viewer.

The process of flowcharting is best understood by reference to a simple example, which is followed through to embrace greater complexity. From this it is possible to see how a hardware choice is made.

In the mid-1980s there was a multi-media show in central London, intended for tourists, called the London Experience. Shows started every 40 minutes, so there was a problem of entertaining those waiting to see a show. An interactive game about London was seen as something that would both amuse and

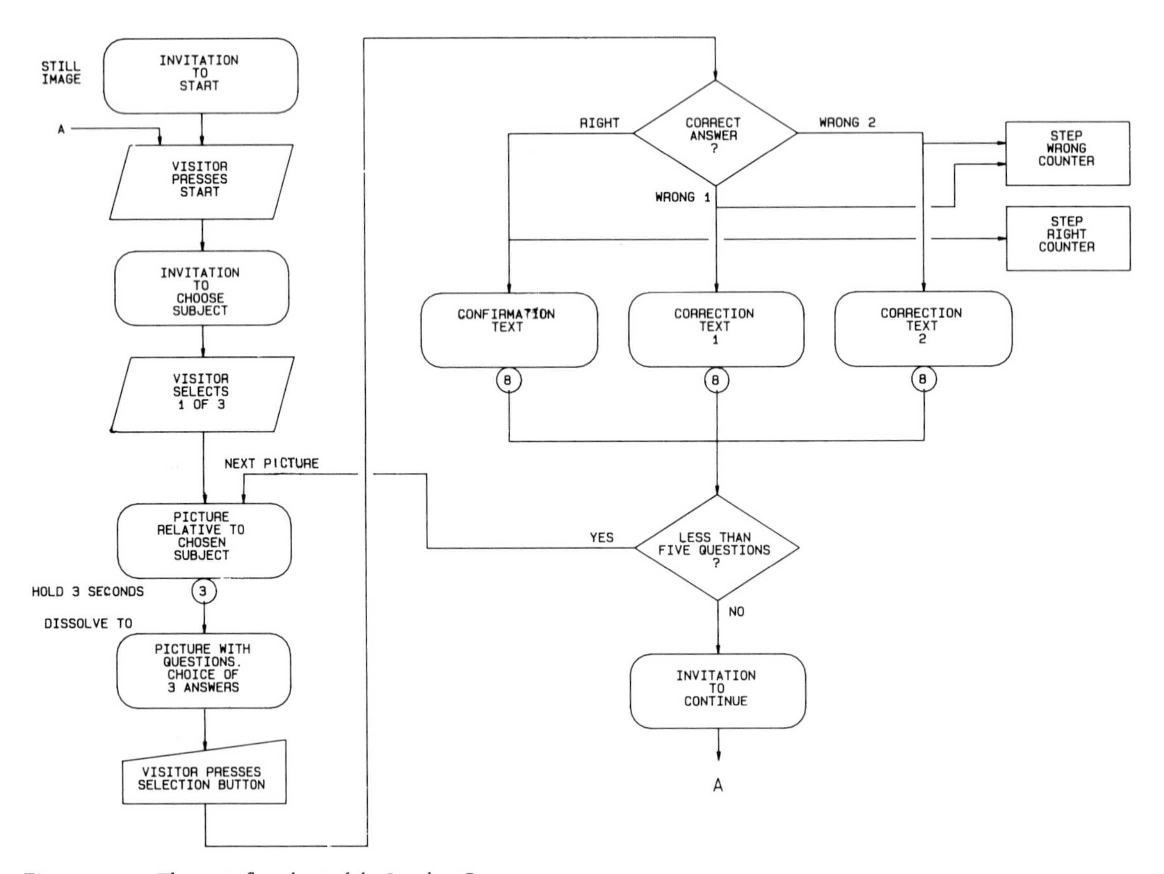

Figure 13.1 The main flowchart of the London Game

educate visitors. Before any decisions were taken about the nature of the game, there were some immediate priorities:

☐ It must be easy for the visitor to join in, and it must be quite obvious how it works.

☐ It must be possible for the game to be enjoyed and give information in both short and long sessions. A visitor might play it for thirty seconds or ten minutes.

☐ The response of the system to visitor input must be very fast.

☐ Assuming that (as was the case) three separate systems are put in, any observer should not see the same material repeated often within a waiting session. Equally, there is no point in having more program material than is necessary to meet the objective.

What could form the basis of the game? Something based on London sights obviously commends itself, so the London Game is a simple quiz game with detailed answers to help visitors learn something about London.

Figure 13.1 shows the game flowchart:

☐ The rest state shows a text image telling the visitors what to do. In this case, a start button must be pressed.

☐ Pressing the start button brings up a choice of three subjects. The visitor chooses by pressing one of three buttons (choice is deliberately limited to ensure a fast response).

☐ An image appears relating to the chosen subject. For example if the subject chosen is *London Squares*, an image of Berkeley Square might appear.

☐ Then, next to the image appears a multiple-choice question with three possible answers. Visitors express a choice by pressing the button corresponding to what they believe to be the right answer.

☐ If the answer is correct, text appears confirming that this is so and a score indicator shows a score of one correct answer.

☐ If the answer is wrong, text appears saying why, and giving the correct answer. A wrong score indication is shown.

☐ After a pause to allow text to be read, the procedure is repeated with a new question.

☐ At the end of five questions a pause image appears, inviting the visitor to press the start button again if he wishes to continue.

Various housekeeping additions have to be made to the flowchart to deal with the following questions:

☐ How does the game finish?

☐ How to make sure there is sufficient variety of material?

☐ What happens if the visitor fails to respond?

In this case there need be no set finish to the game, the idea is that the visitor can go on playing until the next performance of the main show starts, or until he loses interest. In fact, the game does stop if one of the score counters reaches 99!

Although there is an initial choice of three subjects, this group of three is, in fact, one of six such groups

What can you see here:

A. Music Hall?

B. Experimental Theatre?

C. Restoration Comedy?

Figure 13.2 A typical question in the London Game (but since the first edition of this book the Players Theatre has been rebuilt!)

(making eighteen groups of five questions in all). Each time the start button is pressed the system presents a group at random. This ensures that the average visitor is unlikely to see the same set of questions twice. If the visitor fails to respond within a set time delay, the score counters go back to zero, and the rest state image re-appears.

The game as described met the need at the London Experience. It is easy to embellish the game, although in that particular location there was no need to. The embellishments that could have been added include:

☐ A prize given for reaching a particular score.

☐ A printed card issued on request that gives directions to get to a tourist site.

☐ A speeding-up of the presentation of questions as score increases.

The emphasis in the London Game was speed of response, and giving information in an interesting way. There was no *need* from the owner's point of view to find out whether the visitor had retained the information given to him.

The flowchart for a training application can look completely different. A USA automobile manufacturer wishes to train aspiring automobile mechanics in the servicing of car electrical equipment. Here, there is no question of needing to achieve results within the first few seconds. The interactive program is designed to be used over several hours of study and, furthermore, it is to test knowledge of the student and be the arbiter of whether he has passed an examination in automobile electrics.

A possible flowchart is shown in Figure 13.6. It is simplified, omitting most of the AV detail, but showing some of the major management differences. It is also broken down into sections to make it easier to understand what is happening. The following points distinguish this type of program:

☐ The system has to keep track of a student's progress.

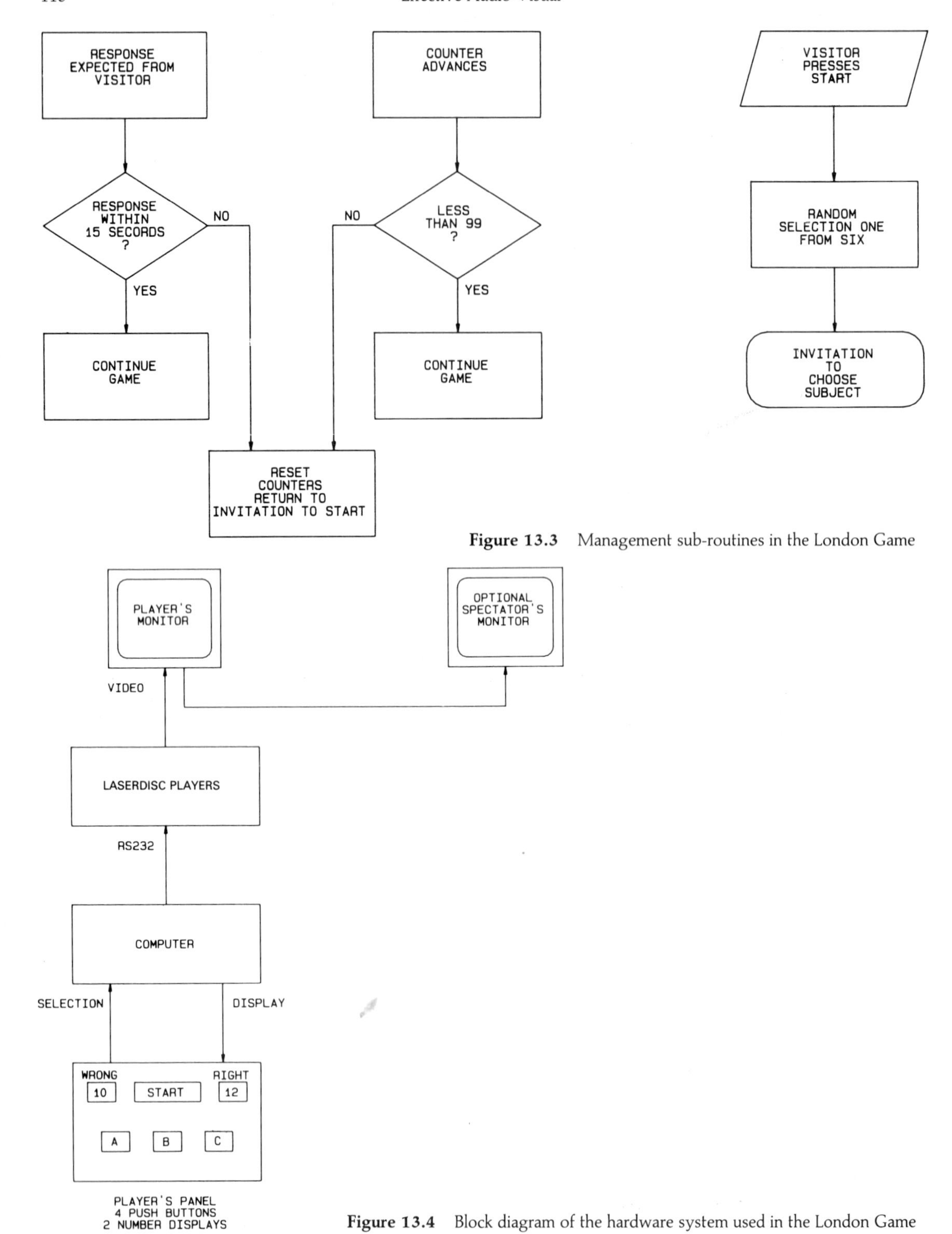

Figure 13.3 Management sub-routines in the London Game

Figure 13.4 Block diagram of the hardware system used in the London Game

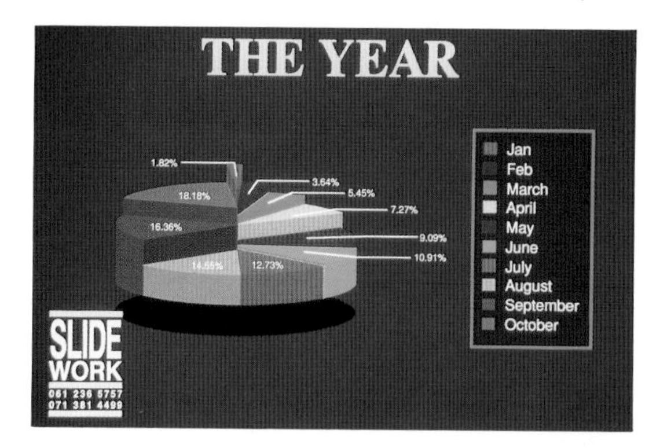

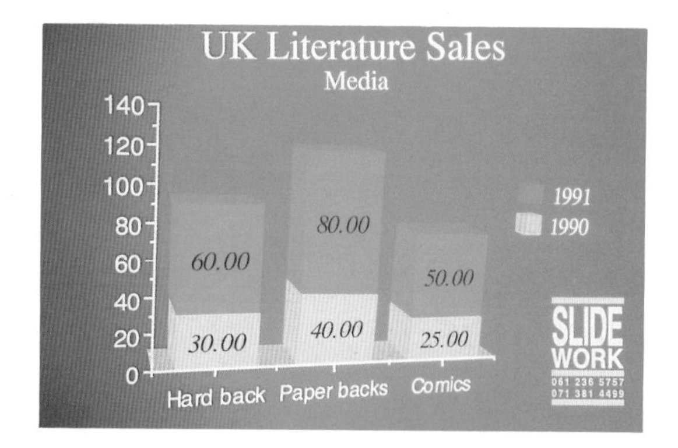

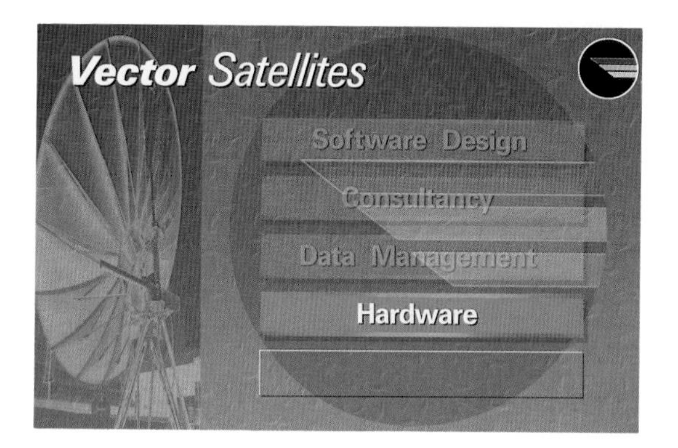

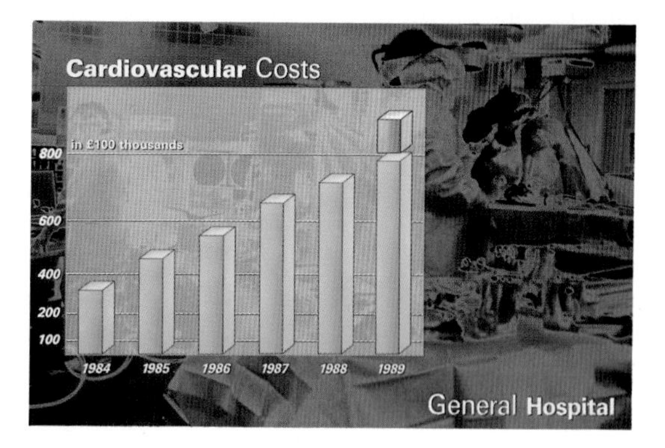

Plate 1 *(above)* Business slides and presentation graphics are now normally made using computers. The two top examples (from Slidework Ltd) are typical of what can be produced on suitably equipped personal computers. More subtle images need dedicated equipment with powerful graphics and effects capability. The bottom two slides were produced on Quantel equipment

Plate 2 Truly portable AV presentations become possible with new LCD screen technology. This portable VHS presenter is from Philips

Plate 3 The History Pavilion at Expo 85, Japan, featured 'Rice and Iron', a multi-image show of great size and impact. It used more than 80 xenon arc slide projectors

Plate 4 *(below)* Major product launches make great use of AV, integrated into the set and overall production. This is the GM-Opel Astra launch in Berlin, seen by 14,000 dealers from thirty countries. It used both multi-image and 35 mm movie. (Production was by HP:ICM, for Adam Opel A.G.)

Plate 5 For the Frankfurt Motor Show moving projected videowalls were used for the Astra's public launch. The projected videowall can operate in high ambient light and occupies relatively little depth. (Photo courtesy HP:ICM and Interactive Television Ltd.)

Plate 6 *(above)* Management information presentation on a big scale. GM Electronic Data Systems Corporation's Management Center uses seven Hughes Superprojectors

Plate 7 The teleconferencing suite at the Texaco headquarters in Houston, Texas. The facilities are controlled by a room controller. (Photo from AMX Inc.)

Plate 8 The multi-image medium is ideal for one-of-a-kind shows in museums, permanent exhibitions and visitors' centers. This is the 'Sounds in Nature' show at the Taiwan National Museum of Natural Science

Plate 9 Modern planetaria combine the use of traditional star projection with multiple slide, effects and lighting control. At the McLaughlin Planetarium in Toronto, 120 slide projectors, 640 effects circuits and 75 dimmers are under computer control

Plate 10 *(below)* Videowalls are not restricted to a TV shape. Pyramids, globes and cylinders are just a few of the possibilities

Plate 11 The videowall principle can be applied to displays of any size. This 144 monitor display was built by BKE Bildteknik for a ZDF (Germany) TV Game Show

Plate 12 The Discovery Theater at Sanrio Puroland theme park outside Tokyo uses animatronic figures, many special effects and projected videowalls in a fully automated show. (Photo from Landmark Entertainment Group of Hollywood.)

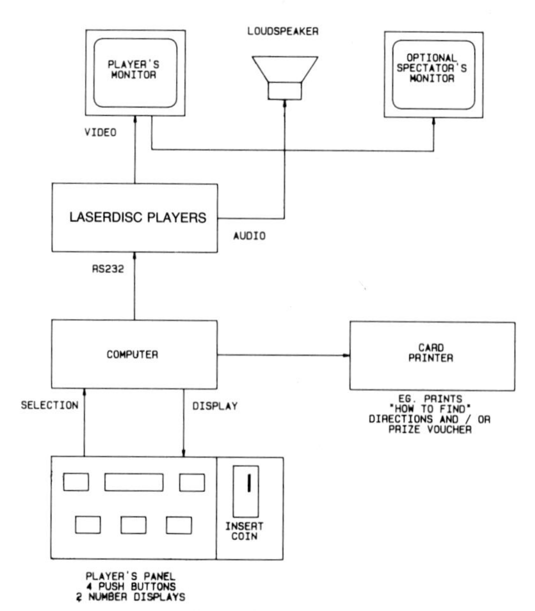

Figure 13.5 Possible embellishments to public information interactive displays like the London Game

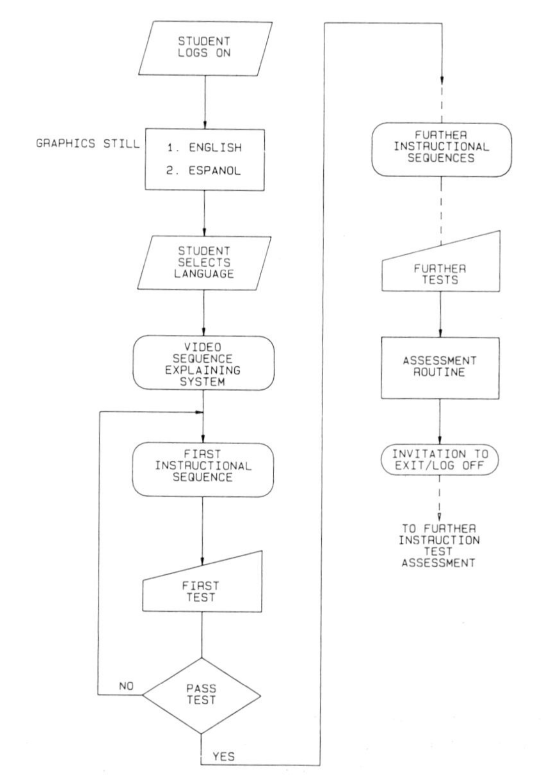

Figure 13.6 The main flowchart of a program for training automobile mechanics

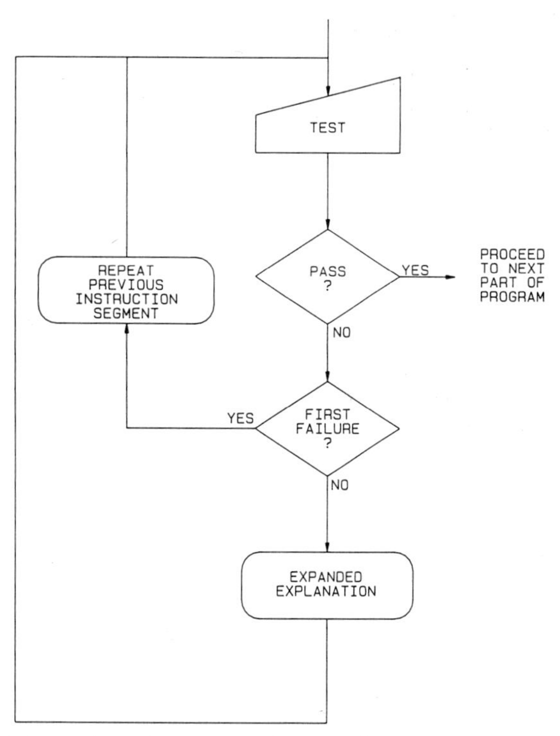

Figure 13.7 An example of a remedial loop

The first time that a student uses the program he must log on. His name can then only be removed from the log of current students by an authorized assessor. Equally, it would be possible to prevent anyone logging on in the first place unless their name is authorized.

☐ The first time the student logs on, a choice of language of instruction can be made. In the example, the choice is between English and Spanish. From now on all audio commentary and text graphics are in the chosen language.

☐ First item in the program is a general explanation of how the program works, and how the student should use the keyboard to respond to questions.

☐ Obviously, for the first-time student the program starts at the beginning. But, if the student has already had one session, the computer starts at the point previously left off. This part of the program needs designing with care to cater for those students who leave off studying in unexpected places. The technique is always to restart at least one segment earlier than recorded as complete.

☐ Most testing of the student is done by multiple-choice questions. Failure to get an answer right usually involves repeat of one or more previous segments, and sometimes involves remedial loops.

☐ Preliminary testing is part of the course and is simply used as a quick check to see whether the student should proceed to the next segment. An option is to keep a record of how students respond to the questions (e.g. how many percent get answers right first time). At this stage the procedure has less to do with assessing the student than with assessing effectiveness of the program.

□ An entirely separate exercise is assessment of the student. The example shown is fairly sophisticated because an attempt is made to ensure that the student really understands, and is not just learning by rote. At set points in the student's progress a series of questions is presented. Importantly, these are retrospective over the *whole* course so far, and they are randomly accessed from a library of questions. This ensures that if a student has to resit the test the questions are not identical.

□ The student's responses are recorded for separate assessment if required. It is then optional whether the student with a satisfactory score is allowed access to the next part of the course, or whether such access is only permitted as a result of an external assessor confirming a pass. If the questions are only multiple choice, there is no reason why the test should not be fully automatic. However, if the test requires the student to answer some questions by text statements, an external assessment is likely to be essential unless statements are extremely simple (in which case there is little point in using them).

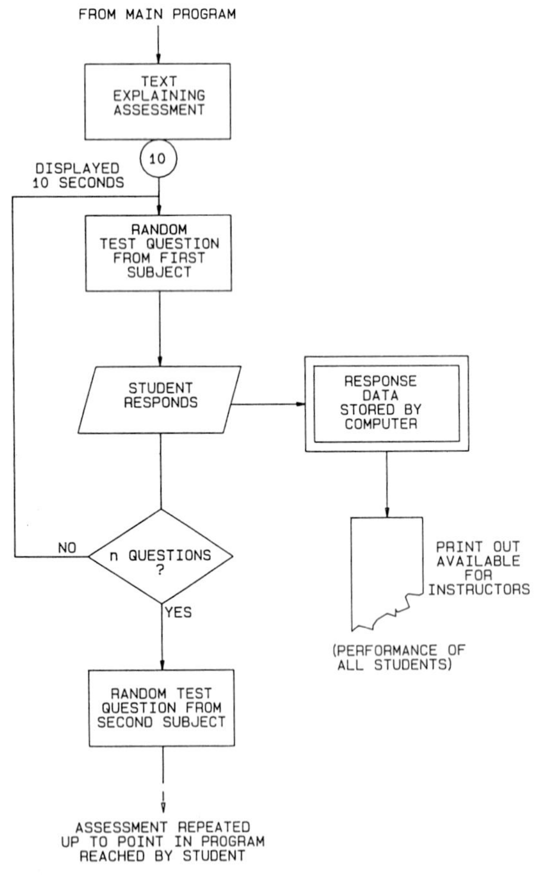

Figure 13.8 An example of an assessment routine

□ There must still be a device to cater for lack of response. Normally a student is expected to log off at the end of a session. This is done with a single key stroke that simply records progress so far and returns the system to its rest state. A similar sequence of events takes place if at any time there is no response from the student within a preset time.

The two examples given are almost two extremes; together they illustrate the possibilities of interactive AV and the importance of flowcharting how an intended program is supposed to achieve its objective. It is immediately clear that training and assessment programs can be extremely complicated, so first-time users are well advised to stick to simple objectives. This usually means breaking the task down into small segments, and getting these right one at a time. This may also reveal that some aspects are better dealt with by separate presentation of linear programs.

The hardware

Much current literature, whether it be journal articles, press reports or advertising, gives the impression that all interactive audio visual is video-based, or conjured by magic out of a computer. However, as Tables 13.1 and 13.2 show, there is a wide range of devices that can be incorporated.

Until recently, mature applications of interactive

Table 13.1 Interactive audio-visual

Still images	Photographs (slides) Artwork Text Charts Computer text Computer graphics Freeze frame from moving picture sequence
Moving images	Movie film Animated cartoon Animated computer graphics
Other visual input	Combinations of the above, e.g. Text superimposed on movie Illuminated displays
Audio	Speech recordings Music recordings Sound effects recordings Speech synthesis Music synthesis Effects synthesis
Other output	Printed text Photographic output Plotted output

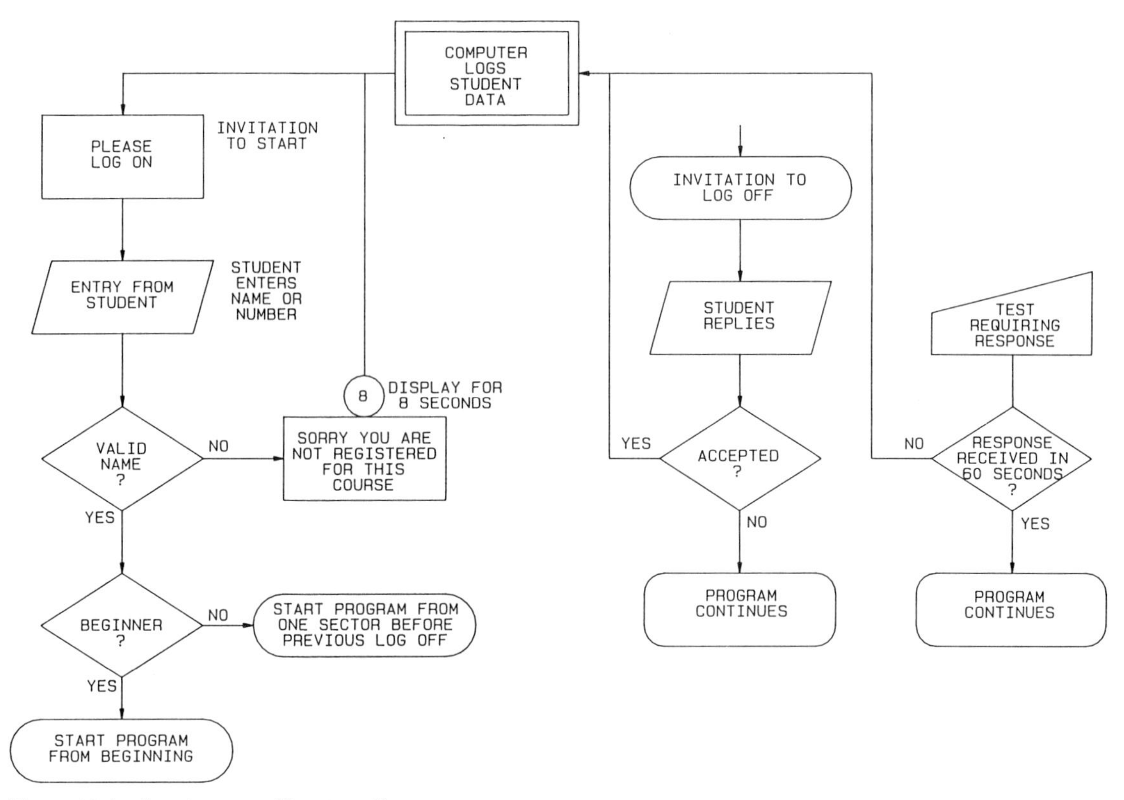

Figure 13.9 Logging on and logging off

audio-visual have been based on the use of analog videodiscs. However, new computer programs and digital disc technology are presenting new options. This chapter, therefore, deals with both non-video methods of interactive presentation, and with video-based methods using tape or analog laserdisc. Chapter 14 reviews the impact of newer hardware.

For reasons which will become clear, many applications benefit from videodisc, especially those requiring multiple users. However, there are many stand-alone applications, and even some multiple-user applications, better served by different or hybrid technologies. These are best explained by example.

Besides deciding on the best method of *delivery* of the interactive program, the intending user has to seriously consider the method of program *creation*. For most applications cost of delivery systems can be seen as reasonable; very often it is the perceived cost of program creation that is the bar to proceeding. Before giving some examples that show the way in which a decision can be taken, performance of some of the available system components is discussed.

Random-access slide projectors are convenient in development of programs using a limited number of photographic visuals. They allow easy editing, but limited capacity (one unit for 80 slides). Random-access performance is reasonable, typically three seconds access time.

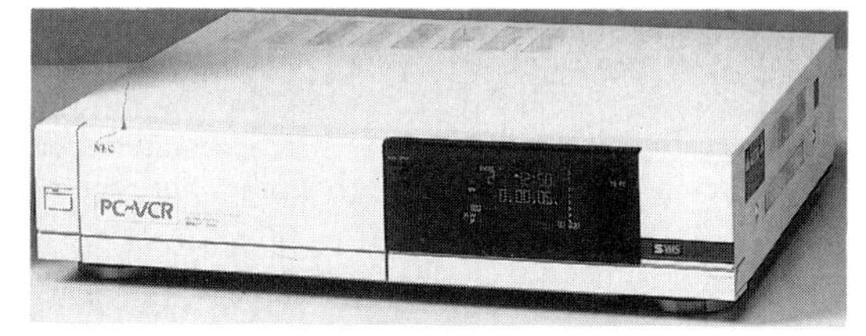

Figure 13.10 NEC have developed the 'PC-VR'. This is an S-VHS videocassette recorder with built-in computer interface, and improved still-frame performance that makes it a useful tool for program development

Videocassette recorders are convenient for program development when the program relies on moving images. They are also suitable for stand-alone applications where equipment is not intensively used. Both U-Matic and S-VHS videocassette recorders are available with computer interfaces that allow accurate selection of individual segments or frames. In fact the

Table 13.2 Interactive audio-visual hardware

Visual display	Cathode-ray tube (CRT)
	Rear projection screen
	Front projection screen
	Illuminated display panel
	Liquid crystal panel
	Plasma panel
Visual source	Slide projector
	Motion picture projector
	Videotape
	Magnetic video store
	Videodisc
	Computer text store
	Computer graphics store
Audio	Combined with visual source
	Cassette
	Synthesizer
	Digital audio store
Participant input	Custom control panel
	Standard keyboard
	Touch screen
	Mouse
	Light pen
	Speech recognition
	Credit card reader
Other outputs	Paper printer
	Card printer
	Graphics plotter
Control	Dedicated controller
	General purpose computer
	Inherent in the display device

mechanical performance of the VCR does give some limitation in accuracy, so if single frames are required it is a good idea to ensure that they are recorded several times. The big disadvantages of videotape are slow random-access performance (many seconds, or even minutes) and poor still-image performance. Thus, if still images are going to consist mainly of text, a solution to the problem is to use videotape for moving images, and computer generated text for stills.

Random-access audio devices are available. The most accessible is the random-access cassette recorder but, as with the videotape unit, access is slow. Digital audio stores based on hard disc are an immediate answer, and these are now easily achieved within a personal computer.

Videodiscs are the ideal basis of many interactive applications. Laserdiscs, in particular, have good still-frame performance. They can easily carry a mixture of moving and still material, have a large capacity (up to 54,000 still frames), have fast random access (sometimes instantaneous, usually between one and five

Figure 13.11 A Laservision™ videodisc

seconds) and can carry a lot of audio information and even additional computer data.

The videodisc has a curious history. It was originally invented for the consumer market where, it was thought, video LPs would be a more acceptable method of distributing entertainment programs than tape. The advent of the low-cost home videocassette recorder and the realization that this could also be used for timeshift recording of broadcast material made the videodisc player very much a second option. The analog videodisc has never really got off the ground as a consumer item, although there is an established market for movie discs in the Pacific Rim countries and the USA. The advent of the high-capacity *digital* disc, described in Chapter 14, will undoubtedly change this situation.

Videodiscs bear some similarity to the long playing phonograph record, but the 'grooves' are very much closer together and information is recorded as pits or variations in the depth of the disc surface. The pit pattern on standard laserdiscs is laid down concentrically as shown in Figure 13.12. The laserdisc surface is reflective and the signal is derived by scanning the surface with a tiny light beam, produced by a small laser. The varying reflected light is detected by a photocell that produces corresponding electrical signals. The pick-up system includes sophisticated elec-

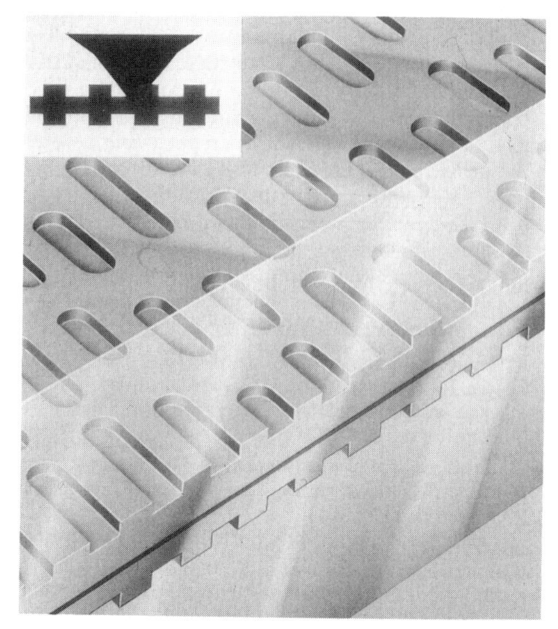

Figure 13.12 The method of recording information on a laserdisc. (Diagram by Philips)

Figure 13.13 The scanning of a laserdisc. (Diagram by Philips)

tronics to ensure that it is directly over the required part of the disc.

The laserdisc is fairly robust. Although it should still be handled with care, the actual playing surface is protected by an outer transparent layer; the laser being focused on the information layer beneath it (see Figure 13.13).

The originator of the laserdisc was Philips, once a major source of players for the PAL market, but who have now withdrawn to concentrate on digital formats (see Chapter 14). All other players originate in Japan, where the principal manufacturers are Pioneer and Sony. The laserdisc is either CAV or CLV (constant angular velocity or constant linear velocity). In CAV players, the disc rotates at a constant speed and one revolution of the disc corresponds exactly to one frame (two fields). In CLV rotation speed varies, but speed of the pick-up over the disc surface is constant. This allows more information to be packed in on the outer parts of the disc, hence a longer playing time.

The result of the one frame per revolution in the CAV discs is that it is very easy to select an individual frame. CLV cannot really be used for still-frame work, and is used mainly for movie films. The standard laserdisc is essentially a replay medium; the producer must make a master tape which is used by the disc manufacturer to make a master disc stamper which is, in turn, used to make as many discs as required. Disc mastering charges may seem high, but with careful planning they are reasonable as a proportion of any particular project, especially because the disc opens up applications of video that cannot be met any other way. Cost of extra discs once the master has been made is very reasonable. Disc manufacturers are able to give a fast turn-round for those users, including most industrial users, who need a few discs quickly. Turn-rounds of a few days are quite normal.

The needs of many one-of-a-kind installations and of those who develop program material, are better met by recordable videodiscs.

Personal computers often form the basis of interactive AV systems. This means that other devices in the system must be of a kind that can be communicated with by a computer. In general this means that they must have an RS232 interface (sometimes now called an EIA232 interface), which is a control connection able to transmit and receive computer data to an internationally agreed format. Random-access slide projectors, cassette recorders and videodisc players are all available with this as a standard feature. However, use of a general-purpose computer is not a *necessary* requirement of an interactive AV system. Often the

Figure 13.14 A touch screen in use on an exhibition stand. (Photo Media Projects)

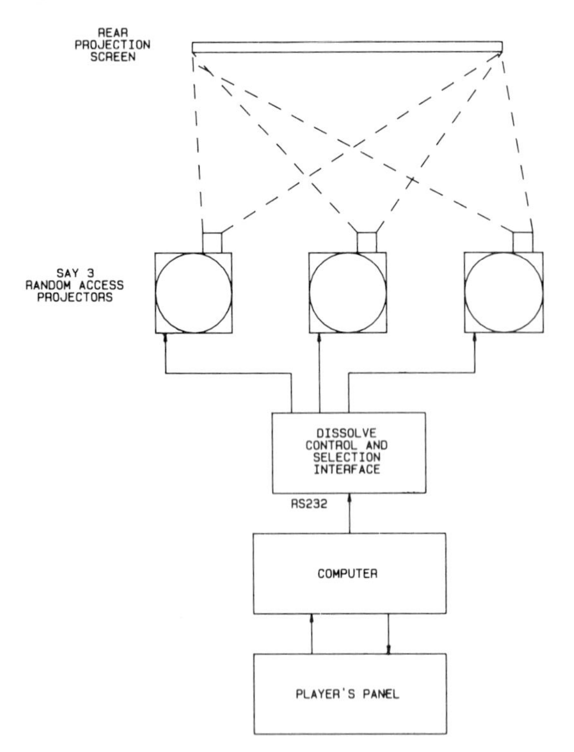

Figure 13.15 Although the London Game was finally committed to videodisc, a display of the same type, required in only one venue, could be achieved with the equipment configuration shown here

final delivery system uses either a special controller or a microprocessor controller within a videodisc player to achieve control.

Touch screens are devices that allow the participant to express responses by touching the viewing screen at a designated point (as opposed to using a standard keyboard or special push-button set). They are usually used over CRT displays, but are equally applicable to flat-screen displays. They work in various ways. Some use a transparent overlay, which includes a grid of printed conductors. Changes in electrical resistance or capacitance locate where the touch is made. Others use an array of infra-red emitters and detectors forming an invisible X–Y grid across the face of the display. They are helpful in public displays and in any application where it is necessary to minimize hardware touched by the participant. They make the preparation of the program slightly more complicated.

Examples of hardware configurations

In order to understand how the hardware configuration is conceived, first consider the London Game application cited earlier.

In its basic form, this game could have been based on two random-access slide projectors and a special controller. The controller could either be specially made, or consist of a standard computer able to control the projectors. There must also be a visitor's control panel fitted with a start button and three choice buttons. This method is recommended for anyone who requires a limited-use single installation, giving highest picture quality, which may need updating.

In fact, the London Experience management elected to use videodisc. The reasons for this choice were:

- There was not space to install a slide-based system. Instead, the participant's panel was fitted with a small monitor, and another larger monitor was placed higher up for other visitors to see.
- The slide system would not have had enough capacity. Actually, at least six random-access projectors would have been needed, which while justifiable in a single installation, was uneconomic for the three complete systems needed.
- The random-access slide projection system would have needed more maintenance than the videodisc system and would have had a slightly slower response. Both these points were, however, only marginal considerations.
- It was possible to use the videodisc prepared for the London Game for other purposes. The game only occupied part of the disc, the rest of it could be used for other material, thus lowering cost of the disc attributable to the game.

The decision in this case was a marginal one. A museum wanting to create its own specialist display, and not requiring such a big image capacity, could well choose the optical projection alternative. In fact, the whole London Game was first made on slides, because this represented the most economical method of production. When the production was costed on a pure video production basis it was so expensive that the project would have had to be abandoned if pure video methods had been used. This simply demonstrates that the potential interactive user must be aware that there may be many ways in which the objective can be met, but only one or two of them will be practical or affordable.

Once the decision had been taken to use videodisc, the London Game *could* have used a touch screen. This was dismissed on the grounds of unnecessary extra cost, unnecessary extra programming, the fact that with only four buttons the input requirements were so simple, and finally because a display with some big buttons to thump was actually more appealing to the visitor than one with just a fingermarked screen.

The automobile electrics training program is the kind of program that needed the arrival of the videodisc as a mature medium to work at all. This type of

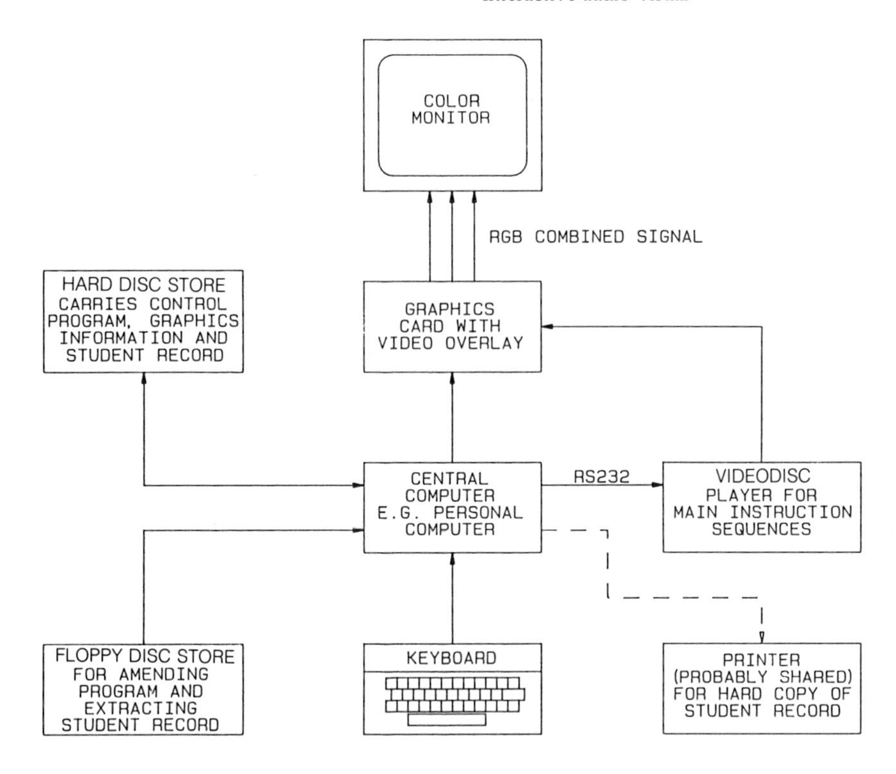

Figure 13.16 A possible configuration for the automobile electrics training program

program would inevitably use both moving pictures (in the form of animated diagrams and film demonstrating procedures) and still images (as diagrams, photographs and text). A possible complete configuration is shown in Figure 13.16.

The computer, as opposed to a small processor in the videodisc player, is needed for this application because it is necessary to keep track of students' performance. The computer keyboard is most likely to be used as the means of student response because this allows students to enter their names, and to answer questions in text format for subsequent retrieval by the assessor. A printer is likely to form part of the system, to extract student performance records in a printed form. This configuration is also ideal where a disc has been prepared that can be used in several different ways, because it allows control software to be changed very easily by the computer. If additional text, not on the videodisc, is required it can be created as a computer text overlay.

This last refinement is not available on all computer/videodisc combinations. It requires that the system has the facility for text overlay, and that the computer can 'genlock' its graphic output in sync with the disc picture. This is usually done with a card that is inserted in the computer that combines graphic overlay facilities with those of videodisc control.

Not all training applications are necessarily best served by videodisc. Teaching of in-house procedures for transaction processing, job progressing, capital equipment movements, etc., may be better dealt with by a computer-based system, or CBT (computer-based training). The UK company Mast Learning Systems specializes in preparing customized programs for small users using the configuration shown in Figure 13.17.

A typical application could be a civil engineering plant hire company. They have equipment rented out at many sites, and themselves have several depots which store the equipment not rented. They have a major problem in knowing where all their equipment is at any time, and whether it is in full working order or needs servicing. The problem is compounded by the fact that any mistakes made by their staff in following the company's documentation and data-logging

Figure 13.17 Computer-based training need only use a standard personal computer. In this photo a random-access cassette recorder is used for voice prompt, but this would now be more easily done by using the computer's own memory. (Photo Tandberg Ltd)

procedures, result in them losing track of individual items.

Therefore, all their staff have to learn the procedures. Because many of these involve entering information on a standard keyboard, and because they are all based on the use of either printed forms or data displayed on a VDU, it makes sense to use a personal computer as the central training device. The program is presented in the form of text and computer graphics. The student responds by entering text on the keyboard, thereby simulating what they will be doing in real life later on, or by answering multiple-choice questions.

The computer on its own is rather dull, and would not give the student the vital sense of participation needed to make interactive AV a success. This is why the system also includes a random-access cassette recorder. This allows an instructor to take the student through the material with a voice commentary and, where a form is described on the computer screen, an arrow or other attention-getting device can follow the commentary. Similarly, the instructor can ask questions and comment on the student's response, verbally, which is much more friendly. The random-access cassette recorder is sufficient for simple programs, but for those with many sound segments it is better to use the computer as the sound store (see Chapter 14).

A feature of programs such as these is that they are highly specific, with a very small audience. They must be easy to modify, and their cost of preparation must be reasonable. Videodisc is too expensive and too inflexible for this type of application, but that is no reason why such users cannot have the benefit of interactive training. Some of these programs also need still images. These can be generated in the form of computer graphics, or can be based on slides which are 'scanned in' to the computer. However, some users find it easier to use the slide itself. It is a simple matter for the computer to control a self-contained random-access slide unit, and these programs should never need more slides than can be held in a single magazine.

Authoring

This last example has exposed the Achilles heel of interactive AV. It is all very well to describe how a system is intended to work, but how does an individual user, even one experienced in conventional AV production, actually put a program together?

For the major corporation requiring a large number of copies of an interactive program on videodisc, there is no great problem. They define their needs and go to a specialist producer who does the whole job for them, and is happy to relieve them of the $100,000 or so that it may cost. For the small user this approach is out of the question, although some specialist help may be necessary to achieve cost-effective results.

The main problem is the need to have a control program in addition to the AV software. The control program may be machine code software buried in a microprocessor or, more likely, it is a program in a higher level language on a personal computer. But even in this case, it should not be necessary for an intending user to have to learn BASIC or C in order to get started.

A number of authoring systems and programs are available that allow users to get straight down to making interactive programs. Representative examples are shown in Table 13.3. They are designed to run in MS-DOS and MS-Windows™ (IBM and compatibles) or Apple Macintosh™ computer environments.

The authoring system itself is nothing more than a computer program that prompts the user to feed in the parameters of his presentation. Flowcharting must still be done, but a good authoring system prevents anything being left out or the creation of dead ends. It also allows the setting up of computer text to augment the video presentation, and allows the setting up of student tests.

Recordable videodiscs

Applications of videodiscs requiring multiple copies or heavy-duty use should use proper pressed discs for the best results. However, there are applications where

Table 13.3 Examples of interactive authoring software

Product name	*Vendor*	*Platform*
Authorware Academic	Authorware	Mac and PC
Authorware Professional	Authorware	Mac and PC
CAL Toolbox and Autographic	Penguinsoft	PC
Course Builder 3	TeleRobotics International	Mac
Crystal	Intelligent Environments	PC
Guide 2.0 and 3.0	Office Workstations Ltd	Mac and PC
HyperCard	Apple Computer	Mac
Hyperdoc	Hyperdoc	PC
IconAuthor	AimTech	PC
Mentor II	Mentor Interactive Training	PC
Procal	Epic Interactive Media Company	PC
Quest 3.0	Allen Communications	PC
SAM	Sandy Corporation	PC
Supercard	Silicon Beach Software	Mac
TenCore	Computer Teaching Corporation	PC
VCN Concorde 3.0	VCN International	PC
Video Builder	TeleRobotics International	Mac

'Platform' is computer on which program runs (PC = IBM PC series or compatible, Mac = Apple Macintosh).
Table prepared from information provided by Videologic Ltd.

the cost and time delay involved in pressing discs is not justified, and it is often convenient to have a test disc made before committing to full production.

This facility is offered by machines and discs made by the *Optical Disc Corporation* of America. Because machines are expensive, it is usual to get ODC discs made by a suitably equipped studio. The discs are compatible with standard Laser-vision discs, so standard players can be used. The service is available for both PAL and NTSC discs and discs are available in both glass and plastic – plastic discs being intended for disc checking work, while more expensive glass discs are for extended use.

Where compatibility is not an issue, then lower-cost disc recorders, such as those made by Teac, are available for in-house use. The players are more expensive than those for standard Laservision, however, and machines are only available in NTSC.

Stand-alone applications requiring highest possible image quality can use the Sony CRV system. This records video as a component, as opposed to composite, signal. It is described in more detail in Chapter 19.

Pioneer market a re-recordable disc system for the professional broadcast market, which may also have application for the advanced interactive program developer.

The equipment described in this chapter records video in an essentially 'uncompressed' form. The new digital recording techniques described in Chapter 14, based on compression, are opening up new possibilities and are making the creation of 'one-of-a-kind' programs a cost-effective proposition.

Point-of-purchase displays

It is difficult to tell whether the biggest non-consumer market for interactive audio-visual will be

Figure 13.18 Price Waterhouse, the international firm of chartered accountants and business advisors, have developed a series of interactive training programs with the Premiere Training Company over a period of several years. These include 'Account Ability', the interpretation of financial statements, and 'TerminalRISK' which is based on discovering the risks that are inherent in computerized information systems. The programs can be bought outright, rented or used on a daily basis at Price Waterhouse's Learning Center shown here. Each program takes between five and ten hours to complete. The programs have required a significant investment, but have proved highly cost-effective because the cost per user is low and the training both effective and fast

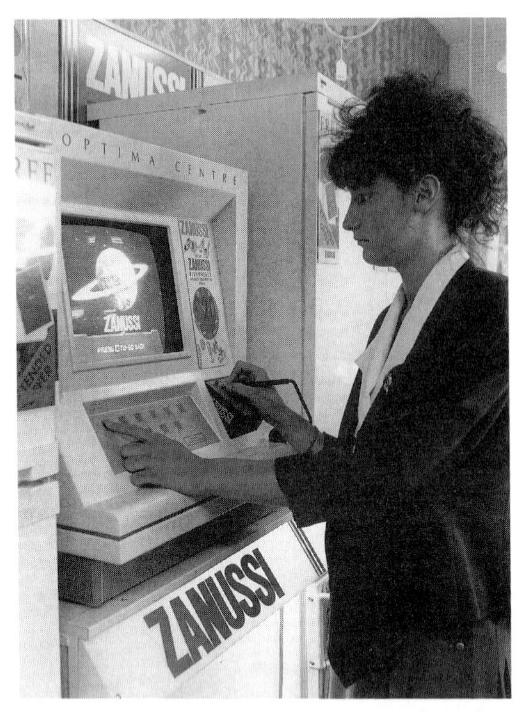

Figure 13.19 The domestic appliance manufacturer Zanussi has installed interactive displays in 500 outlets. The displays give information on 120 different products, and allow users to compare price and specification information. (Photo McMillan Communications)

corporate training, or point-of-purchase displays. In fact, the two applications may well merge together. Multi-branch organizations like AT&T in the USA and Lloyds Bank in the UK already use videodisc systems for staff training and staff information at branch level. It can be expected that they will seek to make further use of their investment by making alternative programs available to the public. Often these programs can use the same disc as the in-company program.

Point-of-purchase systems will see some of the more complex applications of the medium. AT&T offers point-of-purchase systems that allow direct purchase via a credit card reader, but this is where interactive AV is merging with transaction processing. Many users get good results from a simpler approach. Some lessons have already been learnt:

☐ Response must be fast. Programs benefit from having a presenter who is seen on the screen, but long TV-style introductions must be avoided.

☐ Programs demanding a personal input from the user hold interest. For example, in a travel program 'where have you been before?' can be asked, so that other places can be recommended. In a health-food program details of age, weight, etc could be requested to make a diet recommendation.

☐ Generic programs work well. A program on microwave cookers prepared by Convergent Communications for Britain's Co-operative Wholesale Society boosted all microwave cooker sales, not just those of the main brand featured.

☐ While true interactive programs yield best results, they must be made with care. For many applications where budgets are limited, simple random-access programs can be a cost-effective alternative.

☐ 'Catalog' style programs for the retail sale of consumer products have proved a problem. Some high-quality programs of this kind have been abandoned because of the difficulty of keeping the programs up to date. This illustrates the point that when an interactive program is budgeted, the cost of maintaining it must be taken into account.

Interacting with an audience

Interactivity with an audience as opposed to an individual is possible. Sophistication of these systems usually resides in the complexity of the way in which the audience voting is displayed, rather than in the program itself, which must necessarily be fairly simple. Walt Disney World have an excellent example in their Communicore Theater at EPCOT, where the audience's opinion is sought on many topics, each of which is introduced by an AV sequence. Each member of the audience is equipped with a five-way voting panel, and survey results are shown as live computer graphics. Results are added to those of previous audiences, and are also used to determine how the particular presentation should proceed.

Other examples are special displays at EXPOs, especially in Japan, where big multi-screen video theaters have been created with each member of the audience having a color VDU. These systems all use videodisc as the main video source, with computer-generated graphics for the display of audience data.

Back in 1967 the Czechoslovakian Pavilion at EXPO '67 (Montreal) presented a pioneering and highly entertaining audience interactive show. This used 35 mm film as the presentation medium, resulting in the need for several projectors. Members of the same production team went on, more than 20 years later, to create a similar program first shown at the Gas Pavilion in EXPO '90 (Osaka) and now to be seen at Futuroscope in France. In Osaka, practical considerations dictated that the program be presented from videodisc, but at Poitiers it is shown from film to get the highest possible image quality for a large audience.

Recently Sony have opened an 'interactive theater' in New York, using high-definition television as the presentation medium. Shows of this kind can be highly entertaining; most of the technical problems have been solved, resulting in reliable fully automatic presentations. The real problem is now to find the writers with the imagination and talent to create the stories for the medium.

Chapter 14

Developments in interactive audio-visual

This book is written at a time when there is considerable confusion in the AV market. Computer people have discovered AV and called it 'multi-media'. Like some other words which have been hijacked by pressure groups, this can lead to misunderstandings. Multi-media has, until recently, been used to describe the technique of using several different AV presentation methods such as slides, video screens, sound and light techniques, animatronics, etc., in a show presented to an audience.

According to the new doctrine, on the other hand, multi-media means a personal computer, usually operated on a one-to-one basis, with an AV capability. This computer is able to combine elements of sound, still image, moving image, text and data in a single package. In the early 1990s, proponents of the new multi-media were quick to promise many wonderful things; but the truth was that wonderful things were only available at enormous cost and inconvenience. Some aspects of the new multi-media's performance were and are quite inadequate when compared with 'traditional' methods; even though some traditional methods themselves are high technology, but are mature.

The situation has now changed. There is no doubt that the creation and delivery of AV material by computer is a completely practical and cost-effective proposition. However, those who understand the basis of presentation and program construction from the experience of using older methods are in the best position to take advantage of this. Potential users also have to be very careful to ensure they are using the right method for a particular task, and not be afraid to use apparently old-fashioned techniques which may be more effective.

The truth is that many of these new techniques are not actually needed by the majority of professional AV users. The market has only just started to use properly the conventional laser videodisc, and it took ten years for that to happen. The OHP can sometimes be a much better way of making a presentation than using a computer and video projector. The computer salesman, who has never used either an OHP or a videodisc, has no idea what a storyboard is and would not recognise a slide projector if it hit him on the head, is busy assuring those who thought that they might be ready for interactive AV with laserdisc that everything is out of date except some mysterious acronym (of which there are so many, that it would be invidious to select one).

The problem is a commercial one. There are two pressures. First is from publishers of software of all kinds. This embraces music publishers, film and video makers, book publishers and so on. They are all on the look-out for new media methods for distributing their products. Here it is undoubtedly true that there are some very interesting opportunities. However, for the AV user the interest may be in how to exploit these new methods of publication, as opposed to any wish to join in the creation of work suitable for mass publication. It must be remembered that the target of most new media is the domestic consumer, not the industrial or commercial user. Thus, to take just one example, CD-I may be an excellent method of publication for particular kinds of training program to be used by many trainees; but be an absolute non-starter for single-site users.

Second, commercial pressure comes from computer manufacturers. Two generations of personal computers have filled what is now a mature market. Word processors and spreadsheets form probably 95 percent of that market, and for many buyers the computer is a simple commodity item to be purchased at the lowest possible price. What increases value of the computer, and what prompts people to replace their existing machines? Why multi-media, of course!

Where does this leave the AV user, other than confused? The recommendations are:

- ☐ Wait for the big battalions to fight it out, there is no reason for you to fund their developments.
- ☐ On no account be persuaded to be the first to use a particular method.
- ☐ Check costings thoroughly, there may be some nasty hidden extra that cancels apparent hardware savings.
- ☐ Only concern yourself with the end result.

This chapter starts on a warning note but, in fact, there are already some exciting new possibilities to explore without risk, and others which will reach maturity during the second half of the decade.

The disc disorder

At the root of the confusion is the method of data storage. Storage of simple text and data is comparatively economical, and even the smallest computer can store the equivalent of thousands of pages of text either in its own memory, or on inexpensive magnetic memory discs. Storage of images requires much more memory. Simple graphics can easily be designed to use little memory, but full-color photographs are another matter. Full-motion video, requiring thirty frames of images every second requires, in theory, a huge data storage capacity.

Most modern personal computers have big memories and can, in principle, easily deal with the processing of single images. Mass storage of images requires the use of some kind of disc technology, and this is where confusion starts.

Figure 14.1 illustrates the problem. Disc formats can be categorized according to:

- ☐ Whether the medium is recordable by the user or requires pressing. A pressed disc, for example a CD, may be ideal for mass distribution but less practical for individual users.
- ☐ Whether the recording method is analog or digital.

Images recorded digitally are easily handled by a computer, indeed a standard computer could not deal with images stored in any other way. However, recording of an image requires a lot of memory space and, more of a problem, a very high rate of data transfer.

Leaving aside the high definition videodisc (not at present relevant to the average user), the best quality disc recording of video images is achieved by the Sony CRV system. This records the image in *component* form, instead of the more usual *composite* form. By keeping chrominance (color) and luminance (white light brightness) quantities separate and by recording a wide bandwidth, the CRV disc is able to record video to broadcast quality. Discs contain up to 24 minutes of full-motion video. CRV opens up the benefits of disc technology to a wide range of one-of-a-kind applications, but otherwise operates in a similar way to the Laservision disc (which records a composite NTSC or PAL signal) described in Chapter 13. Another high-quality analog component disc recording system is now available from Pioneer, it uses re-recordable magneto-optical discs of 32-minute duration and quality similar to CRV.

CRV records the component video signal in analog form. Quality of the picture is such that, if it were to be recorded in digital form, it would need a data transfer rate of 200 Mb/s (204,800,000 bits per second). By comparison, the audio compact disc has a data transfer rate of 2 Mb/s or, in its computer data storage format of CD-ROM, only 1.3 Mb/s. But Figure 14.1 shows that CD is being offered as a means of video image storage. How is this possible?

The compact disc as a physical medium has a lot of attraction to software publishers. It is economical to produce, easy for the user to handle and compatible recordable discs are now available. Thus, both the needs of the mass user and the one-of-a-kind user can

	ANALOG	DIGITAL
RECORDABLE	Component Recording Video (CRV) Analog Magneto-optical	Floppy Disc Drive Magnetic Hard Disc Drive Magnetic Digital magneto-optical Digital write-once CD
PRESSED	Analog laserdiscs (Laservision)	CD-ROM CD-I and others CD-Video DVD

Figure 14.1 Disc formats for recording video images

in theory be satisfied by a single disc size. The problem which remains, however, is how to record video images with the limited data transfer rate that arises from CDs' audio ancestry.

The confusion that has existed in this field has largely arisen because of disagreements in how this is done. The way in which audio data is recorded on the CD, and the way in which computer data is recorded on to CD-ROM is defined by a series of standards designed to ensure compatibility between manufacturers. Thus one thing that could not be changed was the data rate. (Although in the computer environment, CD-ROM drives are now available in both 'double-speed' and 'quadruple-speed' versions.)

A full-spectrum digital recording, whether video or audio, records every sample as a numerical value. In the case of video each pixel is measured in respect of its red, blue and green content (or in its chrominance and luminance values). In the case of CD audio a measurement of the signal intensity is taken 44,100 times each second. The obvious question is whether this process is actually recording more information than is strictly necessary. For example, if a large area of the video image is all the same color and brightness it ought to be sufficient to record the value once, and to record information about the size of the area concerned. This idea is called *data compression.*

Data compression is possible precisely because the image or audio signal is stored digitally. Clever *compression algorithms,* or mathematical procedures, are able to review the original numerical data and greatly reduce the amount of data space needed to store an individual image or audio sequence. Although there are some proprietary systems of compression, some of which will continue to have a specialized market, there is a move towards standardized methods of compression to facilitate transfer of image data between different systems and users. Table 14.1 lists only a few of the acronyms likely to be encountered as dinner party conversation pieces among multimedia afficionados. It is possible to construct entire paragraphs along the lines of 'CDROM-XA includes JPEG, ADPCM and the CD-I CLUT; CD-I has MPEG, and DVI's PLV is soon to be compatible with MPEG'! Included in this alphabet soup are the two standard methods of image compression:

- □ JPEG is the method of compressing still images.
- □ MPEG is the method of compressing video moving images running at video frame rate.

Table 14.1 Some interactive acronyms

ADPCM	Adaptive Delta Pulse Code Modulation
ASCII	American Standard Code for Information Interchange
CAV	Constant Angular Velocity
CD	Compact Disc
CD-A	Compact Disc Audio
CD-I	Compact Disc Interactive
CD-ROM	Compact Disc Read Only Memory
CD-ROM XA	Compact Disc Read Only Memory Extended Architecture
CD-WO	Compact Disc Write Once
CD-V	Compact Disc Video
CDTV	Commodore Dynamic Total Vision
CIRC	Cross Interleave Reed-solomon Code
CLUT	Color Look Up Table
CLV	Constant Linear Velocity
CRV	Component Recording Video
DCT	Discrete Cosine Transform
DRAW	Direct Read After Write
DTV	Desk Top Video
DVD	Digital Video Disc
DVI	Digital Video Interactive
GUI	Graphics User Interface
ISO	International Standards Organization
JPEG	Joint Photographic Experts Group
MPEG	Motion Picture Experts Group
PCM	Pulse Code Modulation
PCMCIA	Personal Computer Memory Card Industry Association
Photo CD	Photo Compact Disc
PLV	Production Level Video
POI	Point of Information
POS	Point of Sale
RAID	Redundant Array of Inexpensive Discs
RTV	Real Time Video
SCSI	Small Computer System Interface ('scuzzy')
SMPTE	Society of Motion Picture and Television Engineers ('sempty')
SIF	Standard Input Format
Video CD	Video Compact Disc
WORM	Write Once Read Many

Compression

This section gives a brief description of how video compression works. Data compression can be considered as being of two kinds. So-called 'loss-less' compression aims to store all the data contained in the original. It works simply by eliminating easily identified redundant data, for example a system may be recording data to 10-bit resolution, but if most of the data is in the form 10110010$\underline{00}$ where the two least significant bits are nearly always 00, it is possible to record the great majority of the data in 8-bit format only.

'Lossy' compression, on the other hand, admits that some of the original data will be lost. The aim is to achieve the maximum possible compression with the least subjective degradation of the image (or sound).

Figure 14.2 MPEG compression brings video to the CD, photo showing computer decompression card from C-Cube Microsystems and Kudos Thame

It is a general rule that original material, be it a photograph, a video recording or an audio recording is 'captured' in an uncompressed or lossless form, and that compression is only used for the delivery of the final 'program' to the viewer. This is because the process of cascading multiple compressions and decompressions may have uncertain results. However, rules are made to be broken, and no doubt this one will be as compression methods are refined.

As already mentioned, there are many proprietary methods of video compression; however, there are now recognized international standards and these are the key to successful use of video in computers. The JPEG standard was introduced to compress still images, and the 'video' variant of it, running at 30 images per second, has the advantage that each image is complete in itself. This helps in any application requiring program editing or random access.

Video JPEG can be made to work at variable compression rates, typically in the range 30:1 to 50:1. Image resolution can be up to broadcast (CCIR601) standard, but for lower cost applications may be half or quarter this. The process by which JPEG compression works can be divided into three stages:

- The removal of data redundancy. This is done using the DCT.
- Weighting the data resulting from the DCT in a way which is optimized for human vision.
- Using an efficient coding method (the splendidly named Huffman variable word length encoder) to minimize the size of the final data.

Although complex in themselves, the principles of the three steps in JPEG encoding are easily understood. The difficult one is the DCT (discrete cosine transform). What this does it to look at the image in blocks of 8 × 8 pixels. The 'transform' is a mathematical trick to describe the whole block in terms of changes in

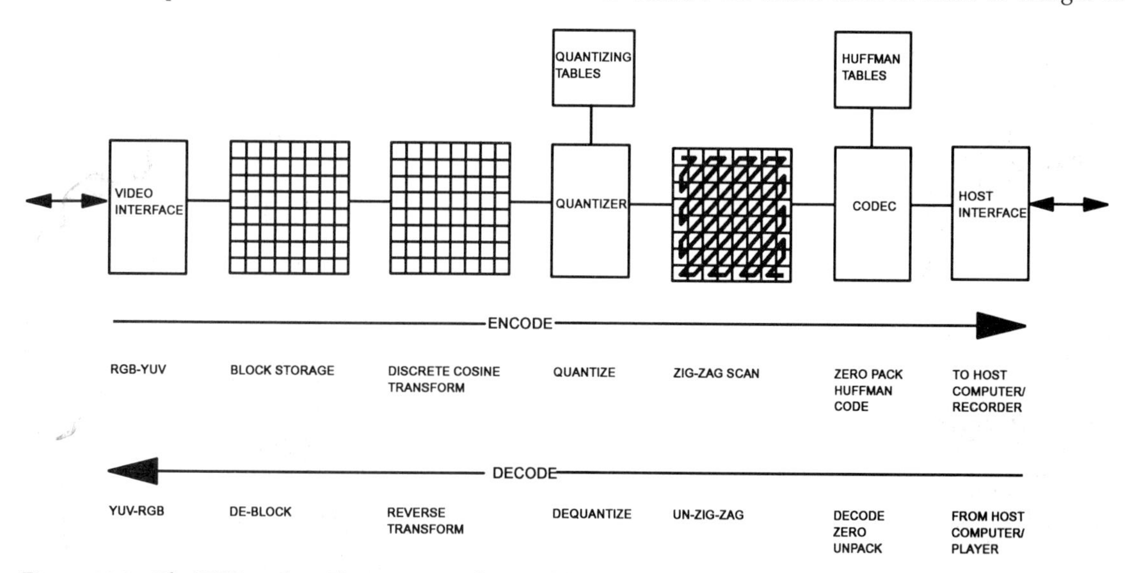

Figure 14.3 The JPEG pipeline. The same principle is used for I picture compression in MPEG. Images are converted to YUV component, and then into blocks of 8 × 8 pixels. The DCT converts this information from the spatial to the frequency domain. The frequency coefficients are quantized, and accumulated in zig-zag order, this increases the number of zeros in the data. Zero packing and Huffman coding then reduces the actual data to be stored to the minimum. Data to be decoded goes through the reverse process

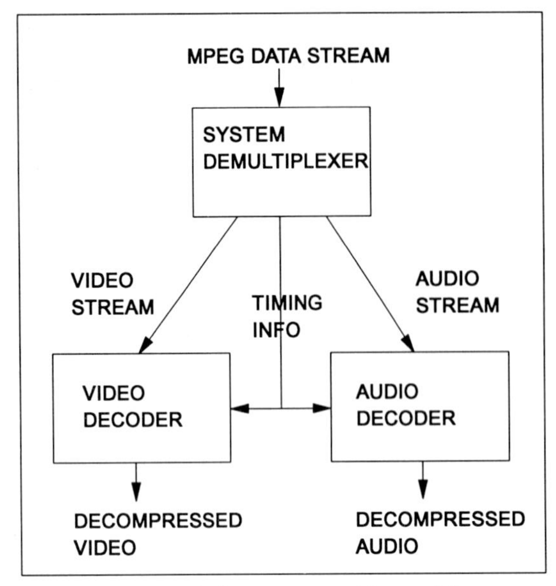

Figure 14.4 MPEG decompression. The serial datastream carries timing, audio and video information all multiplexed together

pixel values. Where there are few changes there are plenty of zeros in the data, which can be eliminated. The amount of data to be stored is further reduced by applying to the DCT factors about human vision. For example, we cannot detect 'high frequencies' corresponding to changes in intensity very closely spaced; and we do not perceive color information to the same resolution as brightness information. The final encoding of the data turns frequently used data patterns into a 'shorthand', in the same way that Morse Code uses the shortest symbols for the most frequently used letters.

MPEG

While JPEG video is useful for some applications, it does not give the highest possible compression rate, and it does not have provision for accompanying audio. MPEG compression addresses both, and it is available in two flavors:

☐ MPEG-1 is optimized for low data rates up to 1.5 Mb/s likely to be available from CD. Its quality is equivalent to VHS.

☐ MPEG-2 is best for data rates between 4 and 8 Mb/s. Image quality is broadcast standard (CCIR601) and can be extended to high definition.

MPEG-2 is 'better' than MPEG-1 only when the higher data rate is available, at low rates MPEG-1 gives better results. An MPEG-2 decoder will also decode MPEG-1 but not vice versa.

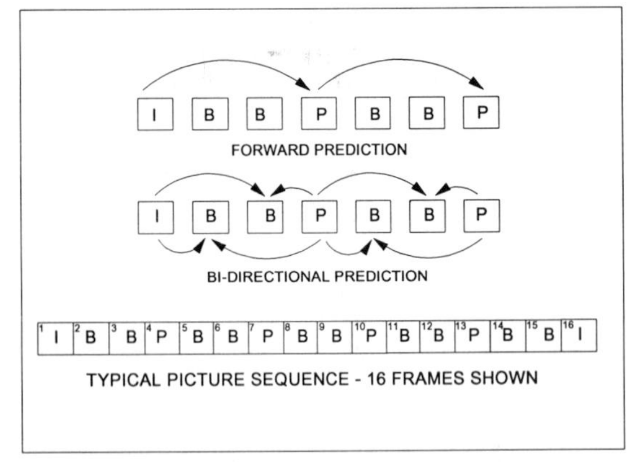

Figure 14.5 In MEPG compression interframe compression greatly reduces the data to be stored. Only the I pictures are derived wholly from data which is *intraframe*. P pictures are forward predicted (top), B pictures are bidirectionally predicted (center). In a typical NTSC video stream, there is only one I picture every half second or so, and usually two B pictures between each P or I picture (bottom)

MPEG-1 uses the 'standard input format' of 352 × 240 at 30 frames per second for NTSC and 352 × 288 at 25 frames per second for PAL. MPEG-2 supports 704 × 480 and 704 × 576, both as full frames or as two interlaced fields to match standard video better. It can also be extended to HDTV. The distinguishing feature for MPEG, compared to video JPEG, is that it gains further compression by *inter frame* compression, which takes advantage of the fact that the differences between one frame and another may be small.

MPEG encodes three different kinds of images:

☐ *Intra* or I pictures. Here the data is based only on information held in the picture itself. These images are compressed using the DCT and coding techniques already described for JPEG.

☐ *Predicted* or P pictures. These are forward predicted pictures based on *motion compensation*, i.e. the movements of blocks of pixels within a series of images are used to predict images. P pictures are constructed from the nearest previous I or P picture. P pictures require only about one third of the data of I pictures.

☐ *Bidirectional* or B pictures. B pictures use both a past and a future picture as a reference. They give the greatest degree of compression, but are not themselves used as references.

The MPEG datastream not only contains the picture data, but also the audio and timing/control information to keep everything in synchronization. The audio is also compressed, using similar data reduction techniques to the pictures – a combination of predictive data reduction and the elimination of information that cannot be heard by the human ear.

The decoding of MPEG reverses the encoding process, but this is altogether easier to do. MPEG decoding can be done with single chips (supported by memory) and inexpensive MPEG-1 decoders are available for computers, or built in to 'set top boxes', or in consumer products like CD-I and 3DO. MPEG-2 decoding equipment is also becoming available as new products and services (like DVD) become available.

MPEG encoding requires extremely fast processing speeds, and is done with special purpose processors or by large computers. Initially MPEG encoding was not done in real time, with, typically, an encoding rate 30–40 times longer than real time. The technique is still used for getting the best possible results on video-CD and CD-I, but is not much use for the AV user. Now reasonably priced MPEG-1 real-time encoders are available, and in due course the currently expensive MPEG-2 encoders will also become more attainable.

For most AV users, the results from MPEG-1 are excellent. The pictures are subjectively much better than VHS, *provided high-quality 'input' material has been used*. The image is remarkably free of noise, provided the source was itself noise-free. The arrival of practical computer-based video has coincided with the introduction of low-cost high-density computer disc drives, with capacities of one, two or more gigabytes, sufficient for hours of MPEG video storage. This gives the user great flexibility. Production, editing and one-of-a-kind applications can all be handled by a computer with big hard disc and an MPEG encoder/decoder card. Applications requiring playback only in multiple locations need only an inexpensive CD-ROM drive with MPEG decoder – both will soon be standard features of everyday computers.

This arrangement is the most flexible for the commercial and higher-educational user because the computer has many other uses. There are, however, other formats which hover between the computer and the consumer hi-fi or video games markets.

CD-I

First there was the CD, then there was CD-ROM. CD-ROM is a standard method of storing data and can be regarded as a standard computer peripheral. It is ideal as a means of distributing masses of data. For example, all postcodes in the UK are available on a CD-ROM published by the Post Office. CD-ROM needs a computer or computer-like device to carry out the search for information required.

CD-I was conceived as a publishing medium based on the data storage principles of CD-ROM, but formatted so that the disc contained not only the 'show' but also the control software. The idea is that CD-I is normally used on low-cost players, similar to standard CD audio players, with very simple controls. Players include special processing chips to decode data into audio and video. Therefore a CD-I disc does not play in a computer fitted with a CD-ROM reader unless the computer is also equipped with the CD-I circuitry.

Figure 14.6 Consumer CD-I uses a simple hand control for interactivity. This program is on golfing technique

CD-I is an audio-visual medium. The way data is stored on the disc is optimized to allow sound to accompany images, whether images are a series of still pictures or a moving image sequence. CD-I stores any required combination of text, numerical data, audio, graphics, still images and moving images. Originally the moving image capability was confined to cartoon-like animation, or to only a part of the screen. Competitive pressures then caused CD-I to embrace full-motion video occupying the whole screen. This was achieved by using MPEG-1 compression.

It must be understood that CD-I is a publishing medium. When programs are prepared for CD-I, the process is carried out in an emulation mode using large hard disc drives on a computer such as an Apple Macintosh or Sun workstation. The completed program is finally copied on to tape which then serves as a master for making the pressed CD-I disc.

The interest in CD-I for the industrial AV user lies in the range of published titles that will become available. It seems very likely that a number of generic training programs will find their way on to CD-I format, and the availability of low-cost playback hardware could increase usage of interactive programs.

The success or otherwise of the CD-I format will depend on the fact that it bridges the consumer and professional markets. The demands of the consumer market for ever more sophisticated games, requiring more and more memory, have meant that the CD has become an attractive 'carrier' for games, because it is far less costly to manufacture than the equivalent memory 'cartridge'. This has resulted in a number of other formats, all of which are in theory usable for audio-visual presentation, but which in practice are not likely to attract widespread support from the professional user. Some examples:

□ The 3DO consortium offer CD-based products suitable for games, video CD and interactive titles. It is a direct competitor to CD-I.

□ CDTV was introduced by Commodore in 1991 in an apparent attempt to pre-empt CD-I with full motion video. It suffered from being offered by only one manufacturer.

□ Sega introduced CD-ROM as an add-on to their product range in 1992. Nintendo are following as they introduce new 64-bit products.

□ Sony have entered the games market with a 32-bit CD-based machine, the 'Play Station'.

Photo-CD

There is another flavor of CD which, while invented for the consumer market, has had considerable success in the professional market, and can be used as the basis of high-quality AV presentations. Photo-CD as a system was devised by Eastman Kodak, working with Philips which provided the 'CD' part. Other film manufacturers also support the format. It can be regarded as a method of storing high-resolution photographic images, and while one way of viewing the images is by a TV set or computer monitor, the points that set photo-CD apart are:

□ It is dedicated to using photographic film as the 'input' medium. The means by which the film is scanned in ensures that, potentially, the full resolution of the original film image is retained.

□ The method addresses the colorimetry and flexibility of film. Thus it can record information which would be meaningless in video environment, such as 'greater than 100 percent white' (needed for highlights) and colors outside the color space recognized by video systems.

□ It is intended for high-resolution picture exchange, recording individual 35 mm images at a resolution much higher than that provided by electronic stills cameras or high-definition television. Images can be output to high-resolution printers giving images of the same quality as a photographic print.

□ For many users photo-CD is an inexpensive way of importing high-resolution images into the computer environment.

A standard photo-CD is derived from 35 mm negative or transparency film, and is usually made by a suitably equipped photofinisher. The images from the film are scanned using Kodak's high-resolution scanner which is based on three CCD arrays able to scan to greater than 3000 line samples. The resulting information is stored in a hierarchical manner which Kodak call an 'Image Pac'. This is shown in Table 14.2.

It can immediately be seen that 'Base' resolution is practically the same as TV image or VGA resolution. In the photo-CD system, 35 mm images are scanned in at 16Base resolution, equivalent to 18 megabytes per image. The RGB signals are converted to 'Photo YCC', luma-chroma signals similar to, but NOT the same as, component TV signals. A workstation 'number crunches' the data so that the Base and lower resolutions are recorded *uncompressed*, and the higher resolutions are recorded as compressed difference signals called 'residuals'. High-resolution images are reconstructed by interpolation of the base image with the compressed high-resolution information. The method does not give as high-compression ratios as JPEG, but does ensure that very little of the original picture information is lost. At the standard 'Photo-CD Master' level, each image needs about 6 megabytes, so one Photo-CD holds about a hundred photographs.

Table 14.2 Photo-CD image structure

Image component	*Resolution lines × pixels*
Base/16	128 × 192
Base/4	256 × 384
Base	512 × 768
4Base	1024 × 1536
16Base	2048 × 3072
64Base*	4096 × 6144

* Only on Pro-Photo CD

The photo-CD itself is made using a 'write-once' CD-ROM recorder. This must be of the 'multi-session' kind, because it is quite possible to store one film on to a disc first, and then add other films later. In order to make the photo-CD format attractive to many different kinds of user, Kodak have initially defined six different types of photo-CD:

□ **Photo CD Master.** This is the standard 'digital negative'. It is written by photofinishers or service bureaux. It stores the full-resolution images up to Base16. There is an optional facility to include text, graphics and audio on the same disc.

□ **Pro-Photo CD Master.** This is the professional version, supporting larger film formats such as 4 × 5, 120 and 70 mm. It records to Base64, and supports pixel level editing, and copyright protection features. It can only be written on a professional photo-CD imaging workstation.

□ **Photo CD Portfolio.** This is created from either Photo or Pro-Photo CD Master discs. It can be written on a computer, or can be written as a service by a photofinisher or service bureau. There is optional text, graphics and audio, and a choice of recording the images at full resolution or at video (Base) resolution only. The resulting disc can hold up to 800 images or one hour of sound, or proportional combination.

□ **Photo CD Catalog.** This again is created from Photo or Pro-Photo CD Master discs, either on a computer or by a

Figure 14.7 The photo-CD system devised by Kodak is an ideal method of getting high-resolution photographic images into the computer environment

service bureau. It includes search software, and images are recorded only at low resolution, allowing 6000 or more images.

☐ **Photo CD Medical.** A special format optimized for medical imaging.

☐ **Photo-CD Print.** Optimized for print, and able to store images in user-defined color spaces and in other formats, for example, CMYK derived from drum scanners.

Photo-CDs can be played back by CD-ROM drives in a computer, and by stand-alone devices. These include dedicated photo-CD players, CD-I players and 3DO players (all of which can also play standard audio CDs). The industrial and commercial user is likely to find the computer platform the most useful. There is then a choice of using photo-CD just as a storage medium, or as a presentation medium in its own right, especially with the Portfolio variant, for which there are both Macintosh and MS-Windows authoring programs. Software is available to convert photo-CD files to other file formats, for example, TIFF, so it is possible to transfer images, or parts of images, to other presentation programs.

Extended audio on CD-I

Philips and Sony, the licensors of the CD system, publish five standards manuals. These are only obtainable by licensees, for example, equipment manufacturers and specialist program developers and are:

☐ The *red book*, which specifies how a standard audio compact disc should be made and exactly how digital data is formatted for the sole purpose of recording audio of the highest quality.

Table 14.3 'Levels' of recording quality for CD-I

Level	*Encoding method*	*Sampling rate*	*Bandwidth*	*Signal-to-noise ratio*	*Audio channels*	*Data for one second of sound*	*Typical playing time*
CD-A	PCM	44.1 kHz	20 kHz	98 dB	1 stereo	171.1 kbytes	1 hr 14 min.
A	ADPCM	37.8 kHz	17 kHz	90 dB	2 stereo	85.1 kbytes	2 hr 28 min.
					4 mono	42.5 kbytes	4 hr 56 min.
B	ADPCM	37.8 kHz	17 kHz	60 dB	4 stereo	42.5 kbytes	4 hr 56 min.
					8 mono	21.3 kbytes	9 hr 50 min.
C	ADPCM	18.9 kHz	8.5 kHz	60 dB	16 mono	10.6 kbytes	19 hr 40 min.

Figure 14.8 Denmark's national estate agent HOME uses IBM's AVC program for effective POS application. Potential purchasers can choose houses or apartments by location, specification and price, and can see full color images of the accommodation

□ The *yellow book*, which specifies the CD-ROM. Here the specification is concerned with the accurate recording of data – it is not concerned with what the data is.

□ The *green book*, which specifies CD-I. This is a specific use of the CD-ROM idea, and it lays down exactly how the different types of data (such as text, images and audio) should be recorded and identified.

□ The *orange book*, which specifies how recordable CD systems work.

□ The *white book*, which defines how video-CD (MPEG compressed video) is recorded.

These books may be of academic interest to the final user, but the important thing is that they ensure a level of standardization. One aspect covered is recording of audio to different levels of fidelity. The full CD audio standard is not necessary for many types of program material, and if audio is to accompany images it is clear that it must give up disc space. Table 14.3 lists different ways in which audio can be recorded. This shows that with only a limited loss in quality it is possible to quadruple audio playing time of a CD. For speech applications it is possible to record nearly twenty hours of signal on a single CD.

The practical applications of this are just beginning to emerge. Within the CD-I regime it allows great flexibility in satisfying multi-language speech requirements. For example, there can be a choice of eight or even sixteen language tracks. It also allows a single program to switch between different standards. A CD-I program on a particular composer could switch between full CD for concert playback, level A for short playback examples and level B for multi-language commentary accompanying images where there was no music required.

Program making

Until a few years ago, the endowment of a personal computer with usable AV facilities was an expensive matter. It needed, by the standards of the day, high speed and a big memory. Useful audio and graphics display facilities, let alone video display, required the addition of expensive plug-in cards. Now, however, the basic facilities required are standard in mid-price or even low-price computers, and it will soon be the case that every personal computer sold will have so-called multi-media capability. This has opened up a huge new constituency for program makers, and has similarly opened up the opportunity for program making to everyone.

Table 14.4 lists a few of the many programs now available for multi-media production and presentation (see also Tables 3.3 and 13.3). The following points should be noted:

□ The acquisition of images and sound at the program production stage may well require additional equipment, for example, a video-compression encoding card, video image 'frame grabber', slide scanner or audio card.

□ It is important to produce programs for the 'lowest

Table 14.4 Some computer programs for multi-media production on computers

Astound	Gold Disk
Audio Visual Connection	IBM
Authorware Professional	Authorware
Autodesk Animator	Autodesk
Persuasion	Adobe Systems
Icon Author	Aimtech
Storyboard Live	Krepec Publishing

common denominator' of delivery platform. This may severely limit the way in which animation and full motion video is used if older computers are involved.

☐ For any complex production it will be the case that several different programs will be used to create the final 'show'. Programs vary considerably in their ability to import different file formats (for example, WAV sound files, TIFF images and Quicktime video).

Examples

An example of the effective use of computer presented programs is given by the computer company ICL. While it could be argued that a computer company is an unfair example, in fact what they did in launching their new TeamServer™ and SuperServer™ products is open to anyone in any branch of commerce or education, and is an object lesson in how to do it right.

They needed several items which traditionally would have been considered separately, but which, when considered together enhanced each other.

☐ They needed an AV program to launch the new products. This traditionally would have been done by a production/staging company on a 'special event' basis.

☐ They needed to reach all members of their own organization, to ensure that they received the same message as the one given at the trade product launch. This is often one of the most difficult aspects of product launches, because it can be impractical to ensure that all staff can see the actual launch presentations.

☐ They needed material which could be used by the company's own staff to launch the products in other markets, and to support ongoing sales. This would traditionally have been done by a mixture of media. For example, videotapes, OHP transparencies and simple presentations on floppy disc. It is difficult to ensure consistent use of such material.

ICL decided to commission a CD-ROM which carried all the required material. Essentially this consisted of the complete launch presentation which itself consisted of eight full-screen video sequences and nine separate slide shows dealing with different subjects and markets. The video sequences varied between the simple 'documentary', for example, a user interview with an electrical engineer responsible for transmission line maintenance, to a complex 'mood' piece with high-production values used as the 'opening module' of the launch.

The important point is that all the video was professionally produced and looked it. The CD-ROM was used as the final show-carrying medium, but the production methods were those needed to ensure the highest quality of presentation. The program material could then be accessed in several different ways:

☐ The various offices of ICL used the CD-ROM as the basis of complete launch presentations of their own, which precisely followed the original. In this case the presenters simply worked through the whole disc.

☐ The disc was presented as an interactive disc, allowing presenters to choose from any part of it. Thus each section of the presentation could be used to support a sales presentation to a particular market.

☐ The 'shows' could be given to large audiences using video projection (the video compression was of excellent quality, few viewers would realize that the video was MPEG-1 from CD rather than videotape) or to individuals and small groups on monitors. The playback system needed for running the presentations consisted only of a standard ICL personal computer, fitted with a CD-ROM drive and an MPEG video playback card. Neither of these two additional items are now expensive, and current multi-media equipped computers would have them as standard.

The CD-ROM format helped to solve the problem of multiple languages. For example, multi-language soundtracks were easily included. The standard graphic 'slides' were created on Harvard Graphics, and this allowed special language versions to be produced locally, while still using the backgrounds of the original. This ensured that the 'look' of the presentations maintained the same high standard wherever they were shown.

The ICL example shows how excellent results can be obtained using a very simple approach. In this case, all the material was produced professionally under the direction of a single production management company, Razor Communications Ltd, working for ICL's marketing communications manager. Razor in turn used the same creative talent which they would have used conventionally to produce the video and slide 'modules', and then augmented this with some overall multi-media programming (to access the material) and specialist sub-contracted services, for example, the video compression and the disc manufacture.

Many users will find themselves emulating ICL's example. Some will have their discs prepared by outside specialists, others will want to do it all in house. In practice, a combination is often best. For example, Electrosonic Ltd is working on producing an annual CD which is arranged as a simple interactive disc allowing users to choose between overall corporate information, and between the products and services of five different business units. Within each subsection,

Figure 14.9 The Hodos route familiarization program produced for British Rail uses DVI

Figure 14.10 Some computers allow the direct recording of audio signals. For standard personal computers it is usually necessary to add an audio card. The Sound Blaster permits sound recording and music and voice synthesis. (Photo West Point Creative)

there is a further choice between product information and applications information. Most of the material is in the form of 'slides' and photographs, but full motion video is also included, for example, a 'house show' corporate video, and several short video sequences showing the company's products in action.

For this CD-ROM, Electrosonic are using outside professionals for video production – and are using the resulting video materials in other ways too – and for the design of titles and backgrounds. However, the text slides, the multi-media programming and CD recording are all done in house which greatly facilitates keeping the disc up to date.

Simulation

While the examples cited so far take advantage of the computer and CD to make programs more flexible and accessible, they cannot be claimed to be doing something which is not possible any other way. The computer does, however, allow completely new methods of training.

British Rail use DVI (a video compression method which preceded the MPEG-1 standard) for train driver route familiarization. The system, developed by Hodos, consists of a mock-up of the train cab with driver controls, a large color monitor and a DVI-equipped computer. The advantage that computer-based video has over the more traditional laserdisc is that speed variation in the image can respond more exactly to driver control.

The Hodos program superimposes the red and green of signal lights on to the video image (replacing those on the original videotape) so each simulated journey can encounter different signalling conditions. The program sounds cab warning bells exactly as heard on a real journey, and alerts the driver to driving errors. Items like advisory speed limits can be shown or not, as required.

Disc wars

For some time it has been realized that the restrictions of the standard CD format, 650-megabyte capacity and limited data transfer speed, could be overcome. 1995 saw the announcement of not one, but two competing disc formats able to hold several gigabytes of data, and therefore able to hold full-length movies with a quality approaching broadcast CCIR601 standard using MPEG-2 compression. Unfortunately this may well be followed by a standards war of the same kind that bedevilled videotape.

Sony, supported by Philips, have proposed a high-density CD able to store 3.7 gigabytes, equivalent to a 135-minute movie. Toshiba are supported by a group which includes Thomson, Pioneer and Matsushita, and have proposed a double-sided disc able to hold 4.8 gigabytes per side, with a 140-minute duration per side. They claim that their variable data transfer rate, averaging 5 Mb/s but able to burst at 10 Mb/s, can give better image quality from MPEG-2.

Are these formats likely to be of interest to the professional AV user? The 1995 prediction, which may well be overtaken by events, was as follows:

- The analog component disc formats (Pioneer VDR-V1000 and Sony CRV) will remain important for broadcasters, professional program development, special applications and as sources for big-screen displays.
- 'Traditional' analog laserdiscs will continue to have a place as reliable video image sources, suitable for interactive

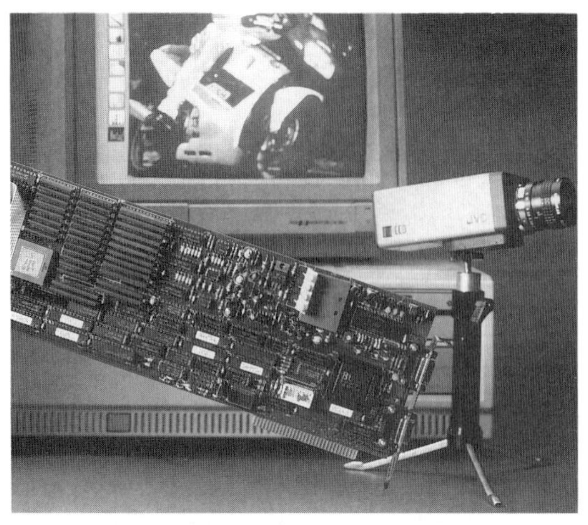

Figure 14.11 Digithurst supply a range of equipment for image capture and processing

Figure 14.12 For advanced interactivity, 'Virtual Reality' puts the participant into the picture. The Virtuality 1000SU from Virtuality Ltd is for standing or moving within a scanned area. The visor provides an interface with the virtual world, and interaction is by a hand-held unit

and public display. Pioneer, in particular, is committed to supporting the format until well into the twenty-first century, so if the format does the job, there is no reason to be concerned about an investment in the medium.

□ CD-ROM, video-CD and related media using MPEG-1 will prove highly effective as program carriers. In the consumer market the emphasis will be on games and multimedia. Movie distribution in this format may be hampered both by the need for two discs to carry a full-length movie, and by the knowledge that new formats may be on the way. In the AV world, the fact that the various CD formats are compatible with computers will make them fit in very well, and their use will be widespread.

□ New consumer formats will take time to become established, with any 'disc war' simply delaying acceptance. By 2000, however, it is reasonable to suppose that MPEG-2-based products will be well established. Until they are established the use of the new formats for anything other than one-of-a-kind applications should probably be avoided by the AV community.

Figure 14.13 The Virtuality 1000SD from Virtuality Ltd is a sit down console which allows the participant to drive his 'vehicle' in a virtual world. While the first public applications of virtual reality are entertainment oriented, it is reasonable to expect that it will be used for serious training and presentation applications

Summary

New methods of AV production on computer, and new possibilities of interactive AV based on compact disc formats are changing people's expectations of how AV can be used. However, the principles of show-making remain the same. Shows do not make themselves, they require hard work whichever method is used. The choice of production method must be tied both to the result required and the abilities and preferences of the producer.

A representative of a major bank in the UK was overheard to say that all training and AV budgets and methods were under the control of the IT (information technology) department. Just as commerce as a whole is getting away from the tyranny of the DP (data processing) department, by virtue of the de-mystification of computers due to the arrival of the PC, here is a company with its own empire-builders determined to hijack AV. A computer is *not* necessary for AV production, and an IT technologist is probably one of the worst people to advise on general AV program production and presentation needs. Nonetheless, computers will create tremendous opportunities for, program makers, and users who approach them on the basis of not being afraid of the technology, and not being afraid to reject inefficient or impractical aspects of their use, will benefit from them.

Chapter 15

Slide presentation systems

A review of AV media may have indicated that slides are the best medium for a particular application; but it may then seem as if the problems have only just begun. What was once the magic lantern of Victorian times seems to have become computerized. In fact, it is the very flexibility of the slide medium that has led to the enormous choice of presentation systems and formats. This chapter examines different possibilities and defines some slide projection technical terms.

Slide projectors

All AV applications of slides use the automatic slide projector. In Europe, especially, there are many automatic slide projectors which use straight slide trays holding about fifty slides. On the whole, these projectors are not acceptable for AV use, although they are fine for the photographic amateur and for simple visual aids work in education.

The standard projector for AV work is the Kodak Carousel™. This range of projectors uses a circular tray containing 80 slides and operates on the gravity slide-feed principle. Two ranges are produced. Eastman Kodak in the USA produce the Ektagraphic III™ range. This is mainly intended for the USA and other 110 volt markets, although there is one model for international use which operates on both 110 and 230 volt. The Ektagraphic is essentially a 'mechanical' projector, with few electronic parts.

For 25 years, the European and 230 volt markets were served by the Carousel SAV series of projectors produced by Kodak AG of Stuttgart. These reliable workhorses give years of service, and although they have been out of production for several years, they can still be seen in permanent installations and in rental fleets. They command a good second-hand price!

In 1992 Kodak AG introduced the Ektapro™ range of projectors. These are high-specification projectors incorporating microprocessor control. The magazine drive is by stepper motor, and slide feed is separately controlled. The top models feature fast random-access selection of slides, a built in triac for lamp control, cordless remote control and an RS232 port for direct computer control.

The success of the Kodak design, the expiry of the basic patent, and the widespread acceptance of the Carousel™ slide tray has resulted in other manufactur-

Figure 15.1 In Europe projectors with straight slide trays are favored by amateurs and are also widely used professionally in education and training. (Photo Leica GmbH)

Figure 15.2 The Eastman Kodak Ektagraphic III projector is the standard in Canada, the USA and other 110V countries

Figure 15.3 The Kodak Carousel SAV series of projectors was a world standard for 25 years, and is still widely used

ers producing Carousel lookalikes, or self-contained slide projection units using the same slide tray. Some examples are the Telex range in the USA, the Elmo Omnigraphic from Japan, and the Simda projector from France. The Telex and Elmo projectors are similar to the Kodak Ektagraphic in that they are 'mechanical' projectors with a single motor. The Simda was the first 'electronic' projector using microprocessor control and separate drive components, and is therefore more like the Ektapro. It should be noted that because some of these projectors have slightly different electrical arrangements, it cannot be assumed that all control equipment designed to work with Kodak projectors will work with other projectors. It is always best, therefore, to purchase projectors and control equipment from the same source, to ensure that only one vendor is responsible for the successful operation of the system.

There is a huge range of lenses available for slide projection, covering every requirement from compact back projection to long-throw projection in large auditoria. There is no excuse for having a projector perched in the middle of the room because the wrong lens is being used. The possibilities are explained in Chapter 21.

Focusing is a common problem affecting choice of slide projector. In an ideal world, all audio-visual users mount their slides in glass and set up their slide projectors properly. They then need to focus them only once, and have no need for either remote focus control or automatic focus.

Any professionally produced audio-visual slide show *must* be delivered in glass mounts. Any professional company providing speaker-support slides for

Figure 15.4 The Kodak Ektapro™ is Kodak's top professional slide projector. It uses microprocessor control to support a wide range of features

Figure 15.5 The Simda projector from France is one of the competitors to the Kodak range. It features a particularly fast slide change. The model illustrated has a 400W lamp for increased light output

Figure 15.6 The Elmo 253ALC is another competitor to Kodak. This model incorporates automatic lamp charge

use in a company meeting or similar occasion supplies the slides properly mounted. Thus, slide projectors used for running multi-image shows and most serious applications of slide projection do not need either a remote focusing capability or an autofocus capability.

However, in the real world, both facilities *are* sometimes required. In a presentation room it is usual to provide remote focus control on projectors used for speaker-support slides, especially to take account of the fact that a presenter may be using slides mounted in different kinds of slide mounts. All the main automatic slide projectors used for AV applications have the facility for remote focus control.

Autofocus is really only needed for card-mounted slides, or for plastic mounts without glass. This is because heating of the slide in the projector gate can cause it to pop and go out of focus while projected. Many mass-produced slides, instant slides and some bureau-produced slides are supplied in glassless mounts, usually for reasons of cost, but also because slides are then easier to mail. A user of these slides has to weigh-up the importance of the occasion; if it is a major presentation, slides must be remounted in glass. If, on the other hand, it is a weekly management meeting and slides are seen once only, or is an informal training session, it is reasonable to leave them glassless. In such a case autofocus becomes desirable, if not essential. Only a few of the projectors widely used for AV work have an autofocus facility.

When slide projectors are used in permanent installations, or in automatic AV presentations, it is best to use projectors with automatic lamp changers. Kodak, Simda and Elmo all make models fitted with two lamps arranged such that when the first one fails, the spare is automatically connected. In the Simda and Elmo this involves a motor arrangement to move the spare lamp into the correct position. The Kodak Ektapro has an elegant arrangement using a moving mirror.

If automatic lamp changers are used in permanent installations, it is a good idea to always ensure that the spare lamp is moved into the No. 1 position, and the new lamp into the No. 2 position. In this way the possibility of the spare lamp itself failing is minimized. Fixed automatic installations in museums and similar environments should have a remote indication that the spare lamp is in use. In this way it is easy to carry out a simple daily check to ensure no lamps need changing, even if the projectors themselves are difficult to access.

More light

A common request is for more light either because the user wants a bigger picture, or to fight ambient light conditions. There are ways of achieving brighter images, but an understanding of the basic problems involved helps to determine whether a high-output projector should be used, or whether there is some alternative solution to the problem. Unfortunately there are cases of users being disappointed by the performance of bright projectors because they have not understood the fundamental problem.

The human eye is good at comparing light levels on a side-by-side basis, but not on an absolute basis. Its response is not linear, so if light output is increased by, say, 30 percent, it is not very impressed. A *doubling* of light output is noticeable, but to create a real impression the light output must be increased by a factor of four.

A standard slide projector can give a good 4 m (13 ft) wide picture in a darkened auditorium. However a large auditorium might need an 8 m (26 ft) screen to ensure correct picture size for the back row of the audience. A doubling of the picture width means a fourfold increase in picture area and, to get the same screen illumination, a projector with four times the standard light output is needed. This application is

Table 15.1 Image sizes from a standard slide projector

Front projection in darkness	5 m (16.4 ft)
Rear projection in darkness	3.5 m (11.5 ft)
Rear projection in low ambient light	2 m (6.6 ft)
Rear projection in medium ambient light	1 m (3.3 ft)
Rear projection in high ambient light	0.5 m (1.6 ft)

This table is for guidance only; actual image achievable depends on site conditions. The dimension is image width on 35 mm or superslides

one where a high power projector is justified, because many users of convention auditoria want to show single slides to large audiences. Tables 15.1 and 15.2 give an idea of different light outputs and image sizes of various standard and special projectors. Some shows might mix screens of different sizes, and could justify a mixture of projectors to achieve uniform illumination on the total screen area.

A point frequently overlooked is that there are

Table 15.2 Approximate light output from some typical slide projectors. All are for standard 35 mm slide, and with 90 mm *f*/2.5 lens except where stated

Kodak Ektapro/Ektagraphic III with EXW lamp (15 hours)	1300 lumens
Kodak Ektapro/Ektagraphic III with EXR lamp (35 hours)	1150 lumens
Kodak Ektapro/Ektagraphic III with EXY lamp (200 hours)	690 lumens
Kodak SAV with over-run 400 Watt lamp and *f*/1.2 lens	2000 lumens
Hardware for Xenon 600 Watt xenon arc projector	4000 lumens
Hardware for Xenon 1000 Watt xenon arc projector	6000 lumens

other ways of achieving subjectively brighter images. A photographer might argue that a particular slide looks best when exposed one stop down, but this

Figure 15.7 Projectors from Eiki based on Kodak Ektographic™ with xenon arc lamps and built-in shutters

halves the light! It is far cheaper to increase light on the screen by judicious choice of slides than to use expensive projectors. Graphic slides with bright colors (no deep blues or dark purples) and selective rephotography with just one stop more can help a lot. On the equipment front: dirty optics, unaligned lamps, low mains supplies and the use of non-multicoated projection lenses can all cause less light on the screen than is really available.

How is more light achieved? One way is simply to use a bigger lamp. For example Simda offer their projectors with a 400 watt 36 volt lamp in place of the standard 25 watt 24 volt. Another way is to over-run the lamp, or use a lamp with a shorter rated life with higher light output (which actually amounts to the same thing). Table 15.2 illustrates this point.

Over-running lamps is acceptable for conference work and other one-off shows, where the procedure can be to use standard or underrun lamps for rehearsal, and a brand new set of lamps at full power for the actual show. The procedure is only recommended when carried out by experienced staging staff.

This type of approach is economic for single conference applications, but is clearly no good for permanent installations, or where more than double the standard light output is required. Here the only realistic choice is the use of projectors with xenon arc lamps. Although expensive, they do give a significant increase in light output, and lamp life is respectable (2000 hours on some models). A xenon lamp cannot be directly dimmed, so it might be thought that such projectors cannot be connected to a dissolve unit. However, projectors are available with motorized dimming shutter or other variable light transmission device that allow the direct connection of standard control equipment.

Figure 15.8 The company 'Hardware' of France specialize in high-power slide projectors, both 35 mm and big format. This is one of their big format projectors

The xenon arc has a different color temperature from the tungsten halogen lamps used in standard projectors. It makes slides a little bluer and subjectively they appear even brighter. It may be necessary to take care when mixing standard and xenon projectors.

The limit to the light output from the projector is the luminous flux through the slide itself. Even with a perfect projector that only puts light radiation through the slide, and rejects all infra-red heat, there is a limit. This is because a slide works by selectively absorbing different wavelengths of light, and this absorption is turned to heat within the slide. The most that a superslide can take is about 7500 lumens.

Slide life decreases more-or-less in direct proportion to light passed through the slide. Users of high-power projectors in permanent installations must remember to allow for extra slide copies accordingly.

In the theater where very large and bright single images may be needed for scene projection, big slides up to 25 cm square are sometimes used in special and expensive projectors giving 50,000 lumens and more. Exponents of the big spectacle, like Jean-Michel Jarre with his mega-concerts use projectors of this kind for lighting building exteriors. The same type of projectors are also used in big product launches, where screen sizes are too big for conventional projectors.

Special format projectors

The standard 35 mm slide and its near relation the 'superslide' meet the needs of most users of slide projection. However, photographers, art directors, advertising agencies and others sometimes require to project slides of a larger format, either because they are seeking the ultimate in quality, or because they wish to project an original transparency.

This has traditionally been done using either

Figure 15.9 The Hasselblad PCP80 Projector for 54 mm × 54 mm slides

Figure 15.10 The idea of dissolve change with slide projection is not new. This splendid Triunial magic lantern, built by J. H. Steward in 1895, has hydrogen and oxygen gas regulators to achieve dissolves. (Photo Doug Lear)

manually-operated projectors, or ones designed for the amateur market. Hasselblad make a professional automatic slide projector for 54 mm by 54 mm slides. It is built on similar principles to the Kodak Carousel, and has many features that make it suitable for professional AV users.

This type of projector can be the basis of very spectacular audio-visual shows. However, there is an important point to bear in mind: the ability of the human eye to resolve detail is amazing, but not so amazing that it can pick out fine detail at great distances. The benefits of this kind of projector are greatest when the viewer is relatively close to the screen, and is able to appreciate the improvement in image quality.

What is a dissolve unit?

Most automatic slide projectors take about one second to advance from one slide to another. This means that there is necessarily a dark interval between each slide. This may be acceptable for simple visual aids work, or for slide AV shows shown to individuals, but it is certainly not acceptable for prestige presentations or AV programs shown to groups.

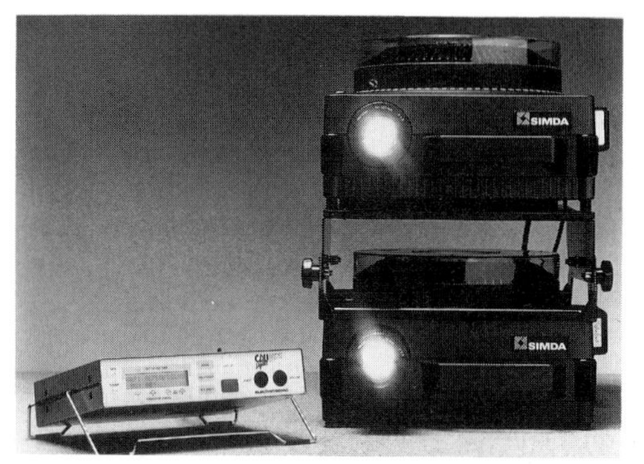

Figure 15.11 A typical modern two-projector dissolve unit. Note the use of a simple projector stacking arrangement

The solution to the problem is to use *two* slide projectors and an electronic dissolve unit. This device controls the projector lamp brightness, and can produce a dissolve from one image to another. The basic idea is not new. No self-respecting magic lanternist of the nineteenth century would have ever given a show without dissolve picture change, although in those days the slides had to be changed by hand and light regulated by turning gas taps.

Modern electronic dissolve units use micro-processor technology to give a wide range of facilities. In simpler applications, slides are loaded into alternate trays and a press of the appropriate button, either on the dissolve unit or on a remote control, gives a choice of cut, dissolve or reverse-cut slide change. The dissolve speed can usually be preset.

Simple dissolve effects are themselves quite sufficient for showing a sequence of lecture slides. However, when a sequence is to become part of an AV show with a recorded sound track, it can be useful to have a wider range of effects. Therefore, many dissolve units allow the show producer to use a lot of different dissolve speeds in one show, to deliberately superimpose slides, and to produce simple animation effects by rapidly alternating lamps without slide advance.

Because all dissolve units need to control the projector lamp, the projector must be fitted with a dissolve socket to allow access to the lamp circuit as well as to the slide advance. Manufacturers of AV slide projectors have tended to fall into line behind Kodak; on 110 volt projectors using 80 volt or 120 volt lamps a seven-pin remote-control socket is used; and on 220 volt projectors using 24 volt lamps a twelve-pin socket is used.

The electronic component that directly controls the

lamp power is called a triac. One of the disadvantages of having this device inside the dissolve unit is that it does dissipate some heat (about 15 watts on the 24 volts, 250 watt lamp) and requires thick power cables from the projector on the 24 volt lamp to prevent voltage drop from causing loss of light output. Therefore, some dissolve unit manufacturers put the triac in plug adaptors, which has the advantage that the dissolve unit can be used with different types of projector just by changing the adaptor (rather than the cables).

A further option is to put the triac in the projector itself. This is done in the Kodak Ektapro projector. The practice is also widely adopted by European projector manufacturers, such as Leitz and Kindermann, which sell mainly to photographic, amateur and educational markets. Dissolve units made for these markets are very compact because the placing of the triacs in the projectors removes both the largest and hottest components.

A reasonable question is: why does nobody make a dissolve projector? In fact such dual projectors *are* available (e.g. Rollei). The reason why they have not found favor in the AV world, as opposed to the amateur market, is that they do not have sufficient slide capacity and do not use the Carousel slide tray. Also, for multi-image work it is often necessary to use more than two projectors, so it makes better sense to standardize on single-unit projectors.

Projector stands

Use of more than one projector on one screen introduces the need for special projector stands. The aim of any stand system is to get projectors as close together as possible, while still leaving access for tray changing.

Although rehearsal of a dissolve slide show is most easily done with projectors side-by-side, the preferred arrangement for actual shows is to stack the projectors, because this keeps the lenses as close together as possible and minimizes the space taken up by projectors.

Shows that are based on unrelated images can use simple stands. For example, if only two projectors are used, a simple tray stand allows one projector to be placed on another. If three projectors are used, a simple triple-shelf stand can be used. However, if multi-image techniques, such as successive reveal or soft-edge masking (see Chapter 16) are used, projectors must be aligned with great accuracy.

The only practical way of achieving this is to use a stand in which the projector is held as an entity, and *not* to use its own rather crude adjustments for levelling. The stand itself allows independent adjustments to each projector for roll, pitch and yaw. Stands are available that take one, two, three or four projectors.

Figure 15.12 A precision stacker stand for professional shows and permanent installations. (Photo Chief Manufacturing Inc)

The stacker stand must be raised to the correct operating height. For travelling shows fold-up stands that pack into a suitcase size are available. Multi-image shows using more than three projectors usually need a substantial shelving system. In permanent installations, the base stand should be as rigid as possible, because the slightest vibration, even that caused by a slide advance on a neighboring projector, causes noticeable picture shake. Also, in permanent installations, the stacker stands should be permanently bolted to the base.

Synchronizing to sound

The principle of synchronizing slide projection to a sound tape is very simple: a separate track on the tape is used to carry control signals. Most AV shows are finally run from the standard compact cassette. This calls for the use of audio-visual tape recorders, because crosstalk of a standard cassette recorder is not good enough to permit use of adjacent tracks. Apart from the fact that use of adjacent tracks prevents use of stereo sound, the control signal is audible. The audio-visual tape recorder has a well-separated control track which eliminates these problems, and gives inaudible control of projection.

Educational use of single-slide projectors linked to sound is sufficiently widespread that there is an international standard for the control signal. This consists of a short burst of 1000 Hz tone. The tone burst is detected by a simple electronic circuit and is used to advance the projector. The same principle is used in self-contained, single-projector slide/sound units used for individual presentation and training, such as the Telex Caramate™. A 150 Hz signal is often used for tape stop or program pause.

A dissolve unit can be controlled by the same kind of signal. However, a single type of pulse signal can only give one type of command. If advantage is to be taken of multiple dissolve speeds, animation etc., it is necessary to have a more complex system of signalling.

For two-projector systems a widely used method is to have a continuous signal of varying frequency that is an analog of the relative lamp intensity. One frequency represents the A lamp full on, with the B lamp off, while at another frequency the opposite occurs. Any frequency in between represents a proportional mix of the two lamps. This kind of system allows any speed of dissolve to be recorded and such recordings can be made in real time using a slider control.

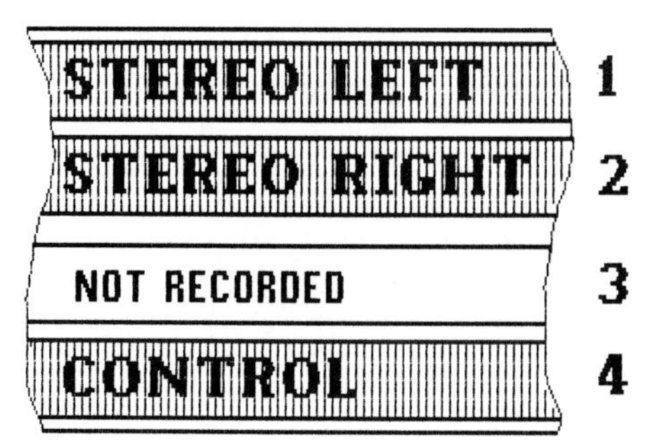

Figure 15.13 Typical track arrangement on a tape cassette used for audio-visual shows. Unlike a normal music cassette the tape only plays in one direction, and must be rewound after each show

Although this method of programming is attractive to photo amateurs and is also suitable for simple in-house production in training departments, professional shows tend to use computer methods for show programming. This results in the need to use complex digital signals to control projectors. The signals used can control more than two projectors, and have the added advantage that they can include slide position data. This means that, in principle, if the tape is started in the wrong place the projectors quickly catch up with the tape and get back in sync. Signals used are described in more detail in Chapter 16.

The single-projector show

Although this chapter is mostly concerned with slide presentation systems that use two or more projectors to produce shows with a professional look, it is important to understand there is a useful role for the slide/sound show that uses only one projector.

One of the great advantages of these systems is simplicity of use. Not surprisingly, the AV cassette recorder with a single-slide projector is widely used both in education and training. It is especially suitable when the material is not a show but, for example, a relatively long verbal description of a number of separate images.

Single-projector shows are also an effective and economic basis of individual AV communication. They give appreciably better image quality and lower production costs than any video-based system. This has led to a demand for special self-contained portable slide/sound units, either consisting of a standard projector in a special carrying case with a folding screen, or an 'AV box' construction.

Why three projectors?

For many users of slide programs two projectors on a single screen are quite adequate. There is no need, in fact it is an obvious disadvantage, to make a slide projection system more complex than is necessary for

Figure 15.14 An AV cassette tape recorder suitable for making simple shows. (Photo Kodak Ltd)

Figure 15.15 A self-contained unit for slide/sound shows, suitable for individual presentation. (Photo Telex Communications Inc)

Figure 15.16 A memory dissolve unit from Arion

a particular application. However, many slide-based AV programs now use three projectors on one screen. Why?

The main market for shows using three or more projectors is in professionally-produced shows where the producer requires the extra flexibility that three projectors can give:

□ Greater slide capacity. Shows made for sales promotion or motivation can be so fast-moving that a 240-slide capacity is needed for a twelve-minute show.

□ Faster slide changing. Three projectors allow more rapid slide sequencing, which may not be necessary for a whole show, but may be useful for a fast-moving introductory sequence.

□ The possibility of dissolve superimpositions, i.e. the ability to dissolve a sequence of slides on another background slide.

□ Simpler slide production. Dissolve superimposition is one of several effects that allow use of simple slide production techniques to produce an elaborate effect on the screen.

□ More animation possibilities, e.g. three-way animation or two-way animation superimposed on a static background slide.

A show that *needs* three projectors is inherently more complex to produce and program than one that uses only two. Two-projector shows are usually programmed in real time. This means that the person programming the show listens to the sound track and simultaneously either moves a slider control or uses push-buttons to achieve required effects. Editing is done by simply re-recording a section.

With three projectors this procedure becomes impractical. Shows using three (or more) projectors require a programming system that allows leisure time programming, so the show programmer can enter commands first, preview sections or the whole of the show, then edit the commands to fine tune the production – usually by making changes to timing of individual effects. The process is described in more detail in Chapter 16.

The memory dissolve unit is typically designed to control two, three or four projectors. Activities of each of the three projectors can be programmed independently, with a selection from a wide range of dissolve speeds and animation effects. In addition, timing between individual cues or groups of cues can be pre-programmed, so visual effects can be precisely rehearsed and, once completed, the show producer can be sure that each effect plays back exactly as intended.

Memory dissolve units can store about 1000 individual cues, which is more than sufficient for the aver-

Figure 15.17 A two-projector audio-visual presentation unit. This unit can both make and play back shows; it is suitable for training departments

age three-projector show, even if all slide trays are full. Completed shows can either be run from the dissolve unit's memory or, more usually, are then transferred to magnetic tape in synchronization with a sound track. The dissolve unit then acts as a playback device, decoding the complex signals from the control track of the tape. For users only needing to playback ready-made shows, dissolve units are available without programming capability.

A dissolve unit does not necessarily control either two or three projectors. Some vendors offer unit dissolvers where there is one dissolve unit per projector; others offer dissolve units able to control four projectors. However, after the simple two-projector dissolve unit, the three-projector unit has become the most common, both because the three-projector stack is the most common and practical multi-image unit, and because it represents an economic use of microprocessor power.

The presentation unit

Most end-users of slide-based audio-visual are more concerned about the successful showing of completed shows than about the mechanics of their production. Even at two projectors, let alone three, the system is already beginning to look a little complicated, especially because it includes separate projectors, dissolve units, AV tape recorder, amplifier, etc. Obviously for most users it is best to combine all the electronics into one box.

This is done with the audio-visual presentation unit. Two types of unit are common. First is a unit intended for training departments and similar users who wish to make their own, relatively simple, shows. A typical unit is a combination of an AV cassette recorder, power amplifier and two-projector dissolve unit.

Second is a playback unit only, suitable for either two or three projectors, and intended for playback of professionally produced shows. These units usually include high-quality stereo sound reproduction with

Figure 15.18 A three-projector audio-visual presentation unit. This one is suitable for permanently installed shows that are given on a regular basis

an amplifier power of at least 2 by 20 watts. They should have some system of noise reduction if they are to be used for large audiences. Most allow for addition of extra dissolve units to run larger multi-image shows.

Presentation units are usually intended to be used for occasional and travelling shows. Not all are suitable for permanent installation, so users requiring such units for fixed or continuous-running shows should check on the suitability of a particular unit for unattended use. Additional features of these units are:

☐ The unit is fitted with a heavy-duty automatic tape deck.

☐ The unit can control projector power for automatic switch-off in unattended push-button started shows.

☐ The unit has built-in routines for continuous running or one-shot shows, with full auto-present routines to ensure that projectors are checked to be at 'home' before start of each show.

☐ Automatic slide/sound shows in museums and other places requiring continuous operation do not use tape as the sound source. They would normally use either CD or solid-state sound replay units (see Chapter 11).

Why multi-image?

It is always a matter of debate as to where single-screen slide/sound stops and multi-image starts. It is usually defined as three projectors or more.

Thus, it is possible that multi-image has yet more projectors on one screen or may, more likely, be multi-screen. The reason for requiring more projectors is simply an extension of the arguments for three projectors; higher speeds, and more elaborate effects. The reasons for more than one screen can be summarized as:

☐ To match the show to the environment. A single standard-format screen can look singularly unimpressive in a large auditorium. The multi-image technique allows creation of big screens to match large audiences.

☐ Low ceilings of many conference sites demand a long, low format.

☐ To allow creation of a total show environment, by using the flexibility afforded by multiple projection.

☐ To produce special show formats, especially in exhibitions and in permanent installations in museums, visitors' centers and tourist attractions.

Because the multi-image medium is so flexible, there are not really any set rules about screen formats. However, it is natural that many users have similar requirements, and there are several popular formats (see Chapter 16). The majority of multi-image shows made for business presentations are in the nine-to-

Figure 15.19 At the Siemens Museum in Munich there is a multi-image show using 84 slide projectors. It gives an effect attainable by no other medium. (Photo courtesy of the show's producer, Kurt Polke (shown here during the programming) and Mietzner & Mattis of Munich who engineered the system)

fifteen projector range, with only an occasional requirement for bigger systems for exceptionally large audiences or sophisticated shows.

Some public entertainment and exhibition shows, on the other hand, use very large numbers of projectors. Shows using 48, 60, 96 and 120 projectors are not common, but neither are they unknown.

Staging slide shows

One of the problems of slide-based AV shows is the number of separate items involved. Despite the underlying simplicity of the slide medium, there can be a tendency to underestimate the effort needed to ensure a perfect show.

There are few problems associated with the extremes of the business. The single-projector show, even the simple dissolve set-up, presents little difficulty. At the other extreme, the large permanently-installed system, provided it has been correctly engineered and receives the recommended routine maintenance, gives reliable continuous service.

The problem is in the middle where, for example, an industrial user has a six-projector show which he wishes to move round several different venues. It can be tempting to understaff the exercise, and involve presenters in the business of staging it. A good rule is: never allow anyone who is part of the presentation to touch equipment!

This suggestion applies to any kind of presentation using AV (not just slide-based), but is especially important with slide and video projection because of the care needed to align projectors; and the know-how to avoid environmental problems in strange hotels. Therefore it is best to have a technician, or at least someone who regularly uses the same equipment set, responsible for staging the show.

Three-dimensional projection

It is possible to project three-dimensional slide images, using the principle of polarized images. The original photograph is taken by a special double camera, or two cameras with their lenses separated by a distance similar to that between two eyes. The two slides are then projected by two projectors, each with a polarizing filter. Polarization of light can be thought of as to making all the light waves move in one plane, like a rope shaken to produce 'waves' but only permitted to go through a vertical set of railings. Thus, light can be horizontally or vertically polarized. If already polarized light is viewed through a polarizing filter, the amount of light seen depends on whether the light is polarized in the same direction as the filter.

Theoretically, nothing is seen if horizontally-polarized light is viewed through a vertically-polarized filter, and vice versa. In three-dimensional slide projection, two slide projectors are used to project each image. One has a horizontal polarizing filter, the other a vertical polarizing filter. The audience view the show through a matching pair of polarizing spectacles, ensuring that each eye only sees the image meant for it.

Three-dimensional slide projection is done with standard projectors, with the sole addition of necessary filters. If dissolve equipment is used it is necessary either to double up on the control equipment, with units working in parallel pairs, or to use dissolve units specially modified to have dual outputs. It is essential to use a screen designed for three-dimensional work, as many types of screen depolarize incident light.

The making of three-dimensional slide shows is, in principle, no more difficult than making ordinary two-dimensional shows. The discipline of planning, scripting, storyboarding and so on is the same as usual. Simple exterior shots are not a problem. However, when it comes to trick effects, especially 'matting' a foreground object into a background, a great deal of experimentation may be necessary to achieve convincing results.

The combination of the additional staging complexity, and the considerable know-how needed to get three-dimensional shows right first time means that there are, in fact, relatively few professional producers with the necessary experience.

Random-access slide projection

Most users of slide projection show slides in a predetermined sequence, so slide projectors simply advance from one slide to the next. For some applications it must be possible to select slides at random from a library, so that any slide can be retrieved in, say, three seconds.

Figure 15.20 The Mast random access projector is built into the Ektagraphic™ projector body. Any slide can be selected in under three seconds. Selection can be by the simple controller shown here, or by computer

Projectors providing this type of facility are called random-access projectors. They require some kind of motor drive mechanism to drive the slide magazine rapidly to the required position. The Kodak Ektapro and the Simda projectors provide the facility as standard, because they both use a stepper motor drive for the slide magazine. In the USA, several vendors offer suitably modified Ektagraphic projectors.

The possible uses of random-access slide projection are many:

- ☐ Display of maps, building drawings, system drawings, plant diagrams etc. 'in control rooms, emergency operations centers, control consoles and other industrial and public service environments.
- ☐ In briefing, presentation and seminar rooms, where slides, charts, maps and so on may need to be shown to groups on demand.
- ☐ In product display applications, both in showrooms and at the point of sale.
- ☐ In simulators and other high-level training applications.
- ☐ In museum, exhibition and visitors' center applications.

Method of selection varies according to complexity of the application. Usually single projectors are operated by a simple calculator-style keyboard, but frequently projectors must be operated by computer. In this case the random-access projector can be supplied with an RS232 interface to allow direct connection to the computer. On other occasions the activities of several such projectors must be coordinated and this can be done either by computer control or by special control panels.

Simple random-access projectors can be purchased

from an AV dealer. However, special applications usually call for an element of project engineering, and may need direct contact with the manufacturer. Some of the many technical possibilities that now exist include automatic display of sequences. For example, a museum display might give a visitor the choice of several slide sequences on different subjects. Pressing the appropriate button immediately selects the first slide in the sequence, and the remainder are shown at a preset interval. At the end of the sequence the screen reverts to a home slide. This arrangement can be extended to include random-access sound, so that a single display is able to give a choice of several mini programs, all with nearly instantaneous start. These systems are not limited to single projectors; both simple dissolve and full multi-image facilities can be added.

Random-access slide projection can be combined with computer data display. For example, a monochrome high-definition data projector can superimpose on a high-quality color photographic image projected by the random-access projector.

Choice of which kind of random-access imaging system always depends on the final application, and on the way in which software is produced. For individual instruction on a semi-institutional basis the optical disc is technically the most attractive method; whereas for storing and retrieving thousands of fingerprint records, random-access microfiche may be the answer. Random-access slide projection succeeds in those applications where slides are the preferred software, for example where:

- ☐ big images are needed;
- ☐ photographic quality is needed;
- ☐ individual images must be easily and inexpensively updated;
- ☐ a single-venue application is involved.

In all these cases a slide system is the most cost-effective and simplest system to install.

Chapter 16

Multi-image programming

This chapter reviews the subjects of multi-image programming, slide mounts and show formats. This could be thought to be too specialized to be of interest to the person commissioning audio-visual programs, however an understanding of basic processes involved is of help in understanding how any AV sequence is put together. Although the discussion centers on multi-image as it is currently understood, based mainly on slide projection, similar principles apply to any video multi-image systems that are now emerging.

Figure 16.1 A range of dustproof slide mounts: 35 mm, Superslide, 54 mm and Idealformat

Slide formats

Although in theory there is a large number of slide formats, in practice the situation is now very simple. The great majority of slides are made on 35 mm film, and it is the dimensions of this film that determine what can be carried on a slide.

Slides are conventionally mounted in mounts 50.8 mm (2 in) square but, of course, produce a rectangular picture. This led to development of the superslide which has an opening of about 38 mm (1.5 in) square – largest possible within the mount. Superslides are made either by cutting down transparencies made on 120 film (i.e. from 54 mm square) or, more usually now, by rephotography onto a special 46 mm film on a rostrum camera. This film is perforated only on one edge and can only be used in special cameras.

There is sometimes a need to project original transparencies of a larger format. For this purpose there are alternative mounts 70 mm (2.75 in) square, which allow direct mounting of 54 mm square transparencies. The application for these slides is mainly in the field of professional photography, top amateur photography and advertising agency work. Even larger sizes are used for special purposes, such as cinema advertising and for some projection in theaters.

All slides should be properly mounted for projection. Although there is an argument for using glassless slide mounts in some high-power projection applications, all normal applications *must* use glass-mounted slides. Glass mounts protect the slide, allow for cleaning when necessary, and ensure that the slide remains properly in focus when projected. There is no place for autofocus in professional slide projection.

Most AV slide mounts use anti-Newton ring (ANR) glass. Newton's rings are colored fringes produced when light passes through two surfaces very close together, which happens when film and glass are placed next to each other. If there is a gap between the two which is of the order of the wavelength of light, colored rings are seen. ANR glass is glass that has been treated to have a roughened surface to ensure there cannot be a significant area of film the critical distance from the glass.

For all normal AV use, mounts with ANR glass are not only acceptable, they are essential. However, purists would point out that areas of film that are projected as saturated white allow the ANR coating to be seen as a mottled effect when viewing the screen from close up. Thus, some super-critical photographic applications call for the use of ANR glass on the film backing side and plain glass on the emulsion side. The argument here is that the emulsion itself can meet the ANR requirement, and, because focusing of the slide is on the emulsion side, the ANR glass is not seen.

The degree of roughening of ANR glass does vary between glass manufacturers, so the recommendation is to carry out a viewing test if it is thought there could be a problem.

Slide mounts are designed to slightly mask the

original image by about 1 mm, to ensure a clean edge and to allow some tolerance in the mounting. An audio-visual show will never use original slides as show copies, both for security reasons and because colors deteriorate with projection. Slide mounts with an extra large aperture are available, therefore, to ensure that when an original slide is duplicated there is no further loss of image area. These are known as duplicating mounts and typically have an aperture of 35.8 mm by 24.4 mm.

Table 16.1 Standard slide formats

35 mm	
Mount size	50.8 mm × 50.8 mm
Image size	36 mm × 24 mm
Available aperture	35 mm × 23 mm
Superslide	
Mount size	50.8 mm × 50.8 mm
Image size	40 mm × 40 mm
Available aperture	38 mm × 38 mm
54 mm or $2\frac{1}{4}$ square	
Mount size	70 mm × 70 mm
Image size	54 mm × 54 mm
Available aperture	53 mm × 53 mm

Note that available apertures are approximate. They represent the masking provided by the mount and vary from make to make. See also Chapter 21

Registration mounts

To simply project a sequence of photographic images on a single screen there need be nothing special about the slide mount, provided it securely holds the film and the complete image is projected.

However, many AV applications have much more critical requirements. For example, successive-reveal slide sequences, or build-up graphics look amateurish if the base image seems to jump around the screen. These sequences, and any multi-screen work where one large image is made up from a number of separate slides, require great precision in slide mounting.

Just getting the mounting right is not, however, sufficient. The whole process of slide production must keep the image in register from the start i.e. the image position on the film always bears a fixed relationship to the position of the film sprocket (see Chapter 2). Finally, the slide must be mounted in a register mount which itself has pegs that accurately locate the film.

There are several reputable makes of registration slide mounts. Lower-cost mounts are two-part and require assembly. Business users prefer register mounts with hinges; although more expensive, they are very quick to use. For ultra-critical applications it is possible to specify that a batch of mounts is all made from the same mold.

Figure 16.2 The Wess Plastic No. 2 registration slide mount is the most widely used mount in professional multi-image presentation

Figure 16.3 Registration slide mounts are also available for large formats. Often these need a viewer/punch to make registration holes in sheet film. (Photo Wess Plastic)

Sometimes it is necessary to register a piece of film that has not been shot on a register camera, or for some reason is out of alignment with other images in the sequence. A viewer/punch is available that allows the user to align a slide then punch round location holes in the side of the film. A special variable-registration mount, with location pins to match these holes (instead of the usual sprocket pegs), is then used. In all other respects the mount matches the normal register mount.

This variable-registration approach is also useful for those who have little general need for registration and are able to do most of their work on non-register cameras. The variable-registration mount can then be used for the few slides in a sequence that are critical.

Slide masks

Slides used in multi-image shows can be a long way from a standard 35 mm photo. The rostrum camera, with its ability to manipulate original material, give accurate multiple exposure and many other tricks, has transformed the humble magic lantern slide. However, not everything is done with the camera. Multi-image shows make frequent use of masks sandwiched in the slide mount with the film. These masks may either be thin metal or high-contrast lith film.

Thin metal masks, made of aluminum or stainless steel, have the advantage that they are completely opaque. Film masks have the advantage that any shape can be inexpensively made as a mask. However, they still transmit some light and before use must be examined for scratches or small imperfections that might give pinpoints of light on the screen. These can be covered using an opaque compound, sometimes known as 'blooping ink'.

One particular kind of mask is widely used in multi-image work to produce large images from a number of slide projectors, without any dividing lines. Normally, attempts to produce this kind of images fail

Figure 16.4 Aluminum masks used for special effects. These have the advantage of being completely opaque, whereas photographic masks on high-contrast film need careful inspection to avoid pinpricks of light

Figure 16.5 An alternative to the sandwiched opaque mask is the ready-masked slide mount. These are especially useful when a large number of slides with an unconventional format are needed. (Photo Wess Plastic)

because it is impossible to line-up projectors with enough accuracy so the viewer is unaware of the picture join. The soft-edge masking technique, however, eliminates the problem.

The technique is best described by example. Probably the most popular multi-image format is an overall screen ratio of 3:1, achieved by projecting two standard slide images side-by-side. How, then, can a single image without any apparent join be achieved? It is done by projecting a third image on the center of the screen, with the center projector projecting half of each picture projected by left and right projectors.

In order to achieve uniform screen illumination each slide is fitted with a soft-edge mask. For slides from the left-hand projector the mask is completely transparent for the left half of the picture, and gradually gets darker towards the right-hand edge where it is opaque. This arrangement is reversed on the right-hand projector. The center projector's mask is completely transparent in the center, and opaque at both edges. The mask sets are matched so that the overall picture is uniformly illuminated.

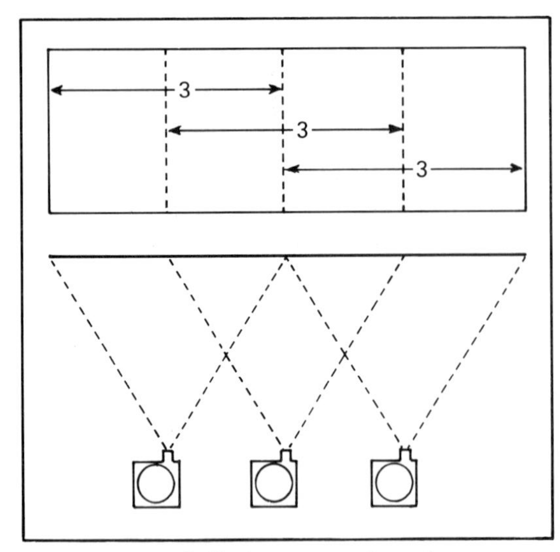

Figure 16.6 Probably the most popular multi-image format uses nine projectors on a single 3:1 ratio screen. Two images side by side with a center-overlap image use soft-edge masking to achieve single panorama pictures

Figure 16.7 A set of soft-edge masks for panorama projection. (Photo RCW Colour Slides Ltd)

This technique only works if projectors are precisely lined up, and if slides are produced in proper register with each other. Even wider screens can be used by repeating the process. The technique has even been done in both vertical and horizontal directions to produce very big images. It is not actually necessary to use a 50 percent overlap, a 25 percent overlap can be sufficient, but 50 percent is often preferred because it gives more flexibility in the use of the projectors.

Show formats

Single-screen shows using slides normally use the standard 35 mm slide landscape format. This gives a satisfying 3:2 ratio. No professional AV show mixes landscape and portrait slides.

Figure 16.8 For many years the Royalty and Empire Exhibition at Windsor, England ran a multi-image show that used soft-edge panoramas

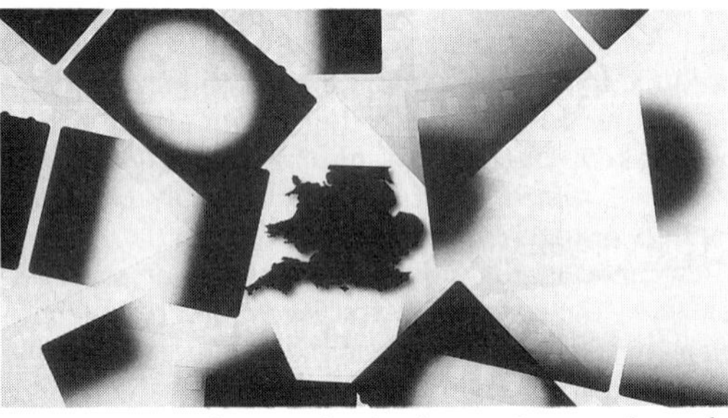

Figure 16.9 The use of soft-edge masks is not limited to panoramas. They can be used for montage and vignetting effects. (Photo RCW Colour Slides Ltd)

Occasionally superslides are used on a single square screen, but this format is not very satisfactory, because if screen height is fixed by practical considerations, usually a low ceiling, the available picture can look smaller. Superslides can, however, make a most satisfactory component of multi-image. Some producers prefer the square image element, and all available light from the projector can be used.

Multi-image formats are a matter for the imagination – anything goes! However, because many business programs must be shown in offices, hotel banqueting suites and other places with low ceilings, there are a number of preferred formats. On the other hand, shows for exhibitions and display do not need to suffer from such restrictions, and the designer can be adventurous.

Although business shows, and shows of a documentary nature used in visitors' centers, make a lot of use

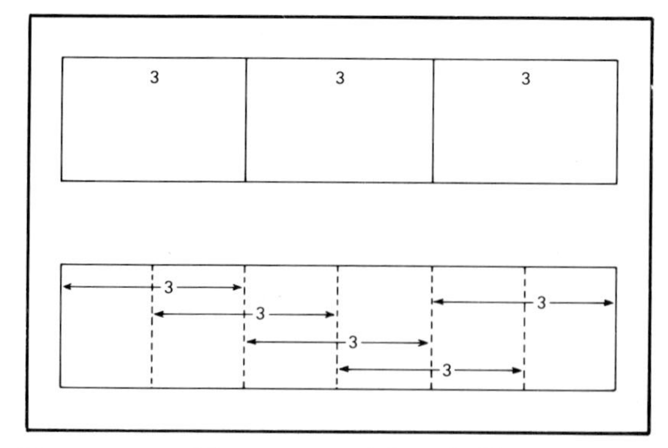

Figure 16.10 Three images side by side. The screens can either be deliberately mullioned, or more projectors can be used with soft-edge masking

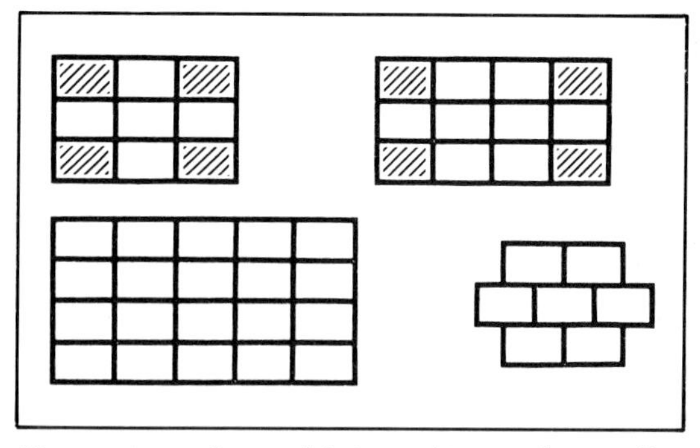

Figure 16.11 Some exhibition multi-screen formats. Here each image would be quite small. The shaded screens are optional

of the single big screen with soft-edge masking technique, exhibition shows often make use of the 'matrix screen' approach. Here, individual images are deliberately divided and the arrangement of the divisions is part of the display design. Usually the screens in these displays are relatively small in order to ensure adequate screen brightness.

Multi-image system basics

Multi-image is such a flexible medium that at first it is a little daunting to try and analyze what is going on. But however complex the format, whether a show is using six projectors or sixty, a multi-image system is completely modular. Thus, in principle, it is only necessary to understand how a small system works to be able to design programs for any size system.

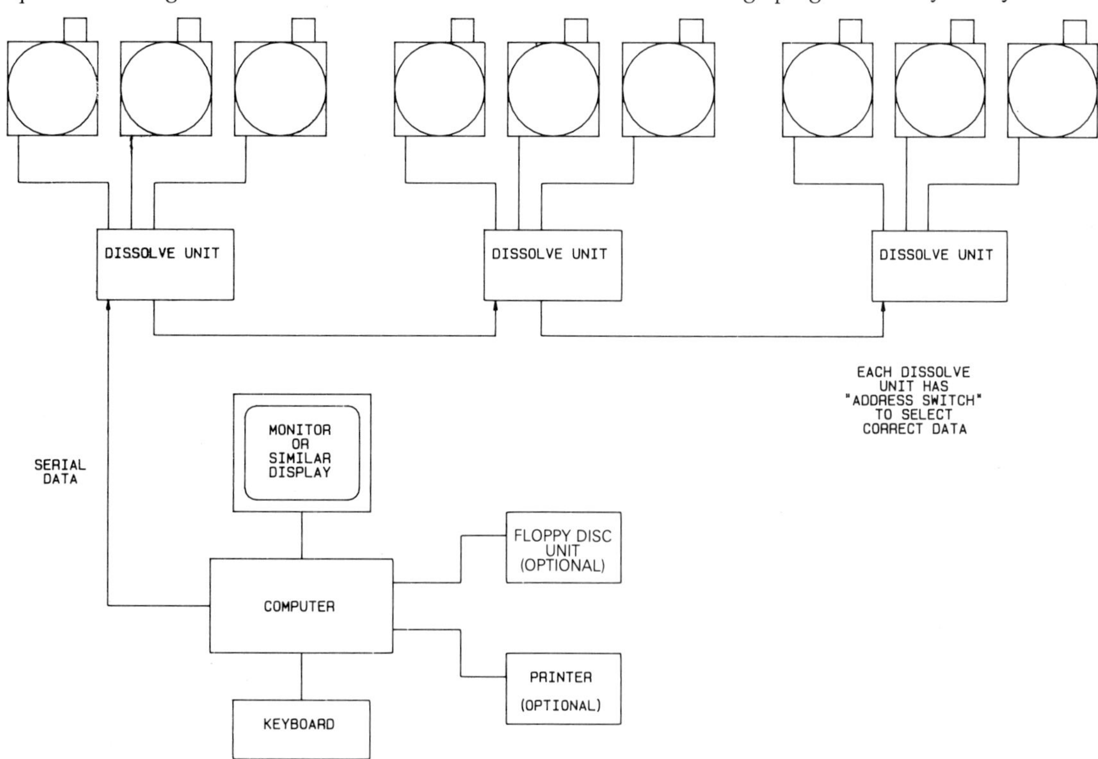

Figure 16.12 Block diagram of a typical programming set-up for a nine-projector show

Consider the nine-projector system shown in Figure 16.12. From a theoretical point of view it does not matter how projectors are arranged. They could each be directed at a separate screen (nine screens, one projector per screen) or even all be directed at one screen (one screen, nine projectors). In fact, they are likely to be organized as three groups of three, because for the reasons explained in Chapter 15 a group of three projectors is able to give a wide range of effects.

Requirements of a multi-image programming system are very simple. Only a few parts of the slide projector can be controlled:

- □ the lamp;
- □ forward slide step;
- □ reverse slide step;
- □ projection of a slide specified by number;
- □ possibly, detection of zero position of the slide tray;
- □ possibly, separate electrical control of the lamp shutter;
- □ possibly, control of electrical power to the projector.

The only one of these items requiring any elaboration is the lamp. A reasonable requirement here is that is be possible to fade the lamp on and off at a choice of speeds, ranging from a quick cut to a long, slow dissolve, and to stop the lamp at some specified intermediate brightness. Lamp flashing and animation effects are simply achieved by a rapid repetition of fast lamp on and off commands.

The job of the programming system is to organize the actions of a group of projectors with respect to time and to each other. All multi-image shows require careful planning. If a proper storyboard is prepared, the task of programming the show is relatively simple. In the old days programming was time consuming, not so much from the point of view of entering the initial program, but more because of the difficulty of editing and making subtle timing adjustments. The arrival of the microprocessor made multi image a more accessible medium, with the side benefit of allowing a wider range of effects.

For small shows using only three projectors it is practical to have the entire programming system in one unit, the three-projector programmable dissolve unit (described in Chapter 15). For more than three projectors it is better to have a split system where there is an interface local to the projector which is, in turn, controlled by a central programmer. This type of arrangement simplifies wiring, and keeps the high-current lamp control local to the projector. Most systems endow their interfaces with considerable local intelligence, greatly reducing the load on the central programmer.

Thus, from commands issued by the central programmer the interface could be expected to:

□ Fade the lamp at any specified speed. Some systems offer a limited repertoire of speeds, but others can offer any speed between 0 and 99.9 seconds specified to 0.1 second.

□ Stop the lamp fade. Some systems offer this as a 'stop fade' command; others allow precise specification of the required lamp intensity e.g. any level between 1 percent and 100 percent specified to the nearest 1 percent.

□ Advance the projector to the next slide. To simplify programming this is usually done automatically when the lamp fades to extinction, with a separate 'no step' command if, for some reason, this is not required (e.g. for an animated sequence).

□ Select a specified slide by number. This may be required at the beginning of a show if one slide tray holds the slides for more than one show, or to allow different versions of the same show e.g. shortened versions or different language graphics.

□ Select an animation routine. An animation routine is set up by causing lamps in different projectors to flash on and off sequentially. This can be done by inserting a 'loop' in the main program, but can also be done by delegating the task to the interface unit. Some systems allow choice from a library of standard effects, others use a system of downloading an animation sequence from the main programmer.

□ Operate the lamp shutter. The filament lamp used in slide projectors (especially low-voltage lamps used in European projectors) has an appreciable thermal inertia. This means that when it is switched off brightness decays over an appreciable time, limiting speed of the visual effect. Some systems allow access to the projector shutter to give an instantaneous 'snap' picture change.

□ 'Home' the projector. In unattended automatic shows it is important to be able to return all slide trays to zero or 'home' at the end of the show. The interface unit should check that trays really are home by means of a microswitch that detects a notch in the slide tray identifying the zero position.

□ Switch power to the projector. Not all systems allow this, but it is a convenient feature, especially for simplifying installation wiring in permanently installed systems.

Some interface units boast additional features that may make them suitable for special applications. These include power fail memory – a system that ensures, in the event of power failure, the unit remembers exactly where the projector is so that when power is restored there is no wasted resetting time; and lamp failure detection, a feature not so necessary now that projectors in permanent installations can be fitted with automatic lamp changers.

Arrangement of interface units varies between systems. At one extreme there is one interface unit for each projector; at the other, one interface unit for every four projectors. The most common arrangement is one interface unit for every three projectors. This arrangement is due to a combination of economics and practicality. The three-projector stack remains the most common component of multi-image configurations. Note, however, the fact that one interface con-

Figure 16.13 A typical projector controller for multi-image. This one controls three projectors independently. It can replay two different signals from tape, or can be directly controlled by a computer via an RS232 link

Figure 16.14 Multi-image programming in progress

trols three projectors does *not* mean they are linked in control terms. Each projector is still individually addressable.

Usually a data cable links all interfaces in a system, and each interface is fitted with an address switch that identifies which unit it is in the chain.

The advent of the 'electronic' slide projector with built-in triacs and microprocessor control, like the Simda and Kodak Ektapro, has introduced the need for an alternative kind of interfacing. While both these projectors can use conventional projector controllers, there are installation advantages in using the RS232 serial control provided by both projectors.

There are some simple programs that allow direct control of the projectors, but best results are obtained using a serial converter which changes the high-speed data from the computer or programming device into the specific data required by the projector. The Dataton system, for example, has a multiple serial controller which receives RS232 data from the computer and converts it into four separate serial streams for controlling four projectors directly. Electrosonic France offer a neat conversion of the Kodak Ektapro that has the interface inside the projector. This results in the simplest possible installation – the computer or programmer output now being fed direct to the projectors with no intermediate equipment at all.

Central programmer

The central programmer is another microprocessor-based device. It can either be a 'dedicated' device, that is designed only for the creation of multi-image shows, or it can be a general-purpose microcomputer. In the former case the unit is fitted with a keyboard that is specific to multi-image programming, a simple extension of the controls provided on a programmable dissolve unit. However, the task of programming and editing a multi-image show is one that is almost ideally done by a small computer, and today most systems are based on the use of personal computers. Typically, these are of the IBM PC or compatible variety.

Multi-image does place some special demands on the computer in respect of communications. For this reason, the computer usually requires addition of one or two accessory cards that plug into spare slots in the computer chassis.

Using the microcomputer as a multi-image programmer is very similar to using it as a word processor. Most people *can* use a typewriter or word processor, although they may be rather slow at it. Similarly it takes some time to become a *skilled* programmer able to work very fast, equivalent to touch typing, although most users do not necessarily have to work at high speed. The most important thing is to be able to think logically how a picture sequence is built up, and to understand both the power and the limitations of the programming system. It is interesting to note that many major producers of multi-image presentations do not have expert programmers on their staff, but rely on the services of highly-skilled freelance programmers when they want a large complex show programmed quickly. This just reinforces the point that acquisition of a basic understanding of programming is accessible to anyone, but that the need to reach the equivalent of '100 words a minute' is limited.

Like a word processing program, the multi-image production program has many editing features designed to simplify the task of initial entry and subsequent editing:

- ☐ Shows are finally stored on floppy disc. They can be stored as whole shows or part shows. Whole shows can be retrieved for editing or showing. Part shows can be retrieved to be added on or inserted into a show being worked on.
- ☐ Cues can be modified, either individually or in groups. They can be deleted either selectively or in specified blocks. Extra cues can be inserted between existing cues.
- ☐ Plain text remarks can be included in a show listing to make it easier to find a particular place. A tab facility is used

to simplify moving to a point in the show whose cue number or timing may have changed as a result of editing.

□ Show listings can be printed as hard copy using a standard printer.

Show principles

The easiest way to understand the process of multi-image programming is to talk through what happens in a typical show. Reverting to the nine-projector example, it is convenient to make the assumption that projectors are arranged in three groups of three, each aimed at a different image area or screen. Image areas may well overlap, but this in no way affects how the program is made.

The program can be considered as a sequence of events, so it is easiest to start by giving these events a number. Thus, a show consists of a number of cues, sequentially numbered, usually from zero. In order to make the show-making program easy to understand, most systems have some limitation in the amount of information stored in one cue. So with some systems it may be necessary to use two or three cues to achieve what can be achieved with one cue in another system. Now that computer memory is so cheap there is no serious constraint on the number of cues available, and most systems allow several thousand cues to a show.

All programming systems use a kind of shorthand to limit the number of keystrokes needed to achieve a particular effect. While it is possible to use a mouse for cue entry, most multi-image programming is done by keyboard entry, and it is easier to describe what is happening by reference to a cue list. The cue list looks as though quite a lot of keystrokes are needed but in practice most systems provide a 'default' facility where, if values remain the same or are simply a standard increment, they are entered automatically.

The listing of the first few cues of a typical nine projector show is shown in Figure 16.15. While the actual command designations are those of one particular system, those of other systems are, in general, just as easily understood. The important descriptors of each cue are:

□ Line or 'line number', the same as cue number.

□ Time. This is divided into two parts, a type of time command and a value. Time value is expressed in standard timecode format, or hours, minutes, seconds and frames. Types of command are:
○ WAIT, which is a 'time from last cue' command
○ AT, which means 'do this at a specified time'
○ HALT, which means stop here until told to release the next cue.

□ Type. This determines the type of command, for example

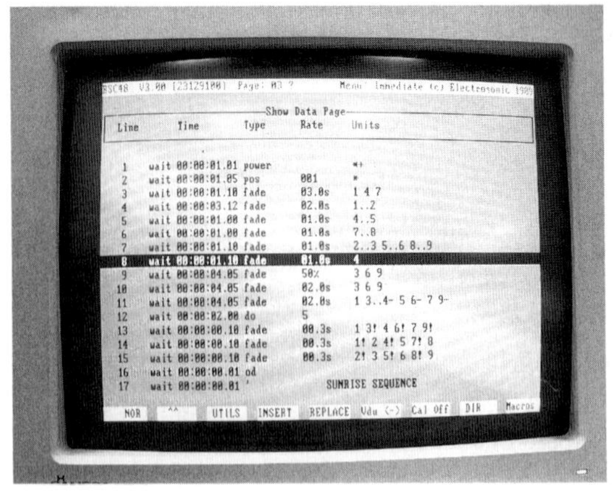

Figure 16.15 An example of the listing of the first few cues of a multi-image program

it may be a projector lamp fade, or operation of an auxiliary effects relay.

□ Rate. If the type of command has a value, for example the time of a lamp fade, or a percentage of light output, it is entered here.

□ Units. This lists the devices affected by the command.

Cues 001 and 002 can be considered house-keeping cues. 001 switches on power to all projectors. In this case the symbols:
 * means *all devices of this type*
 + means on
 − means off
Different systems use slightly different shorthand, but the meaning is usually quickly understood.

Cue 002 has a WAIT value of one second, five frames. Therefore after a little over one second, to allow the projectors to settle down, Cue 002 is automatically released. It is good practice *not* to have a slide in the zero position, because it is easily lost. Thus, the first slide to be seen is in position 1. Cue 002 is a position command, and here the rate column shows the required position (001), and again the * symbol indicates that all projectors are to move there. In a multi-language program with different title slides this command could have been split into two parts, with most projectors going to position 001, but with one or two going to a different slide position.

Cue 003 is the first cue that produces a visible image. This cue is calling for a three-second fade on projector lamps 1, 4 and 7, which has the effect of fading up an image on each of the three screens or image areas.

Cue 004 is released 3.5 seconds later, just after the first fade up is completed. This cue shows a crossfade of two seconds on projectors 1 and 2. Therefore the image on the first screen performs a dissolve change.

Some systems use a letter to symbolize the projector within a group, so instead of projectors 1, 2, 3 the first group might be referred to as 1A, 1B, 1C; but, again, the actual method of designation is not important, understanding what is happening however is.

Cues 005 and 006 perform dissolve changes on the other two screens at one-second intervals. The cumulative effect of cues 004, 005 and 006 is of a 'wipe' fade across the whole image area. Speed and appearance of the effect is clearly dependent on both the WAIT time between cues, and lamp fade time.

Cue 007 is a dissolve change on all three screens. In practice, if the actual projector arrangement is the classic 3:1 ratio 'one over two' soft-edge panorama configuration, this cue corresponds to a single big image dissolve.

In cue 008 there is a superimposition. Status after completion of cue 007 is that projectors 3, 6 and 9 have lamps on to fill all three screens. Cue 008 has brought on projector 4 (a center screen projector) without any corresponding fade-down of projector 6. A superimposition (for example a title) is the result.

Cue 009 prepares lamps of projectors 3, 6 and 9 to fade to an intermediate level, in this case 50 percent of full light level. Cue 010 tells them to fade down to the pre-programmed intermediate level. Now the superimposition slide looks bright relative to its background.

Cue 011 fades out all four projectors that are on and replaces them with images from projectors 1, 5 and 7. Cue 012 is the start of a loop command, saying 'DO the following set of cues five times'. The following three cues correspond to closely spaced quick dissolves (or cuts) giving a ripple effect across the screens. The '!' symbol indicates that the projectors should not advance when their lamps fade down. Cue 016 is shorthand for 'end of loop', and is shown as OD, the reverse of DO. Loops can typically be up to a hundred cues long, and there can even be loops within loops.

Cue 017 is an example of a cue being used for a text description. This cue does not actually do anything other than remind the programmer of where he is in the program.

The above simple analysis of the first part of a program illustrates both the type of command that must be entered, and the mechanical process taking place. Sophisticated multi-image computer programs allow the creation of 'macro' commands to minimize the number of keystrokes used. For example, if a show used twenty groups of three projectors it would be very tedious to enter forty different projector numbers to achieve a dissolve change on each of the twenty image areas. A single command 'ALL DISSOLVE', or even *D might be used. Similarly complex animation routines can be developed on a trial-and-error basis but, once perfected, can be recalled as a single command.

The task of multi-image programming is made easier by the fact that all projectors may be connected up while programming is in progress. As each command is entered, the projectors execute that command, so it is easy to see that the correct slides are being shown. At any time, the operator can go to any selected cue and all projectors back-up or advance to their correct positions corresponding to the selected cue. A program can be run in whole or in part, in which case cues are released manually by the operator, or automatically wherever there is a 'time to next cue' or 'wait' command present.

Status of projectors, and of auxiliary effects relays and other controlled devices present, can be seen at any time. In a small show status is shown on the same monitor screen as the cue listing, but in a big show it may be necessary to have a separate status 'page' to check status. In the case of slide projectors this shows the projector's position and whether the lamp is on or off. The best computer programs allow the user to format the status display to match actual equipment layout. Figures 16.16 and 16.17 show some possible layouts.

Synchronizing to sound

There are some applications of multi-image which only require the program to be put into the computer; cues are then manually released using a remote cue button or the computer's space bar. Examples include speaker support sequences in sales conferences, and use of multi-image in the theater. Often these applications have automatic sequences where a chain of cues uses the automatic cue release feature, and the computer comes to rest at the end of a sequence to await release of a subsequent single cue or chain of cues.

Figure 16.16 When programming a big multi-image show it can be convenient to have two computer monitors. Here the one on the right is showing the cue listing, and the one on the left is showing projector and auxiliary device status

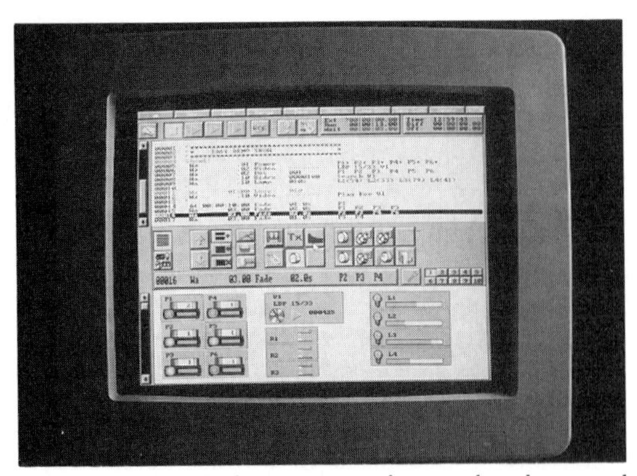

Figure 16.17 The best programs for mixed media control have attractive graphic user interfaces which clearly show the status of the equipment being controlled

For these applications the program should have some special staging features, such as the ability to black-out all projectors in an emergency, and to be able to step backwards through the cue sequence.

Most users of the medium, however, want their multi-image sequence synchronized to a sound track. While it is possible to transfer show data direct to magnetic tape to run in synchronization to the sound (simply by manually releasing cues while listening to the sound track), this method is neither elegant nor precise. The preferred technique is to use a clock track.

Here one track of the sound tape is dedicated to a clock or timecode signal. The nature of this signal is such that time is recorded in a unique code as minutes, seconds and fractions of a second. Thus, the show producer can identify a spoken word or beat of music by its clock time, opening up the way for absolute synchronization between the sound tape and the computer.

The usual procedure is to use a combination of 'clock cues' and 'time to next cue' commands. Thus pre-timed visual effects use the equivalent of the 'wait' command, but initiation of such a sequence, or of individual widely-spaced cues, is by providing the cue with a specific clock time.

The show is reviewed by playing the clock track into the computer, which then releases the cues at the required times. If the timing of individual cues, or even the whole show, looks wrong simple editing commands can adjust individual times, or make offsets to groups of cues or the whole show.

Clock tracks are of various kinds. Some manufacturers of multi-image equipment may have their own proprietary timecode. Generally, this results in a lower-cost system with a code that is easy to copy, and which may typically resolve to 0.1 second. The computer can interpolate commands where a 0.05-second cue spacing is required. Such codes are entirely satisfactory for those making pure multi-image shows.

However, professional show producers and programmers prefer the use of standard time-codes, especially as some multi-image shows may also use video material. Also they may be using multi-image methods to make a video program. In this case the show is first made as a single-screen multi-image show, using anything up to eighteen projectors, then transferred to video using an optical multiplexer. If such material is also required to mix in with conventional video it makes the editor's task much easier if the computer uses the SMPTE or EBU timecode.

Timecodes were introduced into video for the same reasons that they were introduced to multi-image. They make the linking of different sources easier and allow accuracy in synchronization. The SMPTE (Society of Motion Picture and Television Engineers) code works in hours, minutes, seconds and frames, with 30 frames per second for video. The EBU (European Broadcasting Union) code is the European equivalent at 25 frames per second. There is also a motion-picture variant at 24 frames per second.

SMPTE/EBU timecode has become the preferred professional timecode in the audio and video industries. However, the music industry has introduced a timecode signal as part of the MIDI (musical instrument digital interface) system and this may also be encountered in multi-image work.

Control tracks

Some users keep the computer on line to run the actual show, especially where mixed live and recorded sequences must be shown in one presentation. Many users, once they have seen and approved the original programming, transfer the show to magnetic tape, so the one tape carries the sound track and the control data. This allows use of simple replay-only equipment without the need to cart around the computer. Thus, show copies consist of a copy of the audio made directly from the master tape, but have a control track automatically laid down by the computer, in precise synchronization resulting from its reading of the clock track on the master tape.

The nature of the control track recorded on the tape is extremely complex. In all professional systems the control track carries not only the main programming commands, but also a continuous update of slide position information. This means that, in principle, if a tape is started half-way through, all projectors move as quickly as possible to the correct position. Unfortunately, due to the slightly different capabilities of

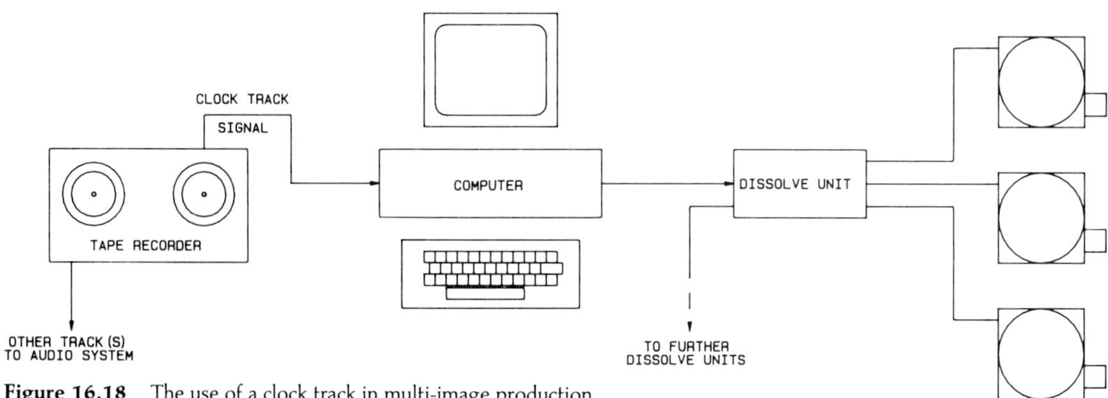

Figure 16.18 The use of a clock track in multi-image production

systems on the market, each manufacturer has its own control signal. This is not such a disadvantage as it may sound, as multi-image is a medium that is applied to single events and single venues. Some manufacturers offer limited compatibility for the replay of shows made on other systems.

A track on standard audio tape is limited in the amount of information that it can reliably carry. There is a trade-off between the number of devices that can be controlled and the speed of cue release. Thus, in theory, a system that can control 15 projectors at 20 cues per second could be offered in an alternative version that controls thirty projectors at ten cues per second. In practice it is easier to limit the system to one defined capability, and some examples are shown in Table 16.2. All professional systems include a complex arrangement of error correcting to ensure that data is recovered uncorrupted; even so, it is best to make show copies as direct first-generation output from the computer, rather than as a tape-to-tape dub.

Many show producers now like to make their soundtracks with digital sound, and play back the shows from a digital source, such as a multitrack digital tape recorder, computer, solid-state sound store, DAT machine or CD. In most cases it is easiest to use a timecode-linked computer for the programming, but for small installations a more compact and simpler playback arrangement can be achieved by putting the show data on to the tape or disc.

In the special cases of DAT and recordable CD, it is a fairly simple matter to encode either SMPTE timecode or, better, the show control data itself (i.e the RS232 serial data that might be directly controlling the projectors) directly into the audio data. This is possible because both DAT and CD have provision for 'user bits' in the datastream, and this provision is quite sufficient for running multi-image and other mixed media shows.

Two examples of equipment able to provide this facility are the DATaDAT device from Electrosonic France, and the SD DAT-Extender made by Stumpfl of Austria. The big advantage of the technique, compared with the recording of a data signal on an analog tape, is that ALL the original computer data is recorded, and there are no compromises arising from the need to use a carrier signal.

Big shows

Modern multi-track tape recorders are inexpensive, so one way of running a big multi-image show is to use several tracks for control purposes. For example, a 72-projector show could be run from three control tracks with each track controlling 24 projectors. However, because computers are now so inexpensive, it is better practice and little more expensive to only use a time-code track on the tape, and retain the computer for control. This completely eliminates any problems of reading the tape, and allows easy modification of the show program if necessary. Some manufacturers, such as AVL and Electrosonic, also offer a means whereby shows can be dumped into non-volatile memories. The show playback equipment is then very simple, with show data being exchanged using a 'credit card-style' memory. This method is recommended for multi-image shows in permanent installations, where shows are run frequently.

What constitutes a 'big show'? For some producers a nine-projector show is big, but shows using over one hundred slide projectors are not uncommon. It is possible to program these using multiple computers, each computer dealing with, for example, 30 projectors. It is also possible to obtain programs that control as many as 192 projectors from a single computer without any speed restrictions. One problem here is that it is difficult for the computer to display status of

Table 16.2 Characteristics of some commonly used multi-image control signals

Manufacturer	*Signal name*	*Number of projectors*	*Cue spacing*	*Bandwidth*	*Aux A*	*Aux B*	*Signal*
Arion	Mate Trac	16	0.05 s	2 kHz	12	12	FSK
AVL	Positrak	15	0.05 s	10 kHz	10	25	FSK
Clearlight	Cue Sentry 20	15	0.05 s	5 kHz	5	33	BP
Clearlight	Cue Sentry 50	15	0.02 s	10 kHz	5	33	MC
Dataton	Syncode II	56	0.02 s	10 kHz	*	224	BP
Electrosonic	Alphasync	24	0.05 s	6 kHz	24	96	FSK

Aux A — Number of auxiliaries available from dissolve units
Aux B — Number of auxiliaries available with extra equipment
* — Dataton permit the substitution of one projector by two auxiliaries
Bandwidth — Recommended − 3 dB response of tape recorder
FSK — Frequency shift keying
BP — Bi-phase
MC — Modulated carrier
Data — In all cases the data on the tape includes slide position information as well as the projection effect instructions. Not all systems permit access to all projectors on every cue

Table 16.3 Characteristics of some commonly used multi-image computer systems

Manufacturer	*Computer*	*Language*	*Number of projectors*	*Cue spacing*	*Timecode*
Arion	IBM etc	Arion	32/512	0.01 s	P
AVL	IBM etc	PROCALL X	30/120	0.01 s	P and SMPTE
Clearlight	Apple IIE IIgs	AMPL 15	15	0.05 s	P
Clearlight	Apple IIE IIgs	AMPL/M (X)	30/(240)	0.01 s	P
Dataton	IBM etc	MICSOFT	56	0.02 s	P and SMPTE
Dataton	Macintosh	TRAX	56	0.02 s	P and SMPTE
Electrosonic	IBM etc	EASY	48	0.033 s	SMPTE
Electrosonic	IBM etc	BSC-192	192	0.033 s	SMPTE

IBM etc refers to IBM™ computers and compatibles using MS-DOS or equivalent

Where two different numbers of projectors are given, the larger number may require extra equipment, or not all projectors can be accessed on one cue

P — Proprietary timecode
SMPTE — Standard SMPTE and EBU timecode

all controlled devices simultaneously. Best programs allow the user to format a number of status display 'pages', each one laid-out to correspond with the actual layout of part of the equipment.

Many shows are now 'mixed media' and it can be instructive to examine a case history to illustrate how multi-image is often now integrated into a show using other media. While the example chosen is a big show, the principles apply to shows of any scale.

The truck and bus manufacturer Scania of Sweden celebrated its centenary in 1991, and as part of the celebrations it made imaginative use of 81 slide projectors, 21 video projectors, and a 35 mm movie projector in a well-paced and highly professional multi-media show.

Over a six-month period more than 45,000 people saw the show at nearly 200 performances. Employees,

Figure 16.19 Programming the 84-projector show at the Siemens Museum in Munich. (Photo of Messrs Mietzner and Mattis, the system engineers and programmers, by Reger Studios of Munich)

dealers, press and customers were invited to visit the Scania headquarters at Sodertalje, where they saw a one-hour presentation in a specially constructed auditorium. A first reaction to the idea of a one-hour multi-media show is that such a length is too long to sustain interest and leave a strong single message. However, the clever way in which resources were used, and division of the show into a number of 'acts', ensured that there were no longueurs and, in fact, the final result was one of the best multi-media presentations to be seen.

The auditorium was about 25 m wide, and comparatively shallow. The decor of the auditorium was a street scene of Sodertalje as it was 100 years ago, and when the audience entered they had no idea of the extent and variety of presentation systems to be used during the show.

Mixed media

The outstanding feature of the show from the AV professional's point of view was the seamless way in which different methods of image projection were blended together. The audience were never aware of switches from slide to video or from video to movie film, yet each medium was used correctly both in terms of production and presentation. Screen illumination from all sources was carefully matched, and there was little perceptible difference between the front and back projected images.

Different sections of the show dealt with various aspects of the Scania story. The first section gave historical background, covering the company's founding and its progress over the years. Other sections dealt with progress in manufacturing techniques, development of new products and the building up of international sales. Others included a dance of industrial robots where two human dancers danced with three ABB robots used in truck production, and a clever illusion where the real last car made by the Scania company in 1929 disappeared into thin air. These helped to pace the show, and clearly divided the different subjects.

Giving continuity to the show was the character of Gustaf Erikson, who was the man who designed the first car made by the VABIS company in Sodertalje. Gustaf first appeared as an animatronic figure seated at his desk and later appeared 'live', played by an actor who, at the very end of the show, disappeared in a well-executed illusion. The use of Gustaf, and other actors, within the show made the audience more involved and definitely contributed to its success.

Multi-language

The show was given in no less than nine languages.

Figure 16.20 Gustaf Erikson, an animatronic figure, with four of the twenty-one moving rear projection screens in the Scania show. (Photo Ljus and AV Teknik)

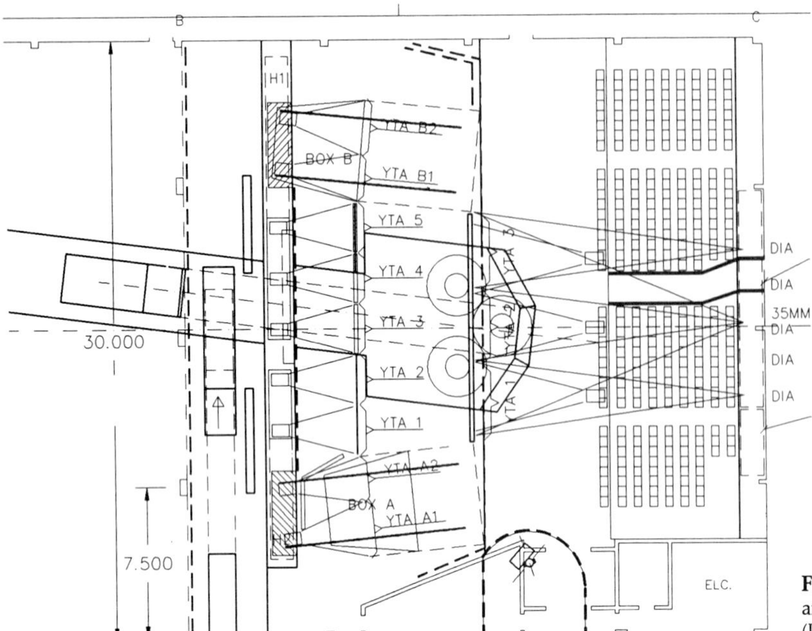

Figure 16.21 Plan of projection arrangement for the Scania show. (Ljus and AV Teknik)

For this reason the actors mimed on the relatively few occasions that they were required to speak. A clever programmed sound system ensured that the sound of the actor's voice always seemed to be coming from the right place.

The show was produced by the experienced AV production company Producenterna of Stockholm. This is a well-established group that has, in-house, professional video post-production, professional audio production, slide making, multi-image production, art-work facilities and so on. This undoubtedly helped them hold together the complex mixed media production, and brought the consistency that might have been missing if different elements had been tackled by separate sub-contractors.

The show format was a mixture of back and front projection. A motorized roll-down front projection screen was normally set to 11.5 m wide by 4 m high (38 ft by 13 ft), but for a full cinemascope movie sequence showing trucks driven on a test track, the screen height increased to 4.8 m (16 ft). In its normal position the screen was served by the movie projector, by three video projectors and 27 slide projectors.

The main screen was flanked by two 5 m by 4 m (16 ft by 13 ft) rear projection screen assemblies. These were each divided into four 2.5 m by 2 m (8 ft by 6.5 ft) screens, with each section served by both a video projector and a group of three slide projectors. These side screen units were on tracks and, in parts of the show they moved back to line up with a 12.5 m by 4 m (41 ft by 13 ft) rear projection screen assembly divided into ten 2.5 m by 2 m (8 ft by 6.5 ft) sections. The resulting 18-screen array was the full 22.5 m (74 ft) width of the stage area.

At the climax of the show, the side screen boxes moved forward again, and the center screen assembly parted in the middle, with its corresponding projector support system moving at exactly the same speed, so it was the screen images that appear to part. In a shower of pyrotechnics, strobelights and clouds of smoke, seven tons of the latest Scania truck rolled towards the audience. (Stopped, of course, by the precision programmed movement system!)

Multi-computer

The question of how to program such a mixture of equipment was addressed at the start of the production process. It was realized that it would be difficult to program everything using only a single computer program as while theoretically possible, in practice it is easier to program the different elements using separate programs each optimized for a particular task. This arrangement also allows different aspects of the show to be pre-programmed concurrently, by programmers who are familiar with the particular medium.

The ABB industrial robots have their own programming system. Everything else was controlled by three

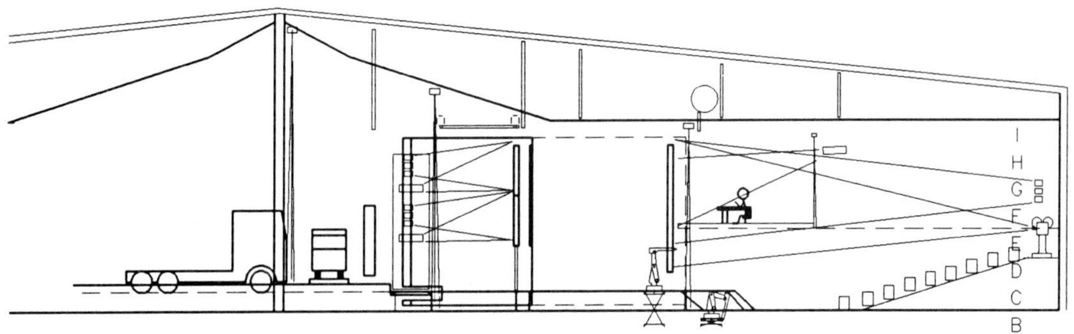

Figure 16.22 Section of the Scania show. Notice the motorized front projection screen. (Ljus and AV Teknik)

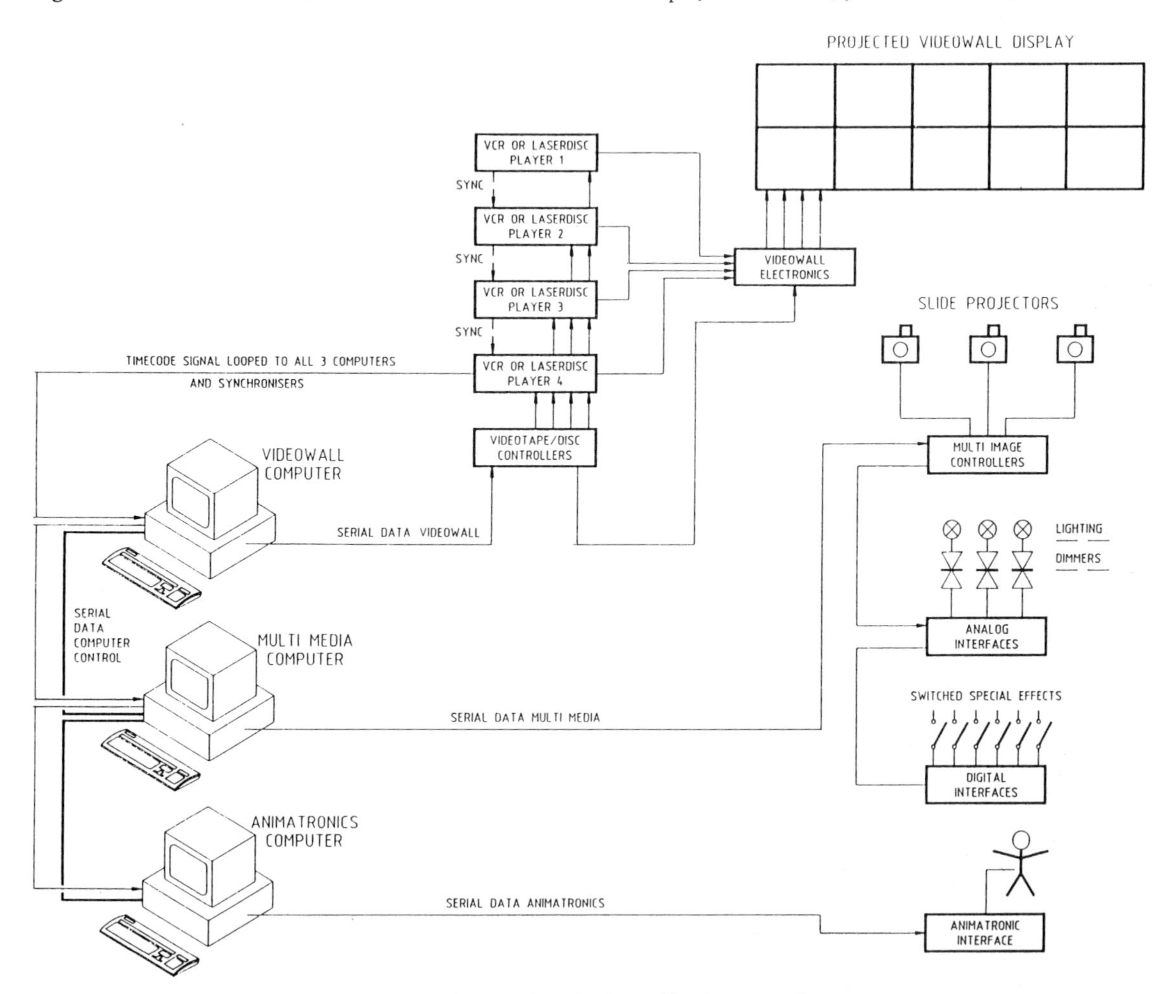

Figure 16.23 The use of multiple computers for mixed media shows like the Scania show

separate computers, all locked to SMPTE/EBU time-code. A 'master' computer ran a mixed-media control program optimized for multi-image. It controlled the slide projectors, the movie projector, the 96 channels of lighting control, the audio routing, and all screen movements and switched effects. It also provided a 'hard black' signal to the video projectors.

A second computer ran a program optimized for

videowall programming, and was used to control the four high-band U-matics and the video projectors. All the video projectors were equipped with videowall image processing equipment, allowing a full range of videowall effects. While videowall programming is similar to multi-image programming, it is not the same and it was therefore faster to use the dedicated program than to modify the multi-image program to also operate the videowall equipment.

Finally, a third computer ran a program optimized for animatronic figure programming, and was only used to control the animatronic figure. Those who program animatronic figures prefer to program various movements in real time, and the procedure is quite different from that used for normal AV programming.

The show required the interlock of no less than six tape transports. Because of the length of the show, and because of the relatively few performances (compared with, for example, a permanent visitors' center show) the video material was run from tape. The high-band U-matics were fitted with computer interfaces, and the tapes were dubbed with framecode. This allowed disc-like computer control of the video sources. In fact the tapes ran the full length of the show, and the tape players were locked to a common sync signal. The master videotape player had one 'audio' track dubbed with SMPTE/EBU timecode. This was fed to the computers and various synchronizers.

Figure 16.24 The reason for using multiple computers is that different styles of programming are used for different media. Videowall programming, for example, benefits from a graphics user interface

Multi-track

Two 16-track tape recorders were locked to the timecode and carried the sound. Twenty-six audio tracks were used, but in any one show only ten were played. Eight were always the same music and effects tracks, and two were voice tracks carrying commentary and acting voices in the selected language.

All audio channels were fed to a switching matrix and VCA system that allowed the position and level of each track to be programmed from the master computer. The movie film ran for only a part of the presentation, but nonetheless did require 'lip sync'. This was achieved by programming the starts of the projector from the main computer, but allowing a reasonable run-up time to allow it to synchronize to the timecode. It carried timecode on its optical sound track, so a standard synchronizer was used in conjunction with a DC drive for the projector. The light output of the projector was programmed by a solenoid-operated douser (shutter).

Screen and truck movements were all controlled by precision electronic motor controllers with soft start. This allowed screens, screen boxes, trucks, moving projector shelves and so on to be positioned with great accuracy (within 1 mm). The motor controllers were controlled by programmable logic controllers (PLCs), which received position feedback from a rack and pinion encoder on each movement. The PLCs themselves received start commands from the master computer.

The Scania show is an example typical of what is being demanded at the higher end of AV programming. It does not really matter whether such a show is run from a number of computers each operating as a sub-system, or from one overall computer. The choice is made solely on the grounds of cost-effectiveness and practicality. What does matter is that those charged with the responsibility for producing such shows have a clear understanding of what it is that they are trying to achieve. Computers and their programs can help with the realization of a project, but they are no substitute for creativity.

Standard protocols

Many devices which formerly were 'stand-alone' and required manual operation or custom control interfaces are now supplied with serial computer control as a standard facility. The serial control is usually RS232 or RS422/485. The former is standard on PCs and is suitable for short distances, whereas RS485 and RS422 are designed for longer control distances. Devices having such control include program source equipment (disc players, tape players, etc.) video processing

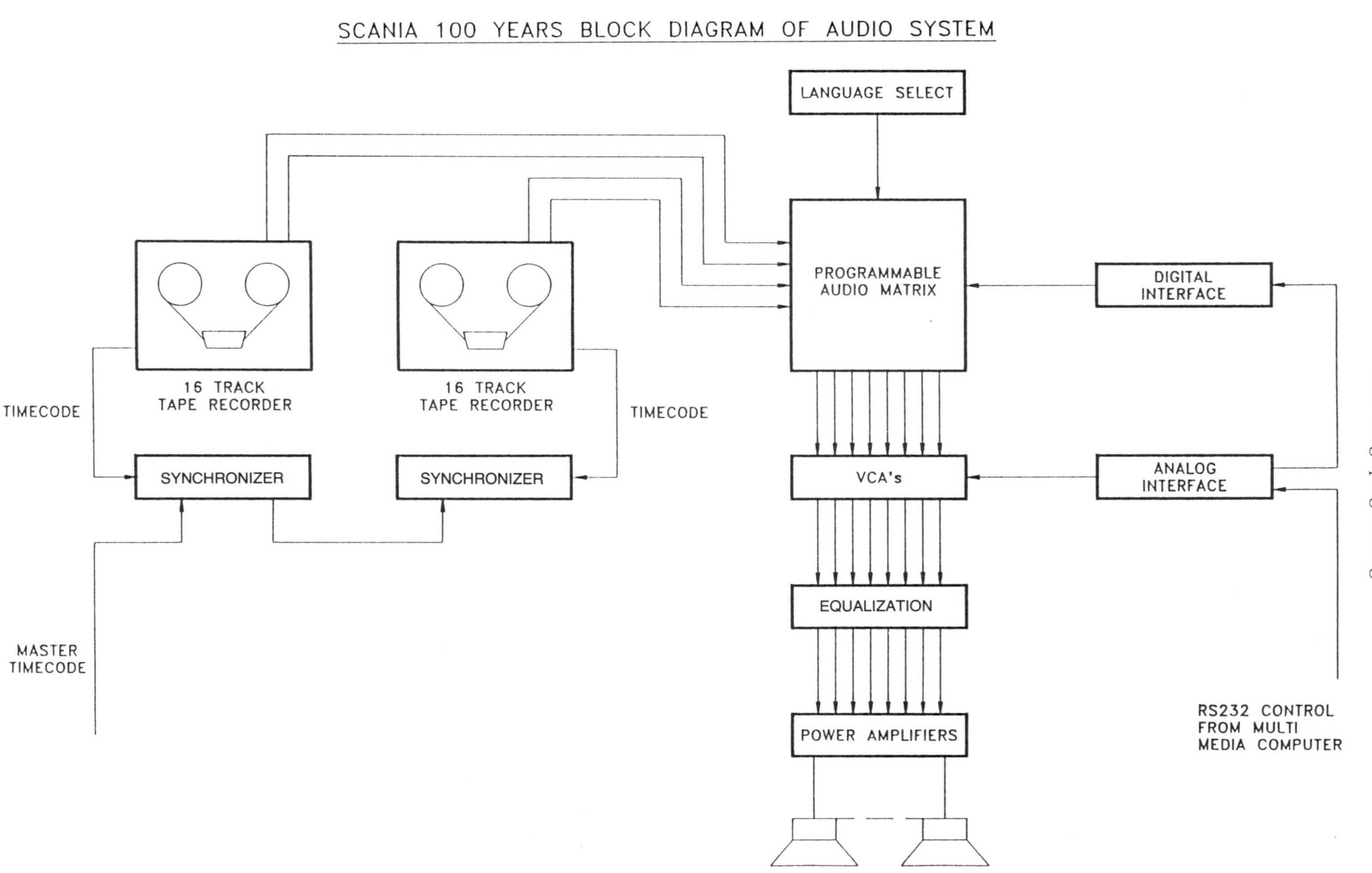

Figure 16.25 The multi-channel sound system used for the Scania show allowed the presentation to be given in ten languages. It included programmed panning of sound to follow actor's movements

Figure 16.26 Finale of the Scania show. The rear projection screens have parted to allow the truck to 'drive' towards the audience

equipment, audio control equipment and lighting control equipment.

It is normally necessary to use customized drivers to control these devices, much in the same way that when using a computer as a word processor it is necessary to install one or more printer drivers to match the printer being used. Thus programs such as Electrosonic's EASY and Dataton's TRAX, while designed mainly to work their respective standard multi-image and effects control equipment, can easily be modified to include drivers for audio processors, video effects machines, CD players, etc.

Apart from SMPTE timecode already referred to, other commonly encountered control signals are:

☐ MIDI (musical instrument digital interface) especially suitable for musical and event-driven applications. The protocol is suitable both for driving electronic musical instruments and for controlling discrete devices through the MSC (MIDI show control) variant. A lot of audio and lighting equipment comes MIDI compatible. MIDI timecode is a timecode similar to SMPTE, but geared to the MIDI data structure and not suitable for recording on analog tape.

☐ DMX. This protocol is the nearest the entertainment lighting industry has to a standard. It is a serial signal for the control of up to 512 circuits of lighting or dimmers, each circuit specified to 8 bits (256 levels). It is carried on RS485 at 250 kb/s. Devices like moving lights and color scrollers are also DMX controlled and sometimes use two 'channels' to get 16-bit resolution.

Video multi-image

There is an increasing demand for hybrid presentations, and for presentations that use the methods of multi-image, but are finally presented by video projection. The new recordable video-discs, for example, the Sony CRV system, allow show sequences to be prepared as a series of video images (very often taken from slides), then presented in the same way as a multi-image show. Producers often have a choice of methods of final presentation, but two points should be made:

☐ those who understand the basic methods of multi-image make the best use of the new possibilities of video;

☐ users should be careful not to find themselves in the situation of using a lot of expensive video equipment, when for much less cost and with a superior or equal result they could be using standard multi-image methods.

Companies such as TVL in the USA and Cadsoft in the UK have taken the multi-image tradition into the video camp and have produced extremely effective computer programs for the controlled presentation of multiple video sources combined with computer-generated graphics and scanned-in images. These programs are mainly used by professional conference staging companies as an alternative to or in addition to optical multi-image.

As a new generation of video projectors with much higher light output comes on to the market, and as increased computer power and high-definition video brings electronic resolution to a subjective par with film, it is inevitable that more large-scale presentations will be electronic instead of optical. However, there will be a place for optical methods for a long time yet, especially where image quality is paramount.

Chapter 17

Movie presentation systems

The movie film is the granddaddy of the AV business. While it could be argued that the magic lanternist with his patter and well-rehearsed routines was the first AV practitioner to use projection, it was the sound film that, at one bound, became the ultimate in audio-visual. Strange, though, that the medium should be so perfect and yet at the same time so limiting.

Perfect, because if a moving-image sequence with sound is required, movie still remains the ultimate method of high-quality program production; limiting, because once made, a film is difficult to adapt, keep up-to-date, and use for different purposes.

Movie film is still the most important method of high-quality program production, wherever the program involves movement and is destined for a wide audience. The industrial AV user's interest may be more limited, but nonetheless movie has an important role to play. This chapter is mainly concerned with the various mechanical possibilities of presenting movie film, but in reviewing these it is possible to define uses of the various movie formats. The comments here are from the point of view of the AV user and do not take into account the amateur market.

Figure 17.1 Modern 16 mm projectors are lightweight, compact and portable. Self-threading and slot-loading models are particularly easy to use. (Photo Elf Audio-Visual)

Super 8 mm and 16 mm

Super 8 mm was, until a few years ago, used as a presentation method, but it has now given way completely to video. It is also the case that 16 mm use has declined greatly, especially outside the education field. This is because most potential users also need video presentation facilities, and on small screens there is little difference in perceived image quality, so there is no point in duplicating the means of showing.

However, 16 mm is still important for the AV user. It is the medium in which sponsored films are made, and is sometimes the preferred production medium even when the program is intended to be finally shown on video. It is the nearest thing there is to a world audio-visual standard. If it is necessary to guarantee that a program can be shown anywhere in the world, then 16 mm is best because a 16 mm projector can be found anywhere and, in contrast to video, standards are uniform.

Sponsored films are usually made by those organizations which need programs for distribution. Thus, shows intended for distribution to schools and training films (especially sales, service and management training films) are best made on 16 mm, even if the same films are also available on videocassette for small audiences.

If a show has been originated on 16 mm, then for large audiences, it should be shown on 16 mm. The impact is significantly greater, particularly because making the effort ensures a better environment for the show.

Most 16 mm presentations are given using portable 16 mm projectors. Present-day portables are lightweight and easy to use. They are available in self-threading versions, which make them little more difficult to use than a cassette tape recorder. Like slide projectors, they are flexible in positioning because a range of lenses is available. Fixed focal length lenses are best, but zooms are available.

The open-gate light output of a standard 16 mm projector with the shutter running is about 700 lumens. This makes it suitable for pictures up to 4 m (13 ft) wide in the dark, especially if a screen of reasonable gain is used (this may not be possible, see Chapter 22). If bigger images are required, then instead of the tungsten halogen lamp used in standard portables, it is necessary to move up to a projector fitted with a xenon arc light source. Portable xenon arc projectors, using 300 watt or 500 watt lamps are available with light outputs of up to 2000 lumens suitable for screens of 5 to 6 m (16 to 20 ft) width.

For permanent installations and other applications requiring big images, pedestal-mounted projectors with 1000 watt and 2000 watt xenon lamps are available, and these can give pictures in widths of 8 to 10 m (26 to 32 ft). These projectors are usually fitted with big reels that allow uninterrupted showing of feature-length films.

35 mm

This is the professional standard which is used for the making and showing of entertainment films. Industrial AV users do not usually come across it, unless they are users of TV commercials, some of which are originated in 35 mm; or because they have the need for a spectacular show (e.g. a product launch) where 35 mm may be essential in order to give adequate picture quality on a big image.

There are 35 mm portable projectors available which are used for preview work. They do not have a significantly greater light output than a 16 mm projector. Most 35 mm projectors, however, are in permanent installations. They are heavy-duty pedestal-mounted machines intended for many years of continuous service.

Because of the varying requirements of different installations, 35 mm projectors are of a modular construction. There is a choice of lamphouses using xenon lamps of, for example, 1 kW, 1.6 kW, 2.5 kW, and 4.2 kW to serve the needs of screens from 6 to 17 m (20 to 56 ft) wide. There is also a choice of sound system and film handling arrangements.

The traditional movie theater uses two projectors. A feature film print comes on a number of reels, each only holding about twenty-minutes-worth of film, so with two projectors there is a changeover at each reel break. A simple system of interlocked shutters ensures that the audience are unaware of the changeover.

Most modern movie theaters, however, run their programs on a single projector basis. This requires

Figure 17.2 The well-known Simplex projector mechanism is available in a form which accepts both 35 mm and 70 mm film. (Photo Strong International Inc)

special film handling equipment, because a reel to carry a two-and-a-half hour program is enormous. Two methods are used. One is the use of big reels, vertically mounted either on the projector, or on a separate stand. Special arrangements are then included for rapid and safe rewind of the film at the end of the show.

A more elegant method, which requires more space, is the non-rewind platter. Here film is stored on a large horizontal plate or platter. Film is drawn from the center, then fed back to a platter on another level. This, of course, fills up from the center, so when the show is finished the next show is re-threaded from the new platter without the need for rewind. Most installations use a three-platter system, with two platters in use for the main show and the third one being available to make up new programs, to strip down old ones, or to carry a supporting program.

Most modern movie theaters are at least partially automated. This means that the film itself is used to initiate the operation of drapes, lighting dimmers and so on. The usual method is to apply a small piece of flexible metal foil to the film at the required point.

Figure 17.3 Modern projectors are of a modular construction. Often the lamphouse and sound system are built into a console which can be matched to the required projector. This is the Strong HighLight console

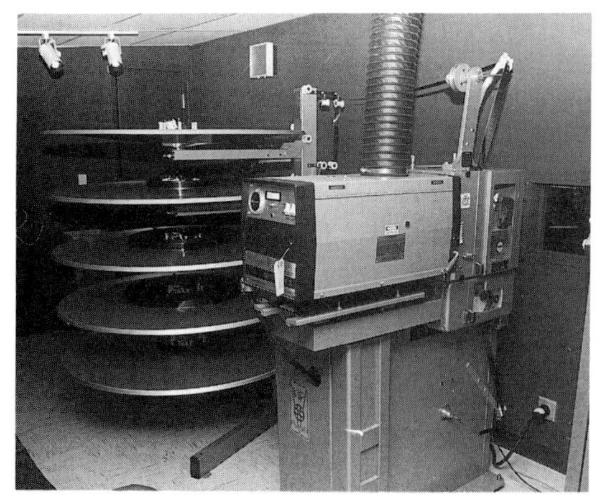

Figure 17.4 A typical 35 mm movie projector installation using a pedestal mounted projector and lamphouse. Here a five-platter non-rewind system has been installed to allow easy changes of program

This is then detected by a proximity detector or infra-red photocell, which cues the automation equipment.

35 mm remains a highly economical and reliable method of presenting large, bright images of high definition. It is now the case that the new generation of LCD and LC lightvalve high-definition video projectors (see Chapter 19) can produce images up to 7 m wide in the dark which are subjectively as good or better than 35 mm projection when viewed from a reasonable distance. For industrial AV users HDTV may be a more practical way of presenting 35 mm film material because the same projection system can be used for graphics and standard video; or, in the case of museum and visitor center applications, it may yield a lower running cost.

At present, any application requiring bigger images on a single screen must use film. The commercial cinema will, for some time yet, find that 35 mm film equipment is cost-effective compared to HDTV for all screen sizes. However, the main reason for this is that the storage and playback methods of HDTV are very expensive and impractical for long programs. This situation must change, and in due course the all-electronic cinema will take over the 'small' screen (6–10 m wide) market.

70 mm

Unless there is a requirement to go into the public entertainment business 70 mm is of even less relevance to the industrial AV user than 35 mm! However, it is the key to spectacular big-image projection.

It was originally introduced as an entertainment film format for big pictures, and film was both shot and shown in wide gauge. In recent years it has become a showing medium only for normal feature films, because the quality of negative films used for shooting the original film has improved so greatly that today's 35 mm negative film can give as good resolution as yesterday's 70 mm. However, when it comes to making the show copy, better presentation quality is obtained on 70 mm, so blockbuster feature films are often transferred for use in prestige movie theaters.

Special-effect shows to be found in theme parks and other tourist entertainments may use 70 mm, both as an origination and showing medium. These sometimes use lenses of very short focal length to give the audience the impression of being in the picture. In these shows it is important to use projectors with good picture steadiness so that viewing is not uncomfortable.

Without doubt, the most spectacular 70 mm films to be seen are those on the Imax process. In this Canadian system, films are made by running 70 mm film horizontally, and the actual show uses a projector working on the same principle. This ensures an original that is far bigger than that achieved in 35 mm or

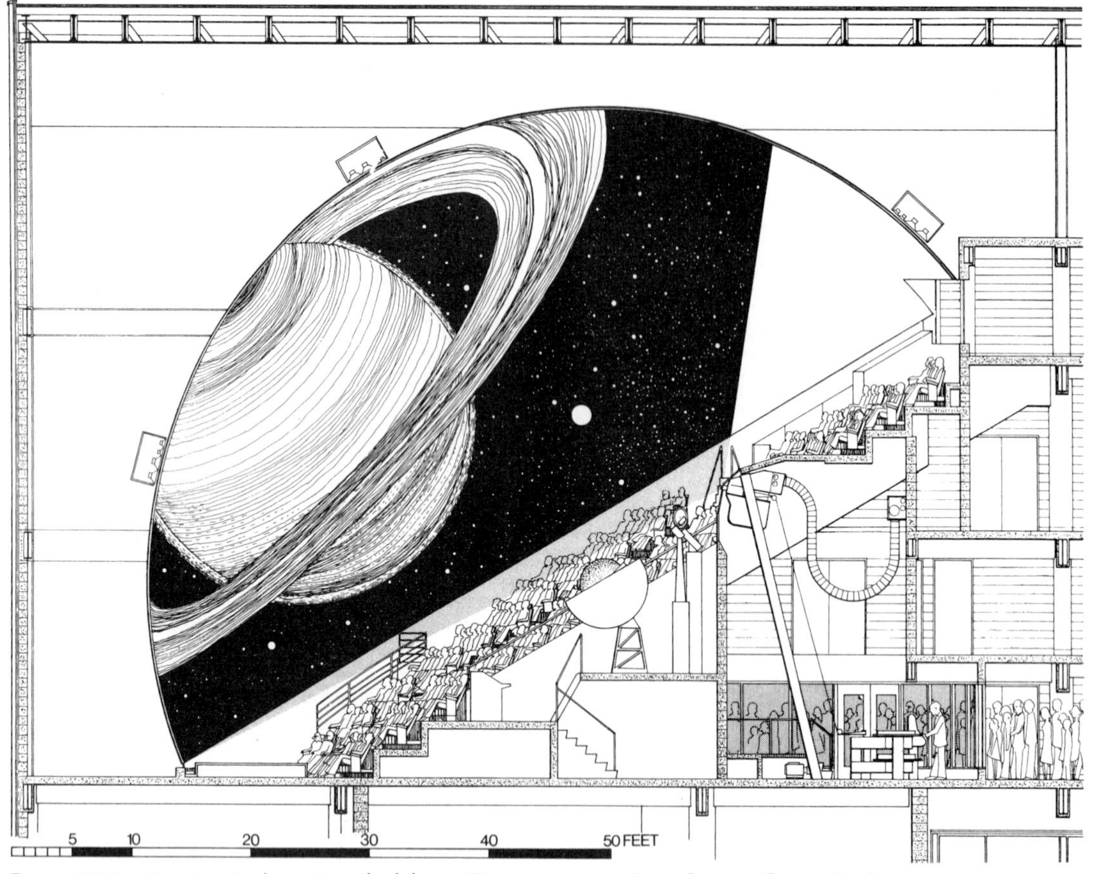

Figure 17.5 Omnimax is shown in a tilted dome. (Diagram courtesy Imax Systems Corporation)

even normal 70 mm production, resulting in exceptional quality. The showing system is one where the audience are quite close to the screen, so the picture must be rock-steady. This is achieved by pin-registering each frame as it is projected and using a rolling-loop system of film transport to prevent tearing of the film.

Imax requires a special auditorium. The screen may be as big as 30 m by 22 m (89 ft by 72 ft). Sound is run from a separate interlocked sound system. Imax is only suitable for short subject films. It has a near relation, Omnimax, which uses the same type of projector with a fisheye lens to fill a whole hemispherical dome, usually a planetarium dome that has been titled. Omnimax is a suitable support for mixed-media shows being given in a dome, but is a less satisfactory medium than Imax. Imax gives huge pictures of superb quality, with excellent sharpness, contrast and saturation. Omnimax suffers from the use of a dome as a screen where cross-reflection reduces both contrast and saturation, and where it is difficult to get a lens that maintains perfect focus over the entire picture.

Figure 17.6 shows the relationship of the 'Imax' image size, 15 perforations of 70 mm film, with 'standard' 70 mm projection where each image is only five perforations. Now that the original Imax patents have expired, there are new vendors of '15 perf' equipment coming on the market. Improvements in film emulsions also mean that intermediate formats are suitable for many applications.

For example, the '870' format, where film travels conventionally (vertically) but uses eight instead of five perforations per image, has the same aspect ratio as Imax. For images up to 18 m (59 ft) wide, 870 has become a formidable competitor to Imax, and many short subject films are available in both formats, 870 can also be used for dome projection, and is an economical format for big simulator rides.

The main problem of the big formats is achieving acceptable image steadiness. One solution is provided

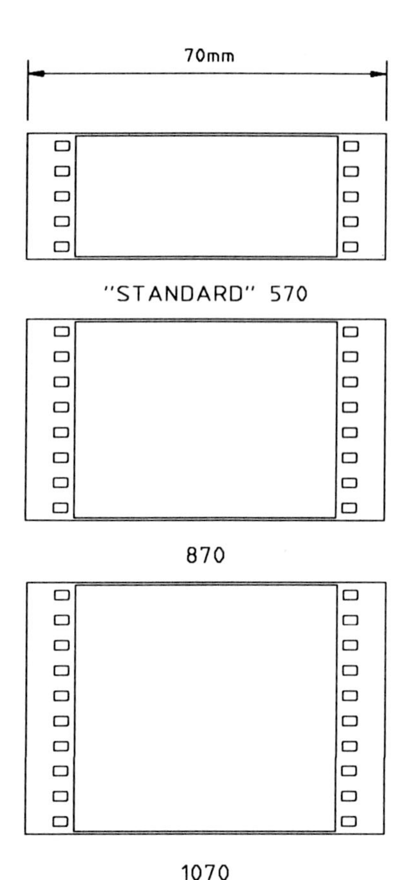

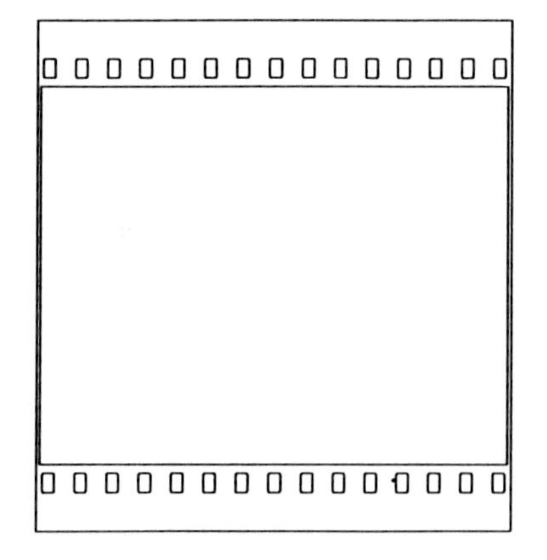

Figure 17.6 70mm film formats compared

Figure 17.7 The linear loop projector from Pioneer Technology Corporation of Los Angeles provides a new method of presenting the 8170 format

by the 'linear-loop' projector, where film is vacuum pin-registered while being projected, and is air-propelled between frames. The elimination of a conventional intermittent mechanism ensures less film wear and greater image steadiness – and also more light, because the film advance time is reduced. The projector construction allows film to be rewound through the gate, which in turn permits a compact installation for the short films used in 'ride' installations.

'Astrovision' uses ten-perforation pull-down to achieve an almost square image. It is also used in Iwerks' Imagine 360 process, where an annular image is printed on the film. A special lens then projects this as a single cylindrical image, giving a 360-degree surround picture from a single projector. While not nearly as bright as the more usual multiple projector image, it is a most elegant system, requiring only a single camera for origination.

Another special-purpose 70 mm movie format is Showscan. This was developed by Douglas Trumbull in Los Angeles as a medium for spectacular shows to be used in theme parks, simulator rides and EXPOs. There are two problems associated with normal movie film which the Showscan process seeks to overcome.

Figure 17.8 A special lens developed by Hughes Leitz to project an image to fill a 20 m diameter dome screen from 870 film

The first is that there is a limit to how brightly a picture can be projected. This may seem odd, as in many projection applications the problem can be getting enough light. However, in the dark, a picture that is too bright suffers from objectionable flicker as a result of a breakdown in the viewer's persistence of vision mechanism, which normally allows him to see the 24 frames per second as continuous movement. (In fact, a twin-blade shutter is used, so the viewer actually sees 48 images a second, each frame projected twice.)

The second problem is that the relatively slow picture rate of 24 frames per second means that, if there is a very rapid movement, the jumps between individual images are large, so the movement has a blurred quality. In addition, stroboscopic effects are badly noticeable on rotating wheels, etc.

The Showscan solution is to run standard 70 mm film at 60 frames per second. This allows a much brighter image to be projected without flicker, reduces stroboscopic effects, and makes rapid movements much sharper. The result is a much more lifelike picture.

Many special movie installations in theme parks run 35 mm or 70 mm films at 30 frames per second to get a useful reduction in flicker, without an inordinate increase in film stock or projector duty.

Movie formats

Although gauges of film are standardized, there are variations in the format of picture shown. Originally all films were made in the Academy ratio, which is 4:3. This is still the format of standard television pictures.

Most industrially sponsored films and all films made for video are still originated and shown in Academy ratio, so AV users usually have to contend with this format which is different from that of slides (3:2).

Entertainment films, however, are now normally presented in a wide-screen format. This means that when making a film the director has to compose his shots to make reasonable sense both in widescreen and in the truncated Academy format. In fact, when films are shown on TV the major broadcasters have a system of automatic panning to ensure a sensible picture composition on the narrower screen.

The simplest method of achieving wide-screen is to use a mask in the projector gate with the different aspect ratio. In the commercial movie theater 35 mm films are shown with an aspect ratio of about 1.85:1. To get a wider screen an anamorphic lens is used to 'squeeze' a wider picture on to the film when taking the film, and to 'unsqueeze' it on projection. This method makes better use of the film area and light available. The final projected ratio is between 2.2 and 2.5:1.

A 35 mm film format which used to be used to secure higher resolution on the original film is 'VistaVision', where 35 mm film travels sideways to provide a film area for each frame which is the same size as a 35 mm slide. The format is now re-emerging as a projection medium, providing a projected film area 75 per cent of that of 70 mm film, with lower running costs. The format is used in IMAX's new 'RideFilm'™ shows, and similar medium-size simulator and entertainment systems.

Movie sound systems

The great majority of 16 mm and 35 mm films are shown with a mono optical sound track. This is the most reliable and simple form of synchronized sound, and is recommended for all standard film presentations. Sound is recorded photographically on the edge of the film as a variable-area optical track; the area of exposed film at any moment is an analog of the momentary signal value. The optical track is scanned by a photocell.

Frequency response of optical sound is limited. However, on 35 mm film it is now possible to have Dolby™ optical sound. This uses a process similar to that used in the Dolby noise reduction system for magnetic tape to increase frequency response and reduce noise of optical sound tracks. The results, in

Figure 17.9 A special-purpose projector for showing 'VistaVision' format 35 mm film, from Strong International

practice, are as good, if not better, than magnetic tracks. Dolby stereo optical sound is able to give 'surround sound' effects by a system of matrixing four audio signals onto the two side-by-side optical tracks.

When there is a requirement for the user to be able to change the sound track on an existing film, or to have extra tracks, magnetic sound is used. Most portable 16 mm projectors have magnetic sound as an option.

Multi-track sound in the movie theater *can* be achieved by multiple magnetic tracks carried on both sides of the 35 mm film where there are three main tracks and an effects track. Generally, though, movie theaters either use Dolby stereo optical sound or, if sound is to be a major feature of a big show, use multi-track magnetic on 70 mm film, also with Dolby noise reduction.

The problem with magnetic sound on film prints is the fairly rapid degradation of sound quality arising both from print wear and, more particularly, head wear. The industry has been searching for ways in which digital sound can be introduced into standard movie presentations, and at least three systems are in contention.

Dolby Laboratories have introduced a system planned to maintain compatibility with existing optical sound. The standard Dolby stereo sound track remains in its existing position, while the digital optical track occupies the space between perforations. It provides five full-range digital audio tracks, a sub-bass track and two control tracks from a single 'SR-D' optical track.

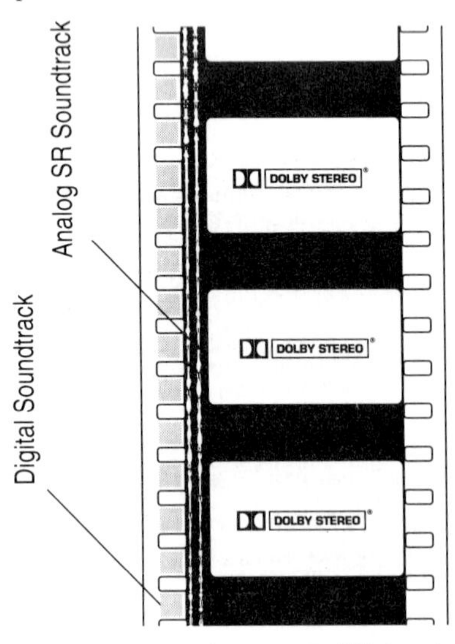

Figure 17.10 The new Dolby™ digital sound-on-film retains the Dolby Stereo SR analog sound tracks in their normal position. The digital sound is stored in the space between the sprocket holes. (Diagram from Dolby Laboratories)

Sony have introduced SDDS, (Sony dynamic digital sound). Here the digital data is printed on the edge of the film, outside the sprocket holes. To guard against film edge-damage the entire information is printed on both sides, providing 100 percent redundancy. The format uses compression technology developed by Sony for their 'MiniDisc'™ system and provides for eight tracks. These are center, left center, left, right center, right, sub-woofer, left surround and right surround. For cinemas with only three loudspeakers behind the screen, the system automatically adapts to six-track working.

Figure 17.11 Components of the Sony SDDS System. On top, the reader which is installed into the projector, below, the processor and remote controller

These systems take advantage of the improvement in filmstock, the resulting possibility of recording high density information, and of modern digital signal processing techniques.

An alternative approach is provided by the LC Concept System from France, and DTS (digital theater sound), the system used for the release of the film *Jurassic Park*. Here the digital sound is carried separately on CD-ROM, and synchronized to the film by timecode printed on the film.

The method has the advantage of flexibility, especially as, for example, the same optical prints can be used for different language releases. It has the disadvantage of 'separate inventory', with the possibility of sound and picture parting company when programs are in transit.

Special multi-channel systems

Applications of movie projection in special displays often require non-standard sound systems, to meet the needs of the display, to simplify production, or for unusual effect. There are several possible methods of proceeding.

In the last few years it has become common to prepare sound tracks referenced to timecode, and with simple and reliable synchronizers it is practical to synchronize movie projectors to standard multi-track tape recorders. This greatly simplifies the production and operation of one-of-a-kind shows.

The simplest systems use the film as the master playback device, while the tape recorder 'follows' the film. Ideally, timecode is recorded on the film's optical track, and a synchronizer compares this with the timecode recorded on the clock track of the tape recorder. In some cases there is only timecode at the film beginning to position the system, then tacho pulses are used from the projector. Here the synchronizer is effectively counting frames to ensure synchronization.

Until recently the tape recorders would have been conventional multi-track analog machines. Now it would be more usual to use compact digital multi-track machines, for example, those of the 8-track ADAT format.

However, it is clear that, because the sound source is only required to 'follow' timecode, there is a wide range of possibilities. Therefore special venue multi-channel film sound tracks are now run from many different sources, including, but not limited to:

- multi-track analog tape;
- multi-track digital tape;
- computer hard disc (DTS system and others);
- CD and CD-ROM (DTS system and others);
- laser videodiscs (these can replay two analog and two digital sound tracks each. In this case, the projector usually 'follows' the disc);
- digital audio workstations (for example, those from Roland and Yamaha);
- solid-state digital sound stores (for example, those based on using PCMCIA flash memory cards, by Alcorn McBride and Electrosonic Systems).

All these systems can give excellent sound quality, equivalent to CD quality. Therefore the choice of which system is the most appropriate will be made by balancing the following considerations:

- show duration;
- number of channels required;
- whether video is also required in the same show;
- most suitable show production method, and production team's preference;
- standardizing with other equipment on the same site;
- minimization of both capital and running costs (including film print costs).

Whatever 'source system' is chosen, it will be followed by a system of equalization and amplification to give the required even response and sound pressure levels in the auditorium (discussed further in Chapter 23).

Mixed-media and multi-projector shows

Film-showing systems used in theme parks, museums and similar venues are often required to synchronize with other media. This is now easily done, usually by a show controller, or show control computer, following timecode derived from the film. In some circumstances, especially where the show uses a mixture of video and film, the film projector itself follows timecode, but this requires a powerful motor controller for the projector.

Anything that can be electrically controlled can be synchronized, using show control programs of the kind described in Chapter 16.

Some shows require multiple movie projectors to be run in synchronization. For example, while it is possible to run three-dimensional films from a single projector using a split frame/dual lens arrangement, most big-show three-dimensional systems use two interlocked projectors. Some systems use electrical interlock ('selsyn' technique) but today it is more usual to rely on the use of electronic motor controllers. These are sufficiently precise that, if two or more films are started simultaneously from a reference point, they will stay in perfect frame synchronization.

Figure 17.12 is a block diagram of a two-projector three-dimensional system used in a simulator show where the audience are seated on individually actuated moving seats, or on a moving platform carrying from 12–60 people. Shows using nine or even more interlocked projectors are quite common. Nine are used in some 360-degree shows based on multiple projection.

Continuous running films

Theme park, museum and permanent exhibition shows require their film systems to run fully automatically, preferably without the need for a full-time operator in the projection booth. Some exhibits require endless operation with no break in projection during opening hours.

16 mm film is no longer considered for endless use, because video projection techniques give equal or better quality and are much more reliable and cost-effective.

For 35 mm and 70 mm films various methods are used:

- Conventional platters with multiple show copies. This requires manual rethreading every few hours, but can be quite acceptable for some theater-based shows.
- Automatic rewind through the gate. This method is only suitable for short shows, with sufficient interval between shows, and only works on special types of projector.
- Endless loop platters. These are a variant of the conventional platter, which ensures that film surfaces do not rub

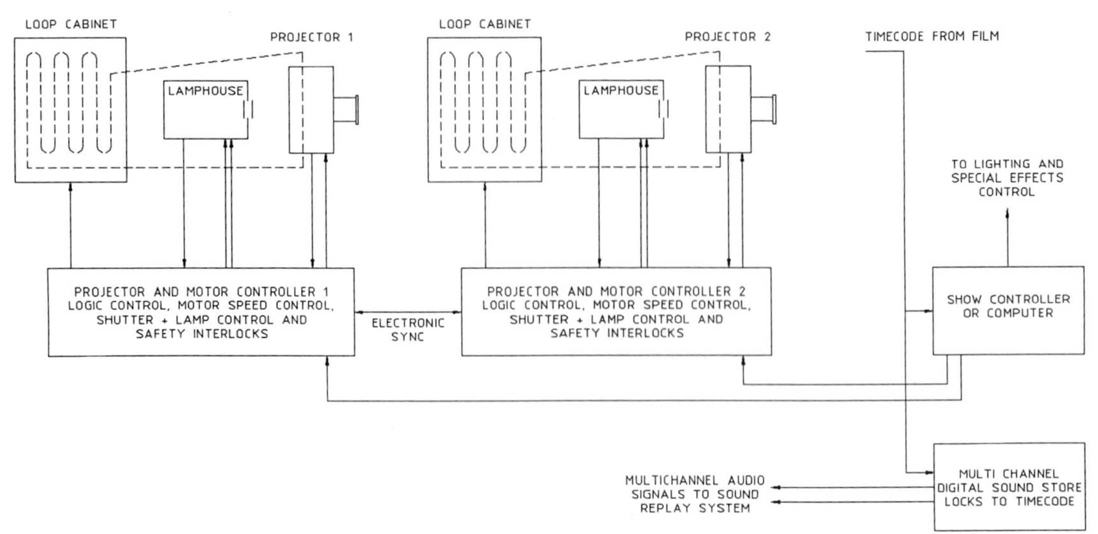

Figure 17.12 Block diagram of a three-dimensional 70 mm system for a theme park. Two 70 mm projectors are electronically interlocked. A multichannel solid-state sound store is locked to timecode read from one of the film prints. A show control computer also follows the timecode

together while on the platter. Minimum program length is about 40 minutes (or multiple copies must be used).

□ Endless loop cabinets. This method ensures the longest possible print life. The cabinets use powered rollers to ensure that the film presented to the projector is not under tension. The rollers only have contact with the edge of the film, and the cabinet usually includes film cleaning rollers and humidity control. The cabinets are BIG. Depending on format and film speed, there can be anything from 90 to over 600 ft of film per minute (27–180 m).

One example: many theaters showing Imax shows run different films in repertory, and here it is best to rewind films between shows, with manual operation. However, in installations dedicated to a single program, which does not change, the installation of loop cabinets is found to greatly increase print life, and to minimize the interval between shows. Imax have an interesting system of 'loops within loops' in their cabinets, which minimizes their size.

Although film presentation can be made to be fully automatic, it must not be thought that no maintenance is required. All automatic film systems require careful daily maintenance. This is mostly concerned with cleanliness, and the single most useful support is a compressed air supply for cleaning sited next to each projector.

Chapter 18

Video production

Most industrial and business users of video are concerned with the *showing* of video programs. Thus, Chapter 19, which discusses video presentation systems, is likely to be of most interest. This chapter covers a few topics on video production which may be of relevance to the person commissioning video programs, or to someone seriously considering establishing an in-house production facility. This chapter is *not* intended as a short primer on how to make a video program. Anyone who needs detailed information of this kind should consult appropriate textbooks.

In principle, the making of a video program is no different from the making of any other AV program. The processes described in Chapter 5 must be gone through in the usual way. The differences with video can be summed up as:

- ☐ The enormous variation in cost between low-quality equipment and professional equipment.
- ☐ The immediacy of the medium, with its ability to replay material as soon as it has been shot.
- ☐ The electronic nature of image storage which allows special effects to be easily achieved by processing electrical signals.

The very high cost of broadcast-standard video equipment has led to a situation where nearly all video post-production (editing and audio mixing) is done at facilities houses. Few program production companies have their own broadcast-standard production facilities, so they usually plan work they need doing, and book the necessary time at a suitably equipped editing suite. Because editing suites may well cost £250 ($400) per hour, planning is all-important if budgets are not to over-run.

There are various levels at which video programs intended for business and industrial use can be made. The choice of which level is appropriate must be made on the basis of the intended application. The levels are:

☐ **Broadcast standard.** This is the highest standard used by professional broadcasters. This standard is usually also adopted by those making TV commercials for public showing, and by those making prestige programs for corporate image use. It is also appropriate to any program that is required in a large number of copies.

☐ **High intermediate.** This standard is only just below broadcast standard. It is used for programs for which wide distribution is expected. The main difference between this and broadcast standard is that cameras are not so expensive, and the range of special effects available are more limited.

☐ **Low intermediate.** This is the standard appropriate to the majority of business, industrial and educational users. The aim is a good commercial quality suitable for copying.

☐ **Training.** This is a polite word for low standard. Here, the equipment used is of good domestic standard. The results are fine for role-playing training applications, experimental work, and limited-interest programs where distribution is not required.

All four levels are relevant to the commercial user. But because equipment prices for complete production systems range from £5,000 to £750,000 ($7,000 to $1,000,000) it is important to decide which level is required.

Videotape recorders

The type of videotape recorder used to make the program is a good guide to the production level aspired to. Unlike movie film where there are only two types of film (16 mm and 35 mm) relevant to commercial production, there are many relevant tape standards for videotape. Apart from the abilities to record a high-quality image and to be able to run at different speeds and give a good still-frame performance, the feature that distinguishes the broadcast machine from the rest of the pack is the ability to maintain quality through several generations of copying.

The process of video editing is completely dependent on copying signals from one tape on to another. It can thus happen that a show copy of a video program is at least two generations, and often three or four, away from the master material. It is the broadcast machine's ability to maintain quality through several generations that sets it apart. Table 18.1 lists the current popular tape formats (see also corresponding entries in the glossary).

Table 18.1 Some professional videotape formats

D1	$\frac{3}{4}$ inch	Component digital
D2	$\frac{3}{4}$ inch	Composite digital
DCT	$\frac{3}{4}$ inch	Component digital
D3	$\frac{1}{2}$ inch	Composite digital
D5	$\frac{1}{2}$ inch	Component digital
Digital Betacam	$\frac{1}{2}$ inch	Component digital
DVCPRO	$\frac{1}{4}$ inch	Component digital
Betacam SP	$\frac{1}{2}$ inch	Component analog
M-II	$\frac{1}{2}$ inch	Component analog
C-Format reel to reel	1 inch	Composite analog
U-matic high band	$\frac{3}{4}$ inch	Composite analog
U-matic SP	$\frac{3}{4}$ inch	Composite analog
U-matic low band	$\frac{3}{4}$ inch	Composite analog
Super-VHS	$\frac{1}{2}$ inch	Separate Y and C
Hi-8	8 mm	Separate Y and C

1 inch C-Format is now obsolescent, although still widely used in broadcast facilities.
U-matic has a huge installed base and, as at 1995, is still available, but for new installations would now be replaced by one of the component formats.

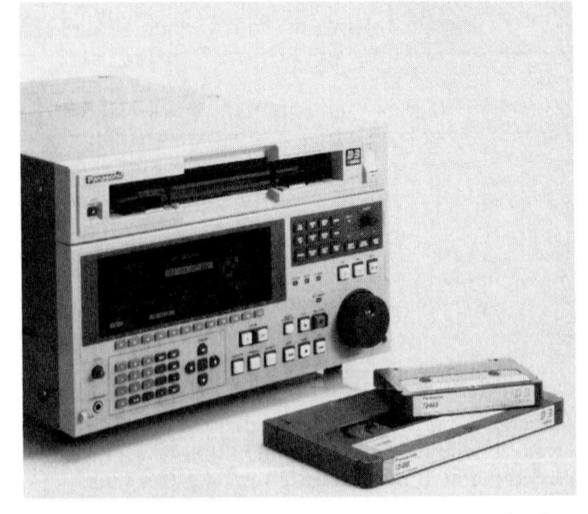

Figure 18.1 The Panasonic AJ-D350 is an example of a modern broadcast videorecorder. It works to the D3 composite digital standard

The evolution of broadcast standard machines was from 2-inch to 1-inch open-reel, and thence to $\frac{1}{2}$-inch cassette. The $\frac{3}{4}$-inch high-band U-matic was introduced as a compact means of program acquisition, but the material was usually transferred to 1-inch for editing. The 2-inch, 1-inch and $\frac{3}{4}$-inch machines all record a *composite* signal (explained in Chapter 19) but newer $\frac{1}{2}$-inch formats record a *component* signal which helps to ensure a higher quality. Sony's Betacam and Betacam SP set the pace for component recording, but these now have strong competition from M-II.

Until recently all video recording was analog, the electrical signal on the tape being an analog of the video image recorded. Now a lot of broadcast origination is done using *digital* video recorders. The idea here is that if the video signal is digitized into signals that only have one of two values (0 or 1), the copying process is much more secure and there is no loss of quality between generations.

First on the scene was D1, using a $\frac{3}{4}$-inch cassette to record an uncompressed component digital signal. This was followed by D2, using a different kind of tape in the same size cassette to record a composite digital signal. D2 cassettes of a given size can record more than twice the duration of their D1 counterparts.

D3 and D5 were designed by NHK to use $\frac{1}{2}$-inch cassettes, which made the formats suitable for use in professional camcorders. Both use the same cassette, but the tape travels twice as fast in D5 to accommodate the component signal.

DCT was designed by Ampex to allow updated D2 cassettes and transports to carry component signals. Similarly digital Betacam was introduced by Sony to have a high degree of compatibility with its analog predecessor. Both these systems use approximately 2:1 compression of the digital signal.

A new contender, especially suitable for news acquisition because of its compactness, is DVCPRO, formally announced in 1995. This $\frac{1}{4}$-inch format uses 5:1 intraframe compression and records 4:1:1 component. While this is not as high a standard as D1, it is equivalent to a broadcast image.

Until the early 1990s, the high intermediate standard of production was served by high-band $\frac{3}{4}$-inch equipment, but is now mostly served by one of the component formats. Similarly the low intermediate standard was served by low-band $\frac{3}{4}$-inch equipment, but is now more likely to be served by S-VHS, or the same class of user has moved up-market to use Betacam.

At the bottom of the pile, systems based on standard VHS are available. The main problem with VHS is that it is not possible to maintain quality when copying through several generations. It may, therefore, be quite suitable for use in a training department, where program copies are not required, but is not recommended for serious production work.

It is important to understand that the way in which programs are made need have no influence on how they are shown. Usually, shows made on high-band equipment are later copied over to low-band or VHS for final showing. The rule is to use as high a quality as one can afford to originate the program, then make copies to match the showing facilities. Chapter 19 explains why the quality of replay equipment is likely to improve, with a trend towards component replay.

Camera

Video cameras come in as wide a price range as video recorders, so usually the two are matched: broadcast cameras with broadcast recorders, domestic cameras with VHS. The aim should be to use as high a quality camera as can be afforded. Until relatively recently all professional cameras used vacuum pick-up tubes, but now virtually all video cameras used in commercial and industrial production are based on CCD (charged coupled device) solid-state technology.

To record a color picture, it is necessary to split the picture information into its red, green and blue components. Broadcast and professional cameras are three-chip (or three-tube) cameras. Incoming light is split into its three color components using dichroic color filters, and each component is fed to its corresponding pick-up chip or tube. All low-cost cameras, and even some professional ones, are single-chip cameras. Here a single CCD chip has a three-color grid filter, and the three color signals are extracted electronically. The problem is that the filter necessarily limits definition; close examination of the resulting image can show a stepped pattern to any color edge.

Camera performance can be measured in terms of the objective quality of the image and in the facilities offered by the camera, e.g. by the type of lens fitted. Apart from the obvious items like lack of picture distortion, picture sharpness, good color rendition, there are attributes that distinguish a good camera from an also-ran:

- ☐ The ability to perform well in low light conditions without introducing graininess.
- ☐ The ability to be precisely set up to match a standard, and to maintain the set-up without further adjustment.

The last point is particularly important in studio applications. Most studios use at least three cameras in order to provide a sufficient variety of shots to give a fluent production. It is clearly vital that the results from each camera are identical. It would be a disaster if, in an interview, the flesh tones of the participants changed as different cameras were used.

Many professional cameras are available in versions that allow the 'docking' of a compact video recorder on to the back of the camera to make a complete camcorder. This has become possible because the new component 1/2-inch tape formats allow production of very compact recorders.

Figure 18.2 The Sony DXC3000 is a 'three-chip' camera. It is widely used for professional, commercial and industrial AV production

Vision mixers

It will already be clear that a video production studio is a considerable investment. This is why only those who are reasonably sure that they can make full-time use of such a facility should consider making the investment. In addition to the videotape recorders, studio lighting, audio recording facilities, cameras and all their related accessories, the assembly of a video program in a studio certainly requires use of a vision mixer.

This device can select from a number of different video inputs, usually the three cameras, plus all sorts of other picture sources such as other videotape machines, caption generators etc. The person using the

Figure 18.3 Professional vision mixing equipment at Yorkshire Television Facilities

mixer can usually see all the material available on corresponding preview monitors, and then selects the required image either by switching or by cross-fading. Modern mixers offer a variety of transitional effects including many kinds of picture wipe. The trend towards the use of component video sources such as Betacam and M-II has meant that mixers must now also work in the component domain.

When all this equipment is hooked together, unless all the picture sources are in sync (i.e. timing of video image scanning is precisely in unison) there are nasty picture rolls as each transition is made. Thus, within a studio, all equipment works to a 'master sync' pulse that is distributed to all the equipment being used. Timing of video signals is sufficiently critical that adjustments even have to be made to take into account different cable lengths.

A particular problem arises when it is necessary to cut to a signal derived from another videotape source. In this case, it is difficult to ensure that the source recorder is perfectly in sync with the rest of the system and even if, in principle, it has been locked to the master sync, there is usually a residual loss of definition. This is put right by the use of a timebase corrector. This expensive piece of equipment is an essential part of the studio armory, and is often used to clean-up material that would otherwise be sub-standard by the time it reached the master tape. Low-band original material is enhanced by its use, so if this type of material must be used as part of a professional production, it can be suitably sharpened up to match the new material.

Many other devices are now considered to be an essential part of the video production studio. One is the electronic caption generator. Captions used to be put up on the screen by pointing a camera at a suitable piece of artwork, or at a roller caption machine. Now digital electronic caption generators allow generation of captions in a wide choice of fonts and in any desired color combination. The low-cost studio can be well served by inexpensive caption generators based on standard personal computers.

The digital processing of video images opens up a whole range of possibilities for the manipulation of image sequences. Now that startling video effects can be seen every day on television, it is not surprising that industrial video users are asking to have similar effects within their shows. The best special-effects generators are still extremely expensive, so most users rent by the hour and, most importantly, obtain the services of someone who knows how to operate the system. However, relatively low-cost systems, with a corresponding reduction in performance, are becoming available for the semi-professional user.

One trick that is provided on even relatively low-cost vision mixing systems in color separation overlay or chromakey. The principles have been known for years, and have formed the basis of many special effects in the movies – electronics just make it easier. The process is used when a studio shot must be married to some other, usually exterior, shot. For example, if a studio commentator wishes to appear to be speaking from some exotic location.

The usual arrangement is for the presenter to stand against a blue background, either a white cyclorama illuminated with blue lighting, or a blue cyclorama illuminated with white lighting. The chromakey system is then fed with two inputs, the studio signal and the remote signal – for example, a color slide of the south of France or a film of a rocket launch. Whenever the chromakey device detects a saturated blue signal, which is easy for it to detect, it substitutes the alternative signal. Thus, the final image is a composite of the two images.

Two problems arise with chromakey. The first is the obvious one that the foreground shot must not include any saturated blue. A blue shirt results in an interesting south of France pattern. The second is that hair and other fuzzy outlines cause the system to dither and to give the foreground a halo or cardboard cut-out quality. The more expensive the system, the less the problem.

Video editing

The editing of videotapes is essentially a copying process. The principles are easily understood, and anyone can learn how to press the appropriate buttons. However, video editing is what puts the polish on a video production, and good editing can make the difference between an excellent and a mediocre show. Creative video editors are a rare breed and, at the top end of the market, the fortunes of an editing facility with millions of dollars invested in equipment can depend on attracting and keeping this talent.

A simple editing suite consists of two machines, the source machine carrying the original material, and the master machine on to which the master tape is being assembled. The simplest method of editing is known as assembly editing and involves sequentially dubbing required sections of the original tapes on to the master machine.

If the material is simply copied across there is a bad picture roll at each transition, owing to lack of synchronization between the machines. The two are therefore linked by an edit controller which allows the editor to select the required section, then command an automatic edit. This process involves a pre-roll, where each machine backs up (usually five or ten seconds), then comes forward using the pre-roll time to get into precise sync. At the designated frame the master machine switches from replay into record and records the required section. Often stopping is automatic because the editor can also designate the stop frame.

Professional video people found a lack of precision in this simple method, particularly in rapid location of material and in being sure that particular sequences were going to fit their allotted slot. Professional editing now tends to be done using the insert editing technique, based on the use of time codes.

Figure 18.4 A professional edit suite. 'Edit 1' at Yorkshire Television Facilities has Betacam SP video recorders as standard equipment, and offers D1 and D2 digital machines as options

Here all material is striped with a unique time code. This is either 'linear' or longitudinal' time code carried on a spare track of the tape like an audio track, or 'vertical interval' time code carried within a spare part of the picture signal. Its importance is that it allows any required section of source material to be located rapidly, and moved to any required place on the master tape. Even a single frame can be placed at a designated frame time.

Not surprisingly, computer techniques are essential to this style of editing. The computerized edit controller not only precisely controls up to four videotape recorders, but also keeps a complete log of all edits made. It can also cue in other devices, such as caption generators and special-effects units, to carry out a set routine over a specified time.

Finally, timecode techniques are the key to a great improvement in audio programming. Because multitrack audio tape recorders can be linked by the same time code system, it becomes possible to bring a much greater sophistication to the audio part of a video program.

Non-linear editing

The means of producing and editing video programs is currently being transformed by the introduction of so-called 'non-linear' editing. These techniques also commend themselves to the industrial user and in-house production facility.

Previous constraints on computer memory capacity have all but been removed, and it is now quite feasible to record useful lengths of video program material on to computer hard discs. Ideally no digital compression is used, with recording to the CCIR601 component digital standard. In practice, most systems use some kind of compression to keep the memory size used to reasonable dimensions.

Once video material is in computer memory, it is very easily and quickly *random accessed*. Instead of winding a tape cassette from one end to the other in search of a few relevant video frames, the computer can call up the required frames almost instantaneously. Editing is much faster, because there is no need to 'pre-roll'; the computer ensures that images are properly synchronized, without having to be concerned with the mechanical constraints of a tape transport with moving parts.

Non-linear video equipment is of two kinds. Dedicated equipment for the professional broadcast studio, and semi-professional equipment based on the use of personal computers as the computing 'platform'.

The professional market is served by vendors such as Avid, Lightworks and Quantel. At the extreme the equipment provides a completely 'tapeless' operation. For example, much of CNN's output is now played out direct from Avid 'Airplay' equipment. However, while it can be expected that tapeless operation of news, current events and commercial programs will increase, tape will continue to be used for some time yet. Its recording capacity, especially when expressed in cost terms, far exceeds that of other media. It will remain the prime acquisition medium, and the prime long program medium.

Non-linear methods have enormous potential for the in-house user. Some of the possibilities are discussed in Chapters 14 and 20. Most practical are systems which combine the flexibility of non-linear with the cost-effectiveness of conventional techniques. It is now quite feasible for a marketing department, for example, to store relevant video material on computer hard disc, and to compile presentations that can easily be used by sales staff.

Until recently the preferred computer platform for semi-professional non-linear video editing and production was the Apple Macintosh, which tended to limit its use to those in the graphics fraternity already familiar with Macs in graphics production. The arrival of the fast PC, and programs to match, has changed

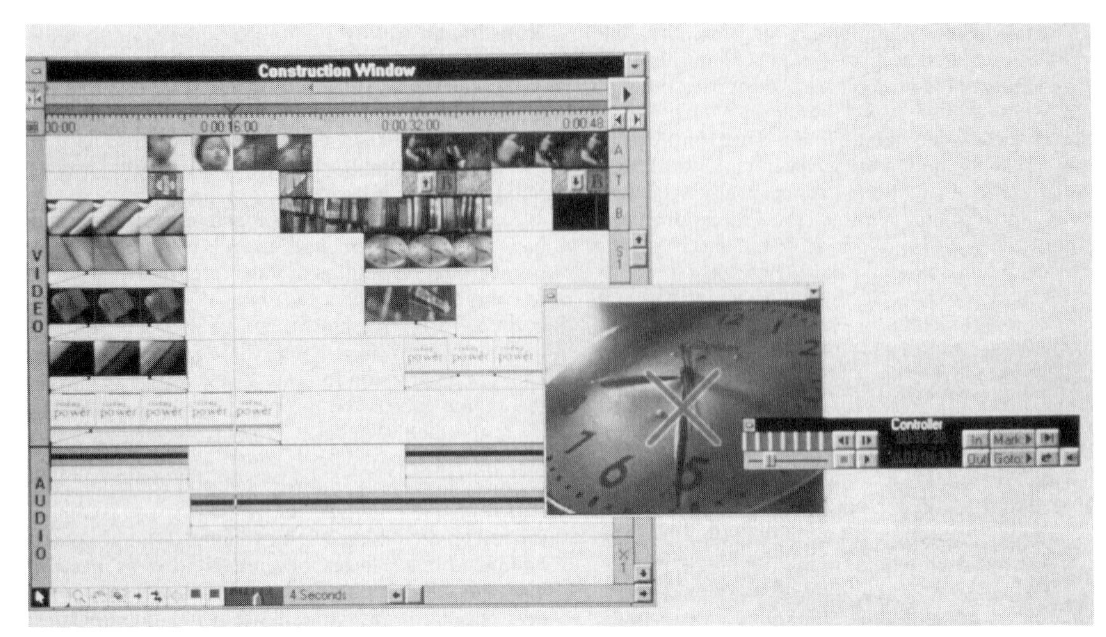

Figure 18.5 The Adobe Premiere 4 program runs on Pentium-based PCs. The 'timeline' display clearly shows what is going to happen. Video and audio are chosen from computer A and B video sources, and transition effects between sources can be programmed

Figure 18.6 Freehand Graphics Ltd use the Fast Electronic 'Video Machine' editing system which combines conventional linear editing with non-linear editing, a cost-effective arrangement for the in-house facility

the situation. One of the market leading computer programs is Adobe's 'Premiere'. The performance of this program is similar when run in its Mac version, or in its latest version for Pentium™ equipped PCs. This is likely to result in much more direct use of non-linear video by business users.

There are systems and products which neatly bridge the gap between traditional and computer methods. One of these is Fast Electronic's 'Video Machine' series of products, which combine the control of traditional VCRs with digital special effects and hard disc digital video recording. This kind of equipment is especially suitable for both intermediate standards of production.

New video sources

Developments in electronics are proceeding at such a rate that it is difficult to keep track of all the new possibilities for originating programs. Some devices which are altering the way in which programs are made, and which have direct relevance to the industrial and commercial user are:

☐ The electronic stills camera. Originally conceived as a consumer device for taking 'photographs' on a magnetic disk for replay on the home TV set, this is now available as a professional unit.

☐ The high-quality scanner. Reasonable cost document scanners that scan the whole or selected parts of a document or color photograph are easier to use than a video rostrum camera.

☐ The recordable videodisc. This is a convenient store for stills and short moving sequences required frequently. Access is fast, and machines integrate easily into an editing set-up. Discs that record a component signal give the best quality.

The small studio

This description of the potential sophistication of video production may seem designed to put off the small user altogether. So, in a way, it is. In general, anyone commissioning a video program should be happy that professional facilities can be used as required by the production. Nonetheless, there is a real demand for small in-house video facilities, especially in the training area.

If in-house use is to be confined to role-playing sequences and limited interest programs, it is probably best to:

☐ Base the system on Betacam or S-VHS if possible, especially if it may be necessary to have programs extended at outside facilities houses. Otherwise, industrial VHS should be chosen.

☐ Choose cameras on the basis of whether the majority use will be on location, or in the studio.

☐ Limit the investment in control equipment to a simple edit controller and vision mixing system.

☐ Consider an investment in a PC-based non-linear editing system and CD disc recorder if material is likely to be required for computer playback.

There are some large users of video who can justify the installation of a complete studio and production complex, but these users should either employ someone with the expertise to specify the system required, or engage the services of a consultant.

There is no reason for anyone running a training department to be afraid of using simple video equipment, provided that video is, in fact, the most appropriate medium for the task. Equally, it is not sensible to over-invest if there is no-one who can make full use of it. There are many horror stories of organizations with cupboards full of unused and out-of-date video equipment, a situation that probably arose out of the mistaken idea that video is all very easy, and that waving a camera in the direction of some activity results in a usable program. Video is no different from other AV media; it requires pre-planning, discipline and a clear idea of its purpose.

Chapter 19

Video presentation systems

This chapter deals with the display of moving video images, as seen on television and from videocassette recorders. The subject of computer data display is covered in Chapter 20.

The television set is part of everyday life, and in recent years the widespread use of video recorders in the home has meant that the idea of audiovisual shows presented on a TV set is easily accepted, and apparently very simple.

There are, however, a few problems. The standard TV set is still a device that is best suited to the scale of the living room, so the AV user has to become aware of the compromises and alternative equipment that must be employed for larger audiences. The domestic user does not have to worry about TV signal standards; the commercial user does. This chapter reviews some of the points about television presentation that most affect the commercial and industrial video user.

TV signal standards

The problem of different signal standards is of most relevance to users who expect to receive video programs from abroad, or intend to send programs abroad. The evolution of color TV unfortunately resulted in the development of a number of different electrical signal formats being used to transmit the color picture information. To some extent, the different standards reflect improvements that were possible in second generation equipment. There are also purely electrical reasons why there should be a difference between Europe and the USA. Unfortunately, additional standards have been introduced for political and financial reasons.

The color television signal starts life as three separate quantities, those of *red*, *green* and *blue*. The highest theoretical quality would be maintained by retaining the three separate signals, but this is uneconomic and, in any case, the task of those designing the original color TV systems was to cram the color signal into the same signal space (or bandwidth) as the old monochrome signal.

In fact the eye does not see color in such detail as black and white. The technical terms are *luminance* representing the total illumination of the image, and *chrominance* representing the color information. As will be described in more detail later the highest quality display and recording systems keep these components separate, but the standard TV signal combines them into a *composite* signal to one of the international standards.

The first system into the field was the NTSC system, introduced in the USA in 1954. The letters stand for National Television Standards Committee or, less officially, 'never twice the same color'. The system is used in the USA, Canada and Japan.

The SECAM (séquentiel couleur à mémoire), also known as the 'system essentially contrary to the American method', comes from France. It is used in France, USSR and most of eastern Europe.

Most of the rest of the world uses the German-developed PAL system, officially 'phase alternating line', but also known as 'pictures at last'. Even within the main standards there are variations. SECAM has two versions, fortunately only horizontal SECAM is now likely to be encountered. NTSC can be NTSC 3.58 or NTSC 4.43. The figures refer to the carrier frequency that is used to carry the color information.

NTSC 3.58 is the official version. The 4.43 modified version is derived in multi-standard videocassette recorders to allow the playback of standard NTSC tapes on PAL monitors. This highlights the awkward fact that usually the whole video chain must be of a particular standard. For multi-standard operation it is not usually sufficient to have just a multi-standard video recorder, the TV set must be multi-standard too.

The AV user fortunately only needs to be aware of the basic video signal standards, because the signal used should always be the simple video signal alone. When a TV signal is transmitted in a broadcast system, the variations between countries are considerable. However, this is a problem for the domestic TV set manufacturer rather than the AV user.

Table 19.1 lists some of the characteristics of the different systems. Chapter 3 described fields and frames.

Table 19.1 Television standards

Characteristic	*PAL*	*NTSC*
Frames per second	25	29.97
Fields per frame	2	2
Fields per second	50	59.94
Lines per frame	625	525
Displayed lines	576	486

Videocassette standards

Commercial users of video only encounter videotape in videocassettes. Unlike the audio business where until recently there has been only one standard cassette, the video business has many different cassettes. Some are now obsolete. Others were once only of interest to the broadcast professional, but because of the demand for higher quality displays are now of importance to the serious AV user.

The majority of videocassette recorders use half-inch tape. VHS™ (Video Home System) developed by JVC has the biggest share of the worldwide domestic market, and there is a compact version of the VHS cassette for portable equipment called the C-Format. Betamax™ is the system developed by Sony, and was second to VHS in world popularity, but for the AV user it is obsolescent.

Figure 19.1 The decreasing size of videocassettes. In size order: U-matic, VHS and 8 mm

The use of 8 mm tape results in a much smaller cassette, and the 8 mm cassette is a popular medium for combined camera recorders. The split between VHS-C and 8 mm varies from market to market.

U-matic is the odd one out. This $\frac{3}{4}$ inch cassette system is not for domestic use. It was originally developed by Sony as a low-cost industrial video format. Over the years its performance has improved to such an extent that U-matic is used as the standard production medium for industrial AV. A variant called high-band U-matic or BVU™ reaches near-professional broadcast standards.

The dramatic reduction in real costs of video equipment and the interest in better quality has led to introduction of new cassette formats. JVC has developed VHS to Super VHS, or S-VHS™. While all the cassettes referred to so far record a composite signal, S-VHS records luminance and chrominance separately. Sony have followed step by introducing Hi-8 as a high-quality 8 mm system.

Figure 19.2 The Sony BVW Series of commercially priced Betacam equipment has brought the benefits of component video to the industrial and commercial user

Figure 19.3 The Professional S-VHS system from JVC is now widely used in the commercial and educational markets

Broadcasters and top professional video production all use systems of much higher quality than those used on the consumer market. Their main attribute is their ability to maintain good quality even when a program is copied through several generations. Until a few years ago the 1-inch reel-to-tape (also, confusingly, referred to as C-format) was king of the studio, but its place has been taken by cassette-based systems that record the signal in component and/or digital form. Sony's Betacam system is the best known.

Table 18.1 lists the current professional videocassette standards. Only the current consumer composite formats of normal VHS and Video-8 are missing.

Out of all this confusion, what is relevant to the normal industrial user? Until recently the answer was easy. All industrial AV work would be done on U-matic, using the standard of the country concerned, for example, PAL in the UK and NTSC in the USA. Now, however, a large proportion of industrial programs are made on Betacam or Professional S-VHS.

Many users are more concerned with the replay of programs than with their creation, so some guidelines are:

☐ Standard VHS is the best medium for widespread distribution, but copies should be made from as high a quality master tape as possible.

☐ Industrial users with existing inventory may be using standard or SP U-matic. This is still a robust and practical format.

☐ New users should consider Betacam or Professional S-VHS as their premium format. This is especially the case if programs are going to be replayed on large projection screens.

☐ The 8 mm cassette is suitable for some portable applications.

Standards conversion

It is possible to obtain both standard U-matic and VHS playback videocassette machines that are multi-standard, and it is a good idea to equip presentation rooms with them. This solves the problem of playing back the occasional tape that a visitor may bring, or which may be sent from one side of the Atlantic to the other. Anyone wanting their tapes to be taken seriously will, however, ensure that they are converted to the local standard.

Conversion of color video signals from one standard to another requires expensive equipment, and it is best to try to avoid the need for it. However, there are studios that provide standards conversion as a service, so tapes received from overseas that will not play on a user's equipment *can* be converted.

There are two types of conversion service offered. One is to broadcast standard. This should be used whenever a program is intended to have wide distribution. For example, if a company in Europe makes a program in PAL, but wants to distribute 10, 50, 100 or 1000 copies in the USA, the best procedure is to take a component copy master to the USA and have it converted to NTSC, to as high quality a format as possible. The resulting copy can then be used as the master for generating bulk copies locally, which may well be on standard U-matic or VHS.

The second service is based on a low-definition standards converter that has the advantage of convenience and low cost. However, its use should be limited to 'one-to-one' copying and dealing with emergencies.

Showing the picture

TV signals used in industrial AV are usually composite video signals that contain all the color picture information, but not sound. They have a bandwidth of about 5MHz. TV pictures received by a domestic TV set are in the form of a modulated UHF signal, in which a color video and associated audio signal have been used to modulate a carrier frequency, typically 600 MHz.

Although it is possible to fit a video recorder with an RF modulator that produces a UHF signal for direct connection to a domestic TV, this is not good practice because quality inevitably suffers. Thus, most industrial video programs are seen on monitors. 'Monitor' is a word that was once strictly reserved for the very expensive sets used by broadcasters to monitor the original picture, but it is now used to describe any set that has a video as opposed to RF signal input. Receiver monitors have both. Modern monitors usually, but not necessarily, also have inputs for Y/C (S-VHS) and RGB analog inputs.

In choosing a monitor the user must decide on:

☐ Size: usually as big as possible for group viewing. Monitors between 25 inch and 35 inch are standard in presentation rooms. Smaller monitors are used for individual viewing and portable equipment. The number of viewers who can easily watch monitors of various sizes is shown in Table 19.2

☐ Loudspeaker: in most fixed installations audio is played through a separate high-quality audio system, but sometimes the loudspeaker must be in the monitor.

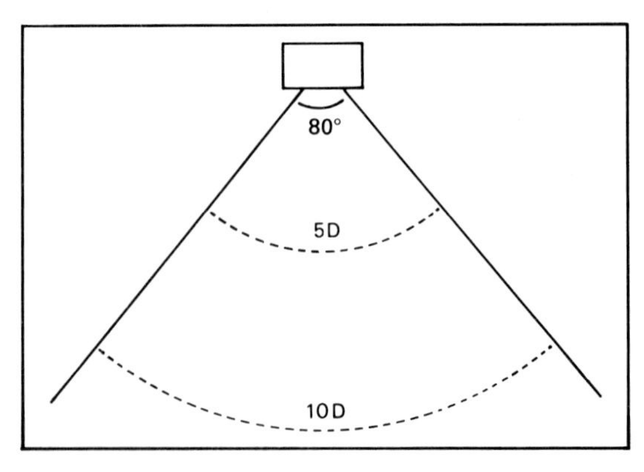

Figure 19.4 A TV set should be viewed at a maximum of 40 degrees off-axis, and at a distance of between five and ten times the screen diagonal.

Table 19.2 Number of viewers of a single TV monitor for general program material

Screen	*Diagonal*	*Number of viewers*
9 in	9 in (230 mm)	4
12 in	12 in (305 mm)	7
20 in	20 in (508 mm)	18
25 in	25 in (635 mm)	28
28 in	28 in (710 mm)	33
32 in	32 in (813 mm)	38
35 in	35 in (890 mm)	45

The figures allow 8 sq ft. (0.75 sq m.) per viewer, with viewing distance being between five and ten times screen diagonal. For viewing computer graphics and text material, viewing distances are between three and six times diagonal, and the numbers approximately halved.

☐ Standards: is single-standard sufficient, or is there a need for multi-standard? 'Quad' standard monitors can deal with NTSC 3.58, NTSC 4.43, PAL and SECAM.

☐ Y/C: it is a mistake to install a monitor unable to accept separate luminance and chrominance signals, as used by S-VHS and Hi-8.

☐ RGB: will the monitor be used for displaying data or RGB video? This subject is covered in more detail in Chapter 20.

☐ Receiver: is a conventional TV receiver for off-air broadcasts required? Sometimes it is convenient to use a receiver monitor. In a multi-use system it can be better practice to use a separate TV tuner, especially when more than one monitor is in use.

Figure 19.5 Barco make a range of presentation monitors suitable for commercial and display use

Standard TV monitors with cathode-ray tube display are the most cost-effective way of showing TV pictures to small groups and individuals.

The largest CRT monitors are very heavy, and at these sizes (35 inch and above) users can also consider the use of compact rear projection units. Units with 40-inch screens are now available which are particularly suitable for displaying both video programs and computer data. While they do not have such a wide viewing angle as the conventional monitor, their screens are flat matt black and do not give specular reflections.

The 'flat-screen' or, more correctly, 'thin-screen' video monitor is now a practical proposition for AV applications. Commercially available flat-screen monitors are based on back-lit liquid crystal technology. The problem has been to get a satisfactory yield from the highly complex manufacturing process, where the failure of only one out of millions of elements of the screen can cause it to be rejected. While more expensive than its CRT counterpart, the LCD video monitor can solve siting problems:

☐ Commercial units are in sizes such as 4-inch, 6-inch and 10-inch diagonal. 14-inch displays are available but expensive. 21-inch displays have been demonstrated in the laboratory.

☐ The LCD video monitor is ideal for incorporating in exhibition displays where there is no depth to accommodate a CRT monitor. It is also widely used in aircraft entertainment systems where the same constraint applies.

☐ It is ideal for installation in lecterns and other 'monitoring' applications.

☐ Note that the viewing angles of backlit LCD displays are limited. This is not a problem when a single passenger is watching a seat-back display, but needs consideration in exhibition applications.

LCD technology is not the only thin-screen technology. Some experts believe that the manufacture of large 'picture-on-the-wall' displays will be easiest done using plasma technology, and displays of 40-inch diagonal have been constructed. 21-inch plasma displays are available as a standard product, for example, from Fujitsu. Plasma displays have a wide viewing angle but are not as bright as CRT displays. Another technology, which has the brightness and wide viewing angles of the cathode ray tube, is the field emission device (FED) which is in pilot production for small screen sizes.

Video projection

Video projection is usually achieved with three separate high-brightness cathode-ray tubes, one each for red, green and blue components. These tubes vary in

Figure 19.6 The largest standard CRT monitor is this 40-inch model from Mitsubishi. It is suitable for data, and has high brightness and a wide viewing angle

Figure 19.7 An alternative to the large CRT monitor is the table top back projection unit. This 40-inch model is the RVP4010 from Sony

diagonal size from 12 cm (5 in) to 23 cm (9 in) and very large lenses are used to collect light and focus it on the screen. Electronic adjustments are made to the scanning signal to get the three colors to line-up and to compensate for off-axis projection.

The most compact free-standing video projection systems use back projection to achieve an image of about 1.15 m (4 ft) diagonal. They use special screens to reject front-incident light, and are convenient to use as they are no more complicated than a large TV set.

The larger presentation room needs a system that can use a conventional screen, and give a specified image size. Free-standing projectors are available, which in the dark can give a picture up to 6.6 m (21.6 ft) wide. They can use conventional front or back projection screens, but benefit from a high-gain screen.

The only way to evaluate image brightness of one video projector against another is to do a side-by-side comparison, with both video projectors showing the same material onto identical screens. This is because objective measurement of video projector light output

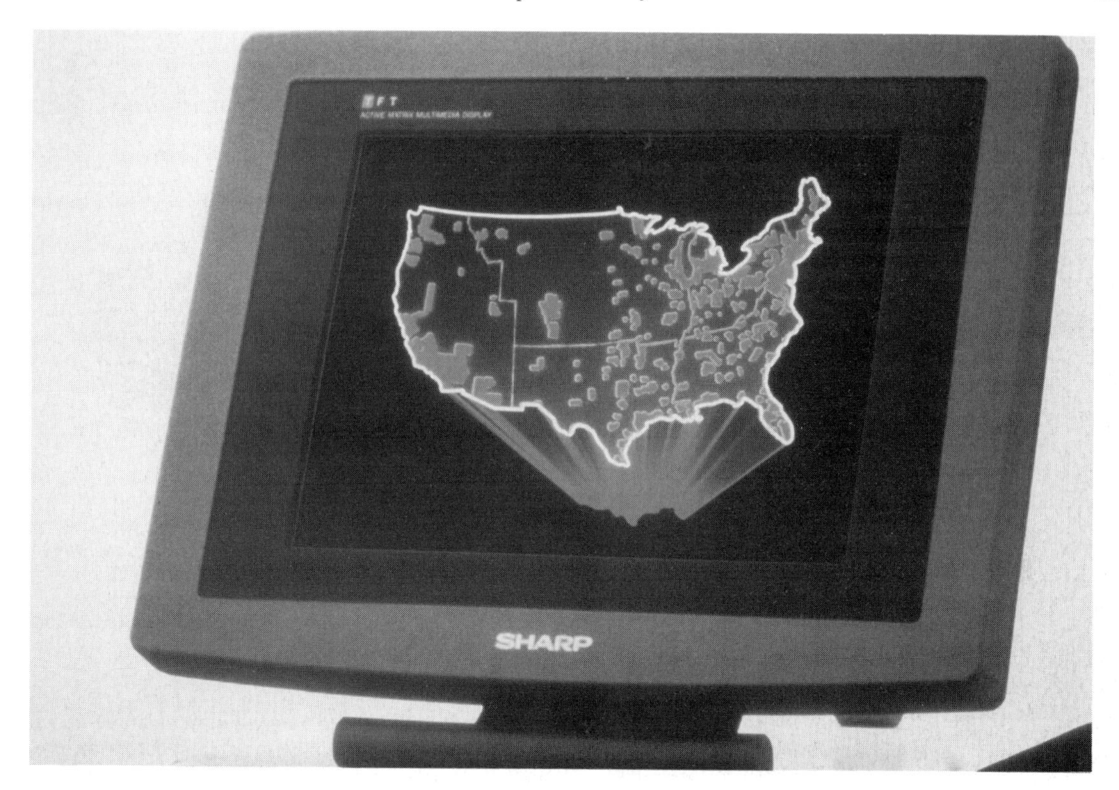

Figure 19.8 A 10-inch LCD thin-screen monitor from Sharp

is not fully standardized. Manufacturers (in particular Barco) give details of test methods, but because the flooding of tubes with 'peak white' light burns them out, they give a light output figure based on typical program material. Actually, current-limiting circuits in projectors prevent tube burn-out on an all-white image, but the resulting lumen figure would be very low, and do not look good in manufacturers' brochures.

Thus, light outputs are usually quoted as lumens for 10 percent or 20 percent peak white. Three-tube projectors used in AV presentation rooms have typical light outputs of between 600 and 1600 lumens expressed on the 10 percent peak white basis.

Three-tube video projectors are only available with one fixed focal length lens system. This means that there is a fixed distance from the screen for a given image size. Fortunately, projectors include electronic keystone elimination, so can be installed on the ceiling or below the level of screen bottom to allow unobstructed viewing. The projector can be made to project an electronically-generated crosshatch pattern in each of the three colors. Usual procedure is to project green first, and to align it so that there is no distortion. The other two colors are then lined-up in turn to match the green.

The latest microprocessor-controlled video projectors have automatic line-up procedures, but it still remains the case that accurate setting up (known as converging) of three-tube video projectors requires specialist knowledge.

Liquid crystal video projectors

The problems of projector line-up are eliminated if there is only a single image source and a single projection lens. This is now possible in a new generation of video projectors using a conventional lightsource with an optical arrangement similar to that of a slide projector. In this case the 'slide' is actually a liquid crystal panel consisting of thousands of miniature liquid crystal cells that act as high speed shutters.

In order to project a three-color picture there are three such panels. Light from the compact arc lamp is split into red, blue and green components by dichroic mirrors, and recombined to emerge from a single lens. The principle is shown in Figure 19.12.

Figure 19.9 Free-standing back projection 'monitors' allow medium sized screens (50–84 inches). This 60-inch unit is from Sony

Figure 19.10 The Barco 1600 is possibly the brightest of the conventional three-tube projectors

Some lower-priced liquid crystal projectors use a single multicolored LCD panel, of a kind similar to that used in OHP overlay panels (see Chapter 20). They rely on color filters within the panel and, as a result, cannot give the same resolution, colour rendering or contrast as the three-panel design. However, Sharp

Figure 19.11 Ceiling-mounted CRT projectors are heavy. It is sensible to install them with a lifting device. This one from Chief Manufacturing locks the projector securely in position when in use, but uses a simple cable suspension for lowering the projector for service

have recently introduced a new design of single-panel projector which does not rely on color filters on the panel. It divides the light from the lamp into red, blue and green using dichroic filters. Each color is directed at the back of the LCD panel at a different angle, and a multi-lens arrangement ensures that each color can only reach a designated set of pixels. The system is analagous to the shadow mask principle used in CRTs.

The portable liquid crystal projector is lightweight and very easy to set up. A wide choice of lenses, including zoom lenses, is available, so projector positioning ceases to be a problem.

□ Low-cost projectors are suitable for video only. Typically they have about 100,000 pixels per color. Their images should be viewed from a distance of at least twice screen width.

□ Liquid crystal projectors cannot give as good a contrast performance as CRT projectors. However, their performance is satisfactory for most applications.

□ Liquid crystal projectors are now available for VGA resolution, using 300,000 pixels per color. This type of projector is the most suitable for business use.

□ It is likely that the liquid crystal projector will supersede the CRT projector in many standard applications. But for the time being, the CRT projector is still best for high-resolution images, and is the most cost effective for some other applications.

With liquid crystal projectors, it is no longer necessary to measure light output in 'CRT lumens' because the projection arrangement is similar to that used for

LCD Projection TV Optical Path

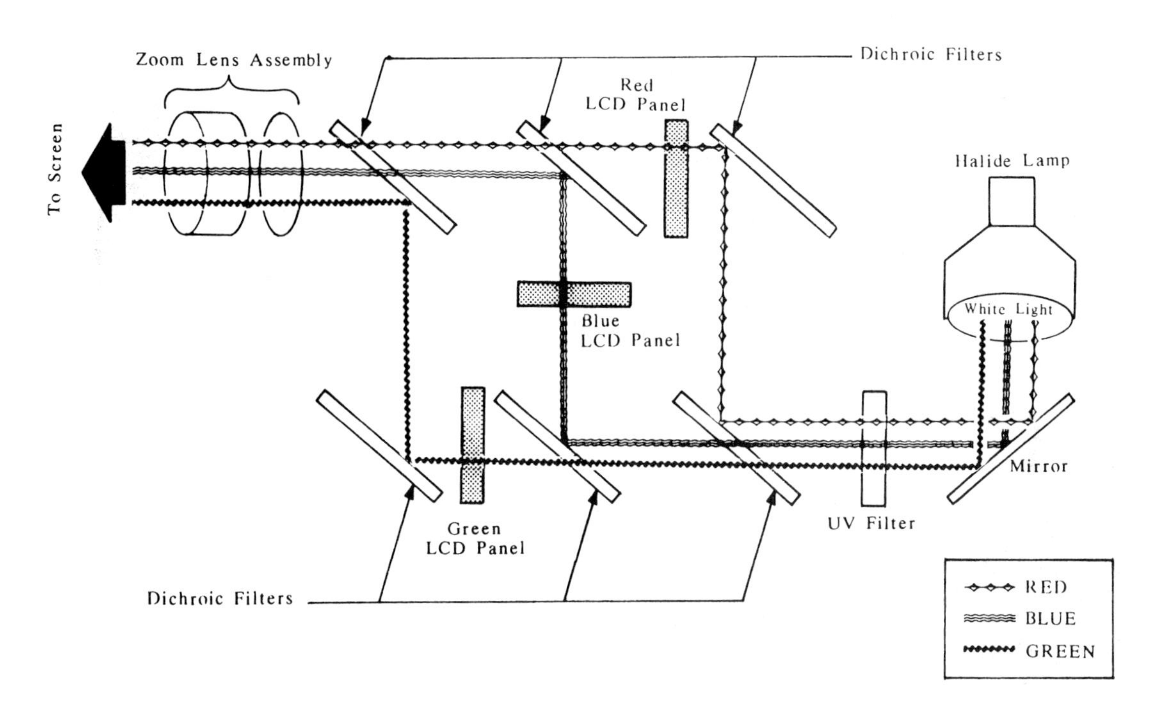

Figure 19.12 (a) This diagram shows how three liquid crystal display panels are used in a video projector. Only one lamp and one lens are needed. (Diagram courtesy Sharp Electronics). (b) A cutaway of a Sharp liquid crystal video projector, where the components corresponding to the above diagram can be clearly seen

slides and film. Light outputs can, therefore, be measured in 'true' or 'ANSI' (American National Standards Institution) lumens. The best of the portable projectors give around 160 lumens, and give results which are subjectively similar to 500 CRT lumen projectors. High-gain screens should be used if possible.

Brighter and bigger images

Many applications of video projection require brighter images that can be achieved from standard projectors. For some applications it is economic to use two or even more standard video projectors on one screen. It is quite easy to line-up two projectors, and the method has the advantage in permanent installations that if one projector fails, there is at least a stand-by image. However for really big images different technologies must be used.

Many years ago it was realized that the best answer to the problem of picture brightness is to harness a conventional light source, such as a xenon arc lamp.

Figure 19.13 This liquid crystal video projector from Sharp is suitable for both video and VGA graphics

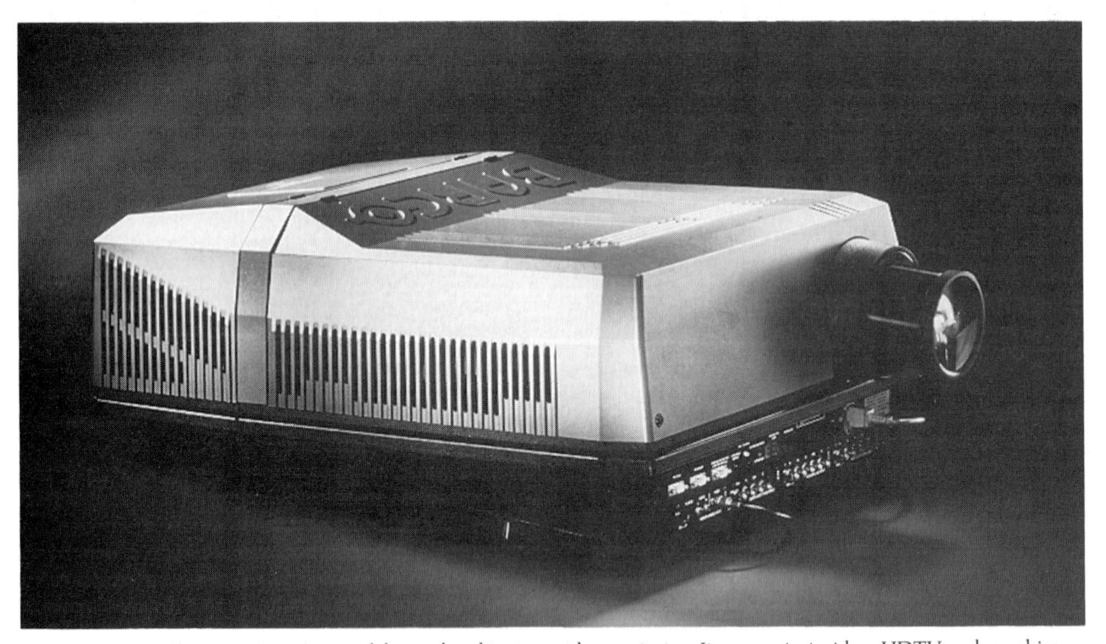

Figure 19.14 The Barco 8100 is a widely used auditorium video projector. It can project video, HDTV and graphics images with resolution up to 1024 × 768, and with a light output of 1,800 ANSI lumens

This led to development of special light valves in which a TV image is constructed by scanning a thin oil layer on a transparent plate. Scanning of the oil layer by the electron beam varies its density, and this produces an image that can be projected like a movie film. The whole process must take place under vacuum, and the oil film has to be continually reconstructed. A very expensive piece of equipment is required to do it.

The biggest example of the method in use is the Gretag Eidophor system. It is a flexible system because, in principle, interchangeable lenses can be used. The largest machines have light outputs of 7000 lumens and give pictures up to 13 m (43 ft) wide.

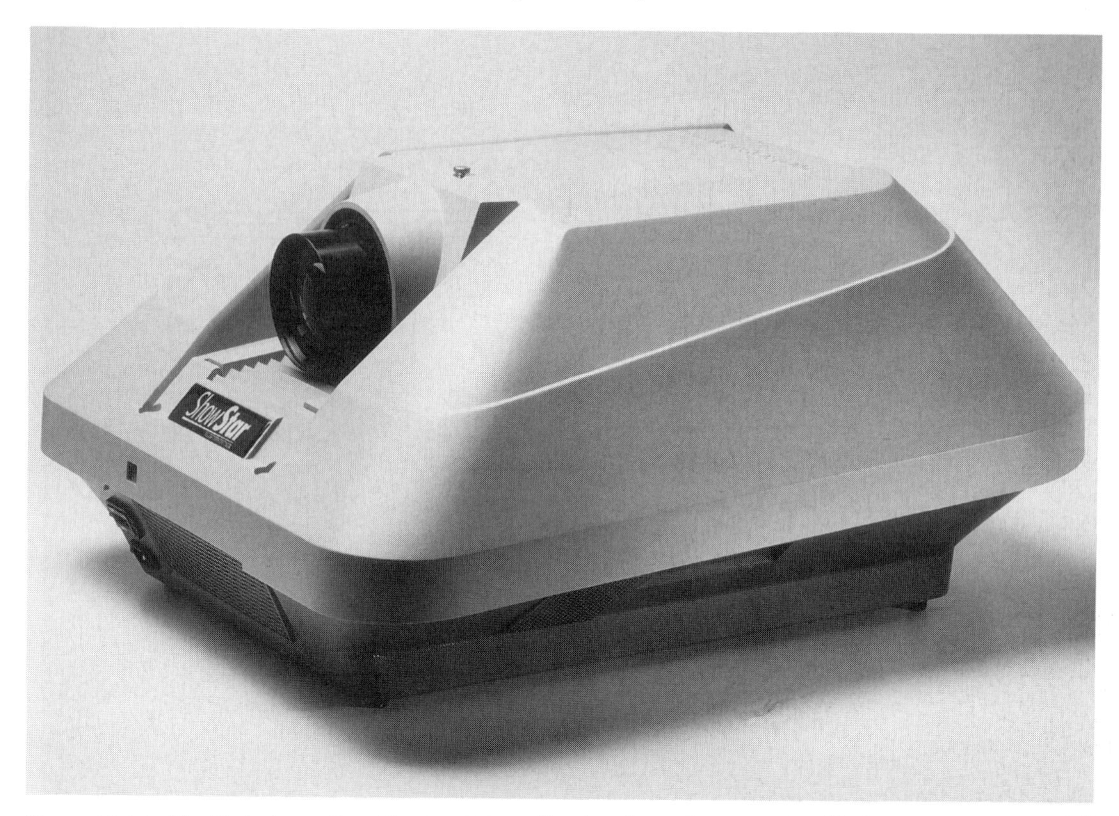

Figure 19.15 The Electrohome Show Star is an intermediate liquid crystal projector suitable for images up to 6 m (20 ft) wide. Output is 1000 ANSI lumens

However, while this method gives the brightest image, it is both expensive and complex to run. Most commercial applications are dealt with by other methods.

First of these is the high-power liquid crystal projector. This is exemplified by products from Barco and Electrohome which work on similar principles to the portable liquid crystal projector already described, but on a larger scale.

The Barco 8100 projector has the following characteristics:

□ It uses 4-inch liquid crystal panels and a special 400 watt metal halide lamp to give 1800 ANSI lumens (claimed to be 8000 'CRT lumens equivalent').

□ Each panel has a resolution of 740 × 568. This number corresponds exactly to the number of pixels in a PAL digital TV image.

□ By using an in-built converter to map the incoming image signal optimally to the available pixels, subjectively excellent results are obtained on standard TV, HDTV and graphic signals up to 1024 × 768. NTSC images are, for example, up-converted to use the full number of lines available in the panel, thus ensuring both the best image quality and full use of the available light.

□ The projector is ideal for auditorium applications requiring images in the range to 8 m wide. Because it is so quick to set up, it (and other projectors in the same family) has become the 'workhorse' of the AV hire business for big screen video.

Both Barco and Electrohome have introduced intermediate projectors with outputs of around 1000 ANSI lumens which are suitable for video and VGA data. It can be expected that models from other manufacturers will join them, and this type of projector will become widely used in meeting rooms requiring good quality imagery under high ambient light conditions.

Even using the conversion technique exemplified by Barco's product, there is a practical limitation on resolution using LCD panels for projection. High-definition liquid crystal projectors are available from Japan, but their light output is very limited and they are extremely expensive. There is, however, another technique which gives high light output and high resolution. This system, exemplified by products from Hughes-JVC and Ampro, uses the principle of 'image amplification'. Confusingly, the main amplifying element is based on liquid crystal, but the way

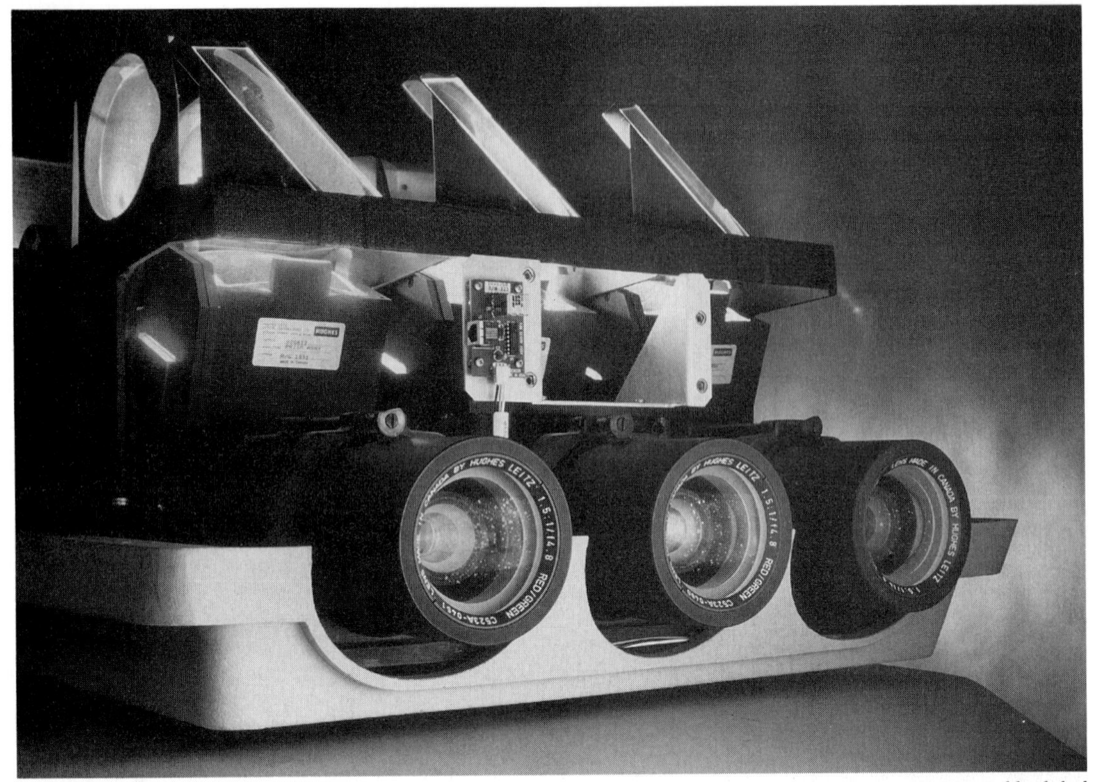

Figure 19.16 The Hughes-JVC ILA series of projectors is suitable for high resolution (up to 1600 × 1200) and high light output (up to 3000 ANSI lumens). The picture shows a projector with cover removed to show the three optical paths

in which the liquid crystal is used is quite different than in the 'transmission' method previously described.

In the Hughes-JVC series of projectors the original image, which can be of resolution up to 1600 × 1200, is displayed on three low-power cathode ray tubes, one each for red, green and blue. On the faceplate of each cathode-ray tube there is a liquid crystal 'light amplifier' lightvalve. This is a multilayer device which works by *reflection*.

Light from a xenon lamp is split into the three primary colors, and each colored beam is directed at the front of the corresponding lightvalve through a polarizer. If no picture is present, the polarization of the front surface of the lightvalve is opposite to that of the incident light, and the incident light simply gets absorbed.

If, however, any part of the liquid crystal layer is activated, it lets the light through, where it then hits a reflective (mirror) layer. This reflected light is then captured by a conventional optical system for projection on the screen. The activation of the liquid crystal layer is done by a photovoltaic effect. Light on the surface of the cathode-ray tube stimulates an electric charge, which in turn activates the liquid crystal layer to change its plane of polarization. From the diagram it can be seen that the reflective layer also acts as the electrically charged layer, receiving charge from the photoelectric effect and applying it to the liquid crystal.

In the Hughes system the output is via three separate lenses. This is less convenient than a single-lens system, but does ensure the highest light output (up to 3000 ANSI lumens) and the highest resolution. The Ampro system is more convenient in that it recombines the three beams to use only one lens, but is not as bright.

The term 'light amplifier' is used to describe the process because the low-level light on the CRT faceplate is 'amplified' by the reflective liquid crystal lightvalve.

There are other methods of video projection under development, most notably the DMD system being developed by Texas Instruments. This, incredibly, uses microscopic moving mirrors in a reflective system. If the mirror is lying flat, incident light is reflected into a

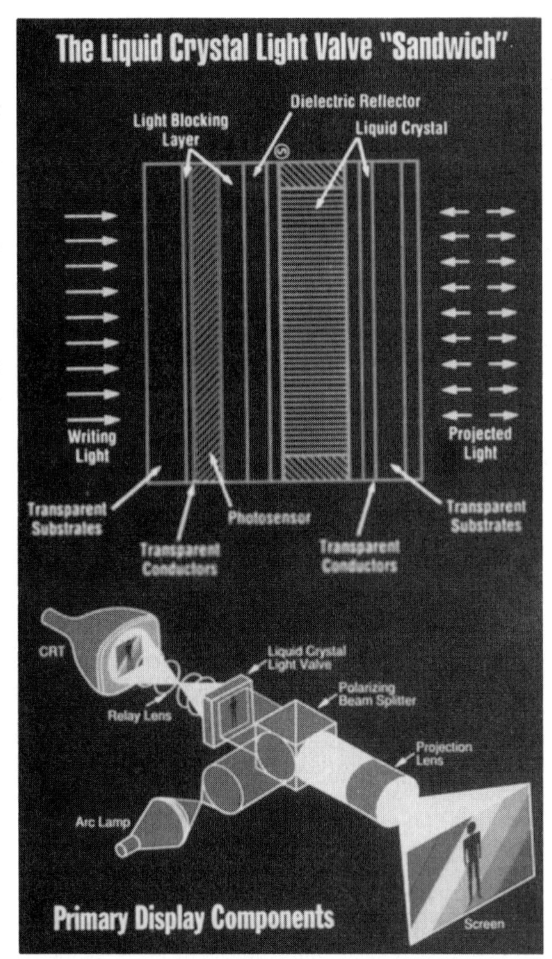

Figure 19.17 The principle of operation of the Hughes-JVC light amplifier projectors. (Diagram courtesy JVC Professional Products Ltd)

light trap; if the mirror is deflected, the incident light is captured by an optical system which images it on the screen. The mirrors are constructed as part of an integrated circuit ('chip') which also carries the driving electronics to deflect each mirror individually – there is thus one controllable mirror for each pixel in the image. The whole array is no bigger than a 35 mm slide.

The DMD system is not yet available in the form of commercial product, but practical demonstrations of both VGA resolution and HDTV resolution projectors have been given. It is reasonable to expect that big screen video projectors based on the technology will come on the market soon.

One other method of video projection is in use for special events. This is the use of a laser scanning system to present the picture. In principle, colored laser beams are directed by electronically-controlled mirrors that cause the beams to scan the picture in a similar way to the electron beam in a cathode-ray tube. The system has the theoretical attraction of being independent of lens constraints, because the scanning arrangement can be programmed to produce the required image size over a wide range of projection distances.

Lasers needed to produce a bright picture are expensive, fragile and require water cooling. Even assuming the present experimental systems are refined to a commercial product, laser-based video projectors are likely to find their main application in specialized permanent installations, or in unusual outdoor night-time events.

High-definition television

Much research is going on world-wide into high-definition television (HDTV). Present television standards build up a picture by scanning an electron beam across a cathode-ray tube face: 625 lines in PAL, and 525 lines in NTSC. In fact, the situation is a little more complicated than this. Each frame consists of two fields: 25 frames, 50 fields a second on PAL; 30 frames, 60 fields a second on NTSC. The fields are interlaced so, for example, on PAL a 625-line picture is built up by scanning the whole picture with half the available lines, then filling in the gaps with the other half. For technical reasons, not all available lines are used; field one carries picture information on lines 23 to 310, while field two uses lines 336 to 622.

In fact, the picture available on standard video is of a very high quality that can only really be appreciated when seen on a broadcast standard monitor. Thus, in terms of producing a picture on a TV set up to, say, 66 cm (26 inch) tube size, quality is limited not by the fundamental standard but by the equipment used. For example, low-cost TV sets are not able to resolve more than about 300 lines, so there is little point in feeding them with a signal containing a lot more information.

Line structure of the video image starts to show up in video projection and is also noticeable on bigger display tubes. A first question is whether it is possible to improve the look of standard video signals without the need to go to a new TV system.

As explained in Chapter 20 the best displays for computer graphics use progressive instead of interlaced image scanning. This requires projectors or monitors able to deal with higher scanning frequencies. For example, the projector must scan 525 lines 60 times a second, instead of only 25 times a second. Pictures presented this way are free of flicker and at normal

Figure 19.18 The Snell & Wilcox 'Supervisor' can up-convert standard video signals either to progressive scan or to 1050 or 1250 lines. Big screen video projection is greatly improved by such devices

Figure 19.19 A high definition monitor from Sony. Notice the 16:9 format. This format is also being used in Europe for broadcast television, but the receivers used are not 'true' high definition because they do not have the resolution of the monitor shown here

viewing distances have much less perceptible line structure.

Modern digital techniques allow a standard TV signal to be converted from interlace to progressive scan, or to have more lines. For example, an NTSC signal can be converted from 525 lines interlaced to either 525 lines progressive or 1050 lines interlaced. Such units are referred to as line doublers or scan converters. The most sophisticated units, made by specialists like Snell & Wilcox (UK), Faroudja Laboratories (USA) and Chromatek (Japan), include sophisticated motion interpolation to iron out the differences between the two fields in a frame. When processing a composite signal, they use special filtering techniques to remove the tendency for the chrominance and luminance signals to interfere with each other, producing undesirable artifacts especially noticeable on patterned images.

There are other simpler products on the market that make a worthwhile improvement in image quality when suitable projectors are used. Any prestige installation using large images derived from present standard video sources should consider using this type of equipment.

The 1125/60 system

Research on high-definition television has not confined itself to just increasing the amount of picture information, but has also addressed the problem of picture format. The present TV aspect ratio of 4:3 was introduced both because it fitted on to the early round TV picture tubes and because it matched the film industry's Academy ratio. It is now accepted that a much more pleasing ratio has a wide-screen look, like that used in movie theaters. Internationally it has been agreed that HDTV systems use a 16:9 picture ratio. This conveniently squares the existing 4:3 ratio.

In Japan, NHK, the Japanese national broadcasting organization, and several manufacturers have had a continuing program of high-definition television development for nearly twenty years. Manufacturers, Sony in particular, offer a complete range of HDTV production and display equipment. For those with stand-alone and closed circuit requirements HDTV is available now. Intending users must be careful to research the total costs involved, as production costs are significantly higher. In some stand-alone applications, production on film using either 35 mm or Super-16 is the most economic approach.

HDTV monitors and projectors give very impressive results, but they do not give as much light as their low-definition equivalents. This means that for HDTV to be appreciated the presentation environment must be properly controlled. The annual NAB (National Association of Broadcasters) show in the USA hosts an excellent exhibition that shows the advances in HDTV presentation. Significantly, though, it is held in very low ambient light conditions.

HDTV for all?

The obvious question is whether and when HDTV will become the standard for broadcasting, and therefore also for AV applications. For the time being AV users are recommended to extract the best they can from existing systems, by using component sources, line doublers, etc., and let the politicians and manufacturers work towards standards and sensibly priced equipment. Those who *must* have HDTV can have it now, but at a price.

The Japanese system uses 1125 lines at 30 frames per second (60 fields). The standard is sufficiently well established that in the USA the SMPTE has recognized it as standard 240M. In 1988 the 1125/60 Group was formed by a consortium of manufacturers and other interested parties to promote the standard as a production medium.

Two things have stood in the way of the use of 1125/60 as a universal standard for production, broadcasting and presentation. The first is financial and practical. The bandwidth required for transmitting the full 1125/60 picture is five times that required for the normal TV picture. This implies a massive investment program, even if there was space on the airwaves for the bigger signal, which there is not. The practical problem of moving everyone to HDTV also implies a phased program over many years, with as high a degree of compatibility as possible.

The second, partly political, partly technical factor is that the PAL/SECAM part of the world has not been too keen to be locked into a system that does not have an easy upgrade path from existing standards, and which is dominated by Japanese manufacturers. A second group has emerged called Vision 1250, promoting a 1250 line, 25 frame per second (50 field) system. This is a convenient doubling of the normal number of lines of the PAL system.

It is now clear that for transmission purposes, the new digital compression techniques will solve the bandwidth problem. It is now possible for a high-definition television signal to be transmitted within the space previously needed by a normal TV signal. In the USA the FCC conducted tests of a number of competing HDTV systems, the end result of which was a 'grand alliance' of the major contenders which agreed to pool their resources to back one system. The grand alliance format can be summarized as follows:

☐ 720 active lines, 1280 pixels per line, progressive scan at 60 Hz or 59.94 Hz. Will also operate at 24 Hz and 30 Hz.

☐ 1080 active lines, 1920 pixels per line, interlace scan at 60 Hz or 59.94 Hz. Will also operate at 24 Hz and 30 Hz progressive scan.

☐ Recommended use of MPEG-2 (see Chapter 14) for video signal transport.

☐ Recommended use of Dolby AC-3 384 kb/s for sound signal transport.

The use of digital techniques makes it permissible to support two different scanning standards, which are equivalent in bandwidth terms. It is currently the case that high-definition cameras used for sports events, etc., actually produce a better picture in the interlace mode.

It would be sensible, but, at the time of writing not a strong possibility, if Europe and Japan were to adopt the same standard meaning that at last there would be a world TV standard. The actual introduction of HDTV to the consumer market is likely to be a slow process. This is because, as already pointed out, there is no real reason for it until big pictures become the norm. The investment both for the broadcasters and for the public will be considerable, and it may be competing for money with other services like video on demand. Note, however, that the use of MPEG-2 should in theory make any new HDTV system compatible with the other new services.

Video presentation units

Just as audio-visual presentation units are needed for slide presentations, there are some applications that need a video presentation unit.

This need often arises in one-to-one presentations, as discussed in Chapter 9. It is met by compact, portable units that combine a videotape player with a small monitor. While convenient for their intended application, they should never be used for larger groups.

It can be expected that the idea of the compact video projector with built-in videotape player will be developed, but even here users will have to be careful to ensure that the presentation environment and screen are suitable.

Figure 19.20 A portable video presentation unit combining videocassette player and monitor, suitable for one-to-one presentations. (Photo Hanimex UK Ltd)

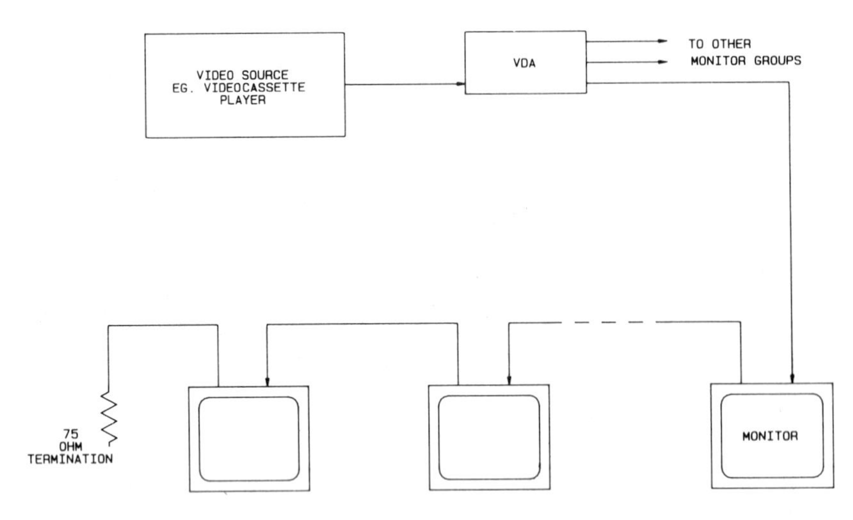

Figure 19.21 When several monitors are to be fed by the same signal they should have a loop-through facility, and may need feeding from a video distribution amplifier (audio not shown)

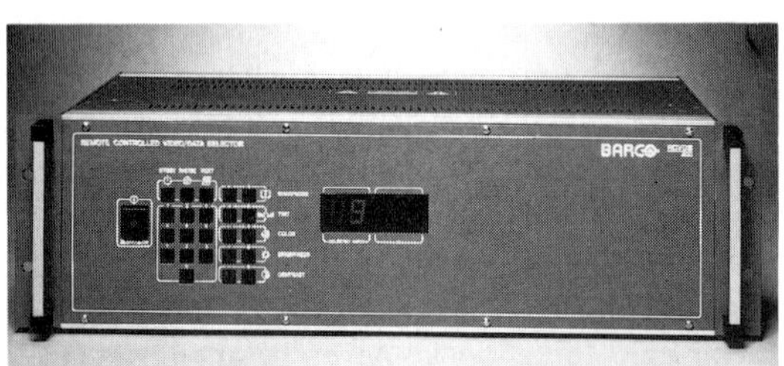

Figure 19.22 A presentation room with many video and computer sources needs remote controlled source switching equipment. The Barco RCVDS is an example of a modular product that meets this need

Video systems

The needs of many video users are satisfied by a simple chain of one source and one display, e.g. a videotape player and monitor. As soon as multiple displays, multiple sources, or both, are needed the user moves into the realms of video systems and usually needs a specialist video contractor to help specify and install the system. Some of the jargon that may be encountered now follows.

If a system uses several monitors, all driven from the same source, a video distribution amplifier or VDA is probably needed. Most video sources can only deal with one or two monitors and the signal must be boosted for multiple monitor use. Some monitors have a loop-through facility that simplifies multiple-monitor installations (see Figure 19.21).

In a multi-purpose presentation room it is certainly necessary for the video monitor or projector to be fed from several different sources; for example, VHS tape, U-matic tape, off-air, and data display from one or more computers. To do this a video switcher is needed to select the source, and often it is convenient if it can be controlled remotely. Switchers are an essential component of the presentation room, but they normally give a rather 'dirty' changeover from one source to another. The achievement of a clean cut without any picture roll is a much more complicated task. It requires all video sources to be synchronized and the use of a switcher that only switches in the vertical interval of the picture scan. These techniques are essential in video production, but only find their way into the highest-quality presentation installations.

Commonly it is necessary to display the output of a computer. Although personal computers used in the home often have a composite or RGB video output that is compatible with a standard monitor, more advanced computers use higher scanning rates to achieve a higher definition image. The special needs of computer display are described in Chapter 20.

Videoconferencing

Sometimes two or more groups of people in quite separate locations use audio and video links to conduct a conference. The process requires specially equipped rooms. The participants must be arranged so that they can be seen by the cameras, and a special sound system must be installed to ensure clear transmission of sound without acoustic feedback or howlround. Arrangements must be made so that visual aid material can be directly fed into the system so it can be viewed by all the participants.

The principles of videoconferencing have been well established for some time. Major corporations such as the Boeing Airplane Company have been using it as

Figure 19.23 British Telecom's Videostream was one of the first European offerings in the video conferencing market. Now their role is much more that of a provider of the digital network as video conferencing becomes a standard procedure

Figure 19.24 Picturetel's model 4800 boardroom system is widely used. It supports both proprietary and standard data compression and operates at data rates from 56 to 768 kbits/s, making it easy to use both on private and public connections. (Photo DCE Video Communications Ltd)

Figure 19.25 Some types of business benefit from desktop videoconferencing. DCE Communications Ltd cite the case or GUI Reinsurance of New South Wales who use equipment like this to conduct face-to-face negotiations of premiums between Sydney and London

part of daily operations for several years. However, until recently, videoconferencing was definitely for the specialist user and for the affluent. This was because of the high cost of renting the links between venues, and high capital cost of terminal equipment. There was also a problem of standards, with little compatibility between systems.

Early videoconferencing systems used broad-band links, that transmitted conventional TV signals, between venues. This was very expensive, and required special arrangements to be made every time a conference was to be held. Regular users could, and still can, have their own private leased circuits, but this is also expensive. The arrival of ISDN (Integrated Services Digital Network) has now made 'line time' a reasonable cost item.

The factors that have encouraged video conferencing are:

☐ Widespread availability of dial-up digital circuits. Now it is possible to make connections only when required, just like the normal telephone.

☐ Advances in chip technology, greatly reducing the costs of equipment and allowing implementation of digital compression to signals.

☐ An agreement on transmission standards.

A modern videoconferencing system uses a 'codec' (coder/decoder) to digitize the combined audio and video signal and format it in such a way that it can be transmitted over the public digital networks. Advances in compression techniques mean that a usable signal can be transmitted with a data rate as little as 6 kilo bits per second (kb/s). A low data rate means that the system cannot cope with sudden or multiple movements; the picture deteriorates during periods of complex movement, and sharpens up for relatively still images.

The digital network offers transmission space in multiples of 64 kb/s, and videoconferencing systems usually operate in the range 64 to 2084 kb/s. All small meetings are well-served by systems using not more than 384 kb/s, but large-screen presentations for big groups benefit from a higher data rate.

Private systems can use any method of coding the signal, provided there is compatibility at each end. However, for the general user there is now an agreed standard called H320, set by the CCIT, which ensures international compatibility. In fact, some codec manufacturers offer equipment that gives higher performance for a given data rate, but they offer their equipment so that it can receive either their proprietary signal or the H320 signal, depending on who they are communicating with.

A videoconferencing system at its simplest consists only of a video camera and microphone, both connected to the codec unit. The codec unit includes a dial-up facility, and special duplex audio circuitry prevents echo and howlround.

The AV user may well be faced with the need to integrate a videoconferencing system into a presentation or board room. This is now relatively easily done. In this case it may be necessary to allow for the following additional facilities:

☐ A means of connecting the system to a VCR so the proceedings can be recorded. This may require some operation if images from both ends of the system are to be recorded.

☐ A means of transmitting 'stills' of higher resolution. The codec normally works on a 256 by 240 or 352 by 288 resolution. While good enough for the conferencing part, this may not be good enough for graphics. It is possible to send stills at, for example, 1024 by 1024 on the same or a separate circuit.

☐ Use of two display screens, one for the conferencing and one for the shared graphic information.

Users of videoconferencing need more than money to make a success of it. They must have a commitment to making the technique work for all participants. This can really only happen when participants are regular users of the system, so its main application is – and will remain – the support of regular management meetings. In theory, therefore, a multinational company can hold a regular simultaneous board meeting in London, Melbourne and New York. In practice, time differences interfere with longer-distance hook-ups.

Another use of videoconferencing that has been identified is that of distance teaching, where one lecturer is able to speak to two or more audiences at once. The extra dimension that the two-way link brings is that the remote audience can have a dialogue with the speaker, and he can see his questioners. Such a method of teaching is likely to be economic and effective in high-level training, where there are only a few expert teachers available, and in adult retraining.

Figure 19.26 As far back as 1985 NEC were demonstrating big screen videoconferencing as a means of distance teaching. The new digital networks make this an economic proposition

Studies have shown that videoconferencing using large screens is more effective than that using the small 'talking-head' approach. The more that the image can be made to look like an extension of the room, the better – especially for infrequent users, for larger meetings and for teaching (see Figure 19.26). The declining cost of wider bandwidth, and the developments in high brightness video projection will both both encourage the use of bigger screens.

Teletext

Teletext is a generic word for text-based information systems available on a public or network basis. Anyone who installs a video presentation system may require one of the teletext variants.

The best known is broadcast teletext. This is a relatively limited information library containing up-to-the-minute public information, including news, weather reports, current events, travel information, stock market prices and even subtitles for the program. The information reaches the viewer via the TV lines of the normal broadcast TV signal that are not displayed.

In the UK, the BBC broadcasts Ceefax while the independent network broadcasts Oracle. They are compatible and only need a teletext adaptor to be added to a TV set, which is best bought as part of the original set. Selection of information pages is usually by an infra-red remote controller. Similar systems are available in other countries.

A more powerful system requires connection of the TV set to the telephone network, giving access to a huge volume of information on central computers. Several 'viewdata' systems are in operation: in the UK, Prestel; in France, Antiope; in Canada Telidon; in the USA NAPLPS, among others.

With these viewdata systems, the user is initially a passive viewer of information, selected in the same way as on the broadcast systems, except that:

- ☐ Cost of the local telephone call must be paid.
- ☐ An extra fee may have to be paid to the information provider.
- ☐ Choice of information available is much greater; in principle, there is no technical limit.

Alternatively, and this is the big difference, viewdata systems allow interactive use. For example, the user can become the information provider or can, in principle, use the system for booking or purchasing.

In practice the features of public viewdata networks are being transferred to new interactive services being offered by the cable and telephone companies, ultimately including video-on-demand services. *Private* viewdata networks are used in the travel and financial trades.

Videodisc systems

In all installations requiring continuous or frequent playing of the same video program, and in installations requiring random-access selection of video programs it is best to use a videodisc as the source. The standard laserdisc is described in Chapter 13, and the reasons for using it in permanent installations are reviewed in Chapter 11.

Analog laserdisc players are reliable, easy to use and require little maintenance. Business and professional users should only use professional players, which are available in a range of prices depending on facilities offered.

Nearly all professional applications require remote serial control by RS232 link, with control commands coming either from a computer or a dedicated controller. There are various other options which are of

Figure 19.27 The Sony LDP1600 laserdisc player has an RGB output. It has an optional synchronization facility

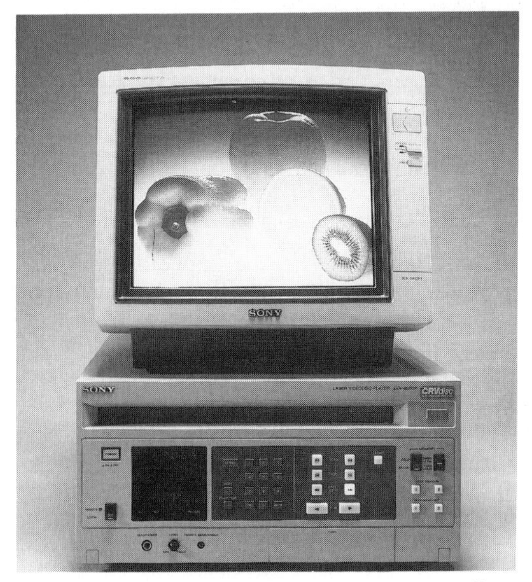

Figure 19.28 The CRV disc system from Sony allows high-quality discs to be prepared in-house. The LVA8000 player includes a framestore, and is widely used in broadcast, medical and big screen display applications

Figure 19.29 The Pioneer VDR V1000 is a disc recorder for *re-recordable* video discs. It records in component format and has twin read-write heads, meaning that it can be playing back from one part of the disc while recording on another

Figure 19.30 The HDL-2000 high-definition disc player from Sony makes HDTV displays a realistic possibility for permanent installations

importance depending on the application. Where budgets are limited, as is often the case in museum applications, for example, there is no point in purchasing a player with more facilities than needed. On the other hand, to try and get away with a machine with inappropriate or inadequate facilities for a special task is a worse mistake. Some of the most important options are:

□ The ability of the player to accept an external sync signal. This is vital for some applications, but does add significantly to the cost.

□ The ability to output an RGBS signal. The disc actually records a composite signal, but those players that provide an RGBS signal do give a better image quality. The facility is particularly useful in combined computer and video displays.

□ The ability to play both NTSC and PAL discs. This is useful for users like planetaria and colleges where a lot of relevant published program material may only be available in either PAL or NTSC.

□ Ability to replay digital sound.

The analog laserdisc gives good quality results at a reasonable cost. Its performance is finally limited by the fact that it records a composite signal. Applications requiring big images and/or a higher image quality should use a component disc system, such as the Sony CRV System. This uses write-once disks, prepared by the user, with a recording quality similar to Betacam SP. Because recording of separate components requires much more disc space, however, a single disc side only carries 24 minutes of material.

The CRV disc is an ideal source for high-quality video presentation. Users such as teaching hospitals find it very convenient to be able to record their own high-quality discs. It is possible to record single frames at a time, as well as complete programs, so there should be minimum wasted disc space. Users requiring playback-only facilities need not invest in the relatively expensive recording unit because simple playback-only units are available (but they are about four times as expensive as a good composite player). CRV is particularly recommended as a source for projected videowall displays, especially where source material is prepared in component form throughout the production process.

The CRV recordable disc has now been joined by the Pioneer *re-recordable* disc. The VDR V1000 machine is designed for the professional market and is widely used in sports broadcasting. It is an ideal source for action replay on big screen video displays in sports arenas and similar venues, and it would also be an economic proposition for any application requiring disc recording features, but where the disc cost of write-once equipment was an important factor.

The VDR V1000 records in component format. Its special features include twin read-write heads so that it is possible to be playing back from one part of the disc while recording on another part. It includes framestores for both heads so that transitions are always clean as well as optimizing still picture performance.

The videodisc is also the key to relatively low-cost HDTV display. For permanent installations required to show the same program on demand it is possible to use a high-definition videodisc player, with final presentation either on an HDTV 16:9 monitor or on an HDTV-compatible projector (or even on an HDTV-compatible videowall). There is now a number of HDTV production studios working with the 1125/60 format, and tapes prepared by them can be transferred directly to the HDTV disc. However, for some

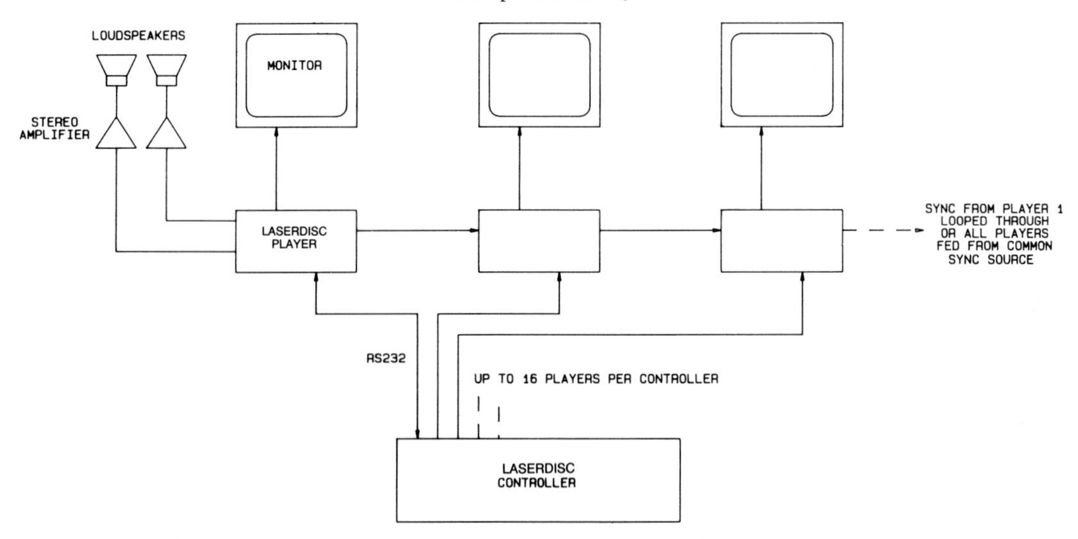

Figure 19.31 The principle of multi-screen video using multiple videodisc players

users the more affordable and practical approach is to use movie film, either 35 mm or Super-16 as the master material. It is then possible to have the film transferred directly to the disc using HDTV telecine equipment. High-definition videodisc players are only about one-third the cost of the most basic high-definition VCR.

Multi-screen video

A widely used display medium is multi-screen video, where a number of monitors operate in synchronization. The best-known example of the technique is the video wall where banks of monitors are disposed in a regular array, usually 3 by 3, 6 by 6 or some intermediate array, but sometimes in groups numbering hundreds of monitors. However, the technique can also be used in more subtle ways, which makes it especially suitable for supporting multi-media exhibits.

A popular requirement is to present a three-screen panorama image, using three video projectors. In order to achieve acceptable image quality it is essential to use three players, one for each projector. Furthermore, the original material cannot be derived from a single video image, but must either be created using three cameras, or derived from movie film.

A single frame of video has less than 600 lines in it. If it is to be stretched across three screens, the resulting image can only use one third of the original, resulting in less than 200 lines. On the other hand a 35 mm film can be scanned in three sections to produce three separate tapes, each with reasonable resolution. If the alternative method of three video cameras is used, there does need to be some experimentation to ensure the horizon lines stay in one piece!

The playback of the three-screen panorama is achieved either by using three VCRs, or by three videodisc players. The disc method is very simple and reliable, and is the only practical method for permanent installation. The disc players must be of a kind that can be synchronized. The tape method, used at conference presentations and so on, requires use not only of synchronizable machines, but also of a device like an edit controller to ensure synchronized start. Sony U-matics are available that accept a computer interface allowing the VCR to behave like a disc player in respect of frame selection and cueing up.

This method of achieving multi-screen video is not restricted to only three players. Systems have been installed using 40 or more synchronized videodisc players. There is also no reason why screens have to be adjacent. Mixed-media shows often use separated screens built into a set. The Canadian Pacific Pavilion at Expo '86 had one of the first shows using the simple idea of placing a video monitor in portrait format as the 'head' onto a model 'body'. The many characters were disposed in about twenty separate settings, and each 'talking head' needed a videodisc source. All the disc players ran in synchronization.

Multi-screen video is the key to the imaginative use of video in exhibition display. It requires careful planning and a disciplined approach to the overall production. Producers with a background in multi-image programs seem to make the most successful use of the multi-screen video opportunity.

Videowalls

In the 1980s the videowall progressed from an expensive curiosity to a standard method of video presentation. The basic videowall is an array of either video monitors or video back projection screens. It is normally arranged with the same number of screens vertically as horizontally, as this results in an overall display that has the same format as a standard TV image. Exciting videowall displays use non-standard formats, but do require special production techniques, and normally require multiple sources.

The basic videowall consists of a video source, the array of monitors, and an electronic 'splitter' that splits the incoming video signal into the requisite number of sections. Splitting is done digitally, and output of

Figure 19.32 (a) Multi-screen video near the North Pole. At the North Cape visitors' center at the top of Norway a five-screen video presentation is the main show. Five laserdisc players serve the five screens. (Photo Husmo Photo). (b) An off-screen photograph of the North Cape five-screen show. (Photo Husmo Photo)

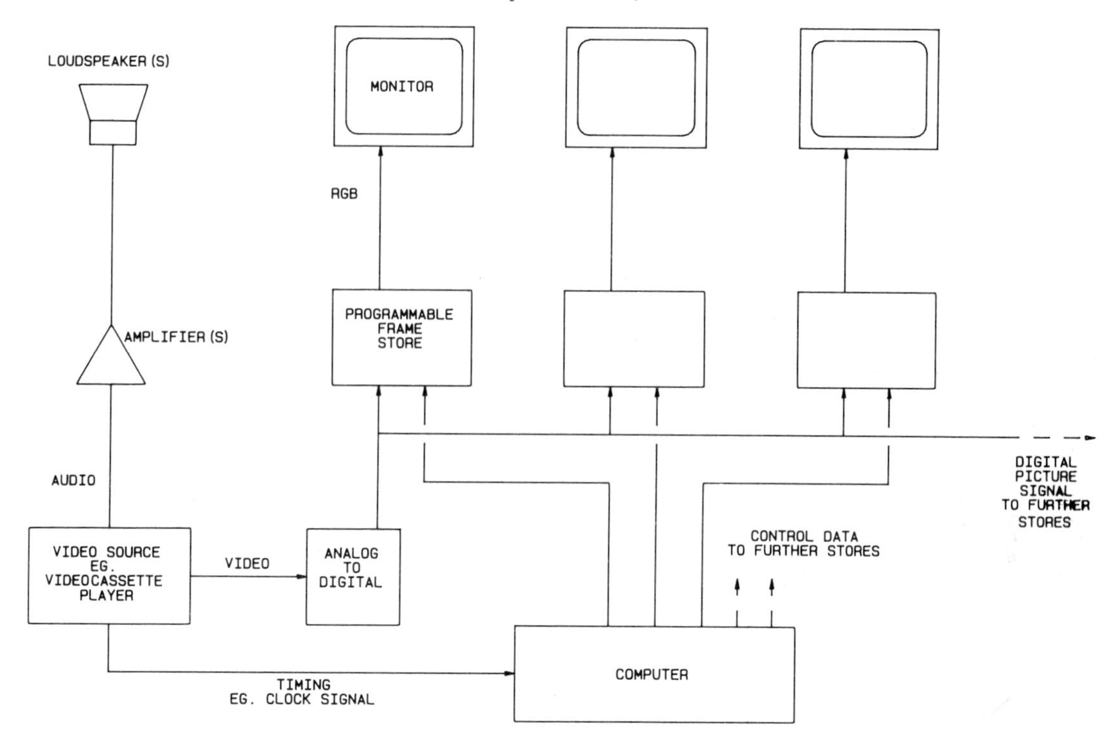

Figure 19.33 The principle of multi-screen video using multiple framestores, as used in professional videowalls

Figure 19.34 The Barco SCM monitor is specially designed for videowall work. It can stack 16 high

the splitting unit is the requisite number of RGBS signals. For small point-of-sale displays and similar applications the splitter is kept as simple as possible, and is dedicated to the particular array size, for example, 2 by 2 or 3 by 3.

Professional videowall controllers allow the videowall to be treated as a multi-image AV medium in its own right. They allow production a wide range of special effects, such as changes in magnification, image freezing, anamorphic effects and digital video effects. The programming of the videowall is then usually done by computer, and is synchronized to SMPTE/EBU timecode.

Although, in principle, any RGBS monitor can be used in a videowall, in practice it is best to use monitors that have been specially designed for the application. These ensure minimum gap between the image sections, and are designed for safe stacking. The 28-inch monitor is the 'standard' for videowall work, but both 21-inch and 32-inch monitors are available in stackable form, and other sizes can be adapted to the task.

The videowall represents the first practical method of achieving a large high-brightness moving image presentation in high ambient light conditions. Furthermore it does this when occupying a comparatively small amount of floor space. As well as the monitor videowall, there is also the projected videowall, where each screen element consists of a back projection unit using a special video projector and a high-gain rear projection screen.

Figure 19.35 Monitor videowalls are often used on TV studio sets, especially on sports, current affairs and game-show programs. (Photo courtesy BBC)

Figure 19.36 A four-sided projected videowall display at the Minneapolis Target Center arena. These displays provide an amenity for the audience, and are a useful means of attracting sponsorship

Figure 19.37 A mobile videowall in use. Morgan Broadcast Ltd offer this 4 × 4 'cube' videowall for hire, here seen promoting Ford cars on Waterloo Station concourse

Figure 19.38 The English Electric Starvision daylight video display is used at sporting events and open-air concerts. (Photo courtesy of Screenco Ltd)

So-called projection 'cubes' form the basis of very effective large-image displays. Individual screen elements are of sizes ranging from 40-inch to 60-inch diagonal. The reason they are no bigger than this is twofold. Any bigger and units are impractically big to transport, and are too deep. Secondly, the videowall is required to have a screen illumination of at least 350 and, preferably above 500 footlamberts, and this can only be achieved on comparatively small screens. Theatrical applications with controlled lighting conditions *can* use bigger screen sections, with correspondingly lower screen illumination and greater projection distances.

Projected videowalls are used in venues such as sports halls, conference auditoria and TV studios for live image presentation, and in venues such as shopping malls for advertising and promotion.

It is a natural assumption that projected videowall is the preferred medium, as the split between image sections is minimized. However, there are many applications where monitor videowall is more suitable. Monitor videowall is not only much less expensive, but it also occupies much less floor space and requires minimum maintenance.

It is also an assumption that videowall is primarily a means of big image presentation. While this may be true in 'live image reinforcement' applications at conferences, videowall can also, in the right hands, be used for highly creative video presentation. This is especially the case when multiple-image sources are used. Videowall then becomes a medium in its own right: different from conventional video in the same way as a 24-projector slide show is different from a single slide projector.

Videowalls are best at presenting standard video program material. The highest possible quality source should be used, as every blemish is visible on a large display. Graphic material with text should be used very sparingly, and only with careful planning and previewing to ensure that the final result is acceptable.

Videowall systems are now available for high-definition television and for up-converted video running at 1050, 1125 or 1250 lines. These give excellent results suitable for viewing at comparatively short distances. Videowalls are also used for high-resolution graphics, as mentioned in Chapters 12 and 20.

Outdoor video displays

Projected videowalls can be used in most places with high ambient light, for example, shopping malls and exhibition halls. They can even be used outdoors, but only under certain conditions. There is obviously no problem at night, indeed the problem then can be that the display is too bright. Equally at dusk or on an overcast day, there is no problem. However, if sunlight is present videowalls are not effective, and can only work at all if they are sited back to the sun.

The big displays used in outdoor sports stadia and similar venues work on an entirely different principle. Here each pixel is a separate light-emitting element, and the displays can achieve a brightness about four times that of a videowall. Various techniques are used:

- ☐ Ordinary tungsten lamps. These displays are very inefficient and are not suitable for full video because of their slow response.
- ☐ High brightness light-emitting diodes (LEDS). These give a bright display, but of very poor color quality. The advent of an effective blue LED may change this.
- ☐ Special lighting elements based on miniature 'flood gun' cathode-ray tubes. These give the required fast response and high light output, and are the basis of displays like the well-known Sony Jumbotron. Other vendors include Mitsubishi Diamond Vision, English Electric Starvision and Omega.
- ☐ Systems using liquid crystal displays.

Because the individual pixels on these displays are usually at least 1-inch (25 mm) square, they are only suitable for long-distance viewing.

Chapter 20

Computers in audio-visual

It is now almost true to say that computers, mainly in the form of personal computers, are an essential or optional component of every aspect of audio-visual presentation and production. Within the business use of audio-visual, it could soon be the case that computers will be the most widely used delivery platform for AV programs.

This chapter reviews some aspects of computer use to augment points raised in other chapters and to draw attention to a few items which do not easily fit under other chapter headings.

Computer data display

NTSC video displays 480 lines, so it is not surprising that, when computer graphics for personal computers were defined, the VGA standard opted for a pixel array 640 × 480. However, NTSC is an *interlaced* signal, and this results in unacceptable flicker for the predominantly still images needed in data display. VGA is a *progressively* scanned signal, where the entire 480 lines is 'refreshed' 60 times a second. The implication is that the display signal has at least twice the 'bandwidth' of the NTSC signal.

As the power of computers has increased, so have the demands for higher image resolution. Now a computer workstation or top-of-the-range personal computer will, typically, display at a resolution of 1280 × 1024 with a refresh rate of 72 times per second. This allows very high-resolution images to be presented on special monitors which give a rock-steady image and are suitable for close-up viewing.

Table 20.1 shows the increasing demands placed on the display device. The table is not complete, and many other resolutions and frequencies are used. PAL and NTSC video are shown for comparison, and these are the only interlaced scanning systems in the table.

A simplistic calculation shows that the 'bandwidth' required for the high-resolution displays is at least twelve times that required for video. It should, therefore, come as no surprise that the display devices are more complex and expensive.

Table 20.1 Image resolutions and display frequencies

Resolution	*Vertical frequency*	*Horizontal frequency*
NTSC 640 × 480	29.94 Hz	15.75 kHz
PAL 768 × 576	25 Hz	15.625 kHz
VGA 640 × 480	60 or 72 Hz	32 or 38 kHz
S-VGA 800 × 600	60 or 72 Hz	38 or 48 kHz
1024 × 768	60 or 72 Hz	48 or 56 kHz
1152 × 882	60 or 76 Hz	64 or 72 kHz
1280 × 1024	60 or 72 Hz	64 or 78 kHz
1600 × 1200	60 Hz	78 kHz

Display devices are loosely categorized as being suitable for 'video', 'data' or 'graphics' in ascending order of performance, but yesterday's graphics are today's data as products have improved. The lowest cost display devices can only show video. Medium performance data display devices will accept signals of up to at least 31.5 kHz and possibly up to 50 kHz. Graphics display devices are those designed for horizontal scan frequencies above 50 kHz.

The lack of a single standard for computer displays means that the display needs to be configured for the image it is showing at the time. Unlike PAL or NTSC which, under normal circumstances, can be expected always to appear in a precisely known format, VGA and its higher relatives can appear in versions with slightly different timings and synchronization arrangements. A typical graphics or data projector has a number of internal memories, so that when it detects a new horizontal scan frequency, it automatically operates in the required manner; but in most cases it is necessary to have set up the display in the first place. Otherwise parts of the image may be missing, or the image may not occupy the full display area.

High-resolution monitors are expensive because of their ability to accept a wide range of input frequencies, and because of their use of tubes with very fine dot pitch. Some are able to sense the input frequency automatically and display accordingly; others must be set up to the required signal. Graphics monitors are not as bright as video monitors because they need to use a very narrow electron beam.

The showing of computer output to large audiences can be done in several ways:

- by very large monitors (up to 39-inch diagonal);
- by LCD panels placed on standard overhead projectors;
- by video, data or graphics projectors.

Large monitors are convenient for permanent installation, and give a bright image. They are, however, generally limited to VGA resolution in sizes above 21 inch, and are very heavy.

For most group presentations VGA resolution is, in fact, quite acceptable provided the images have been designed for group viewing and obey the sizing rules set out in Chapter 2. Very often the need for high-resolution display devices is determined by the source computer, not by any real need.

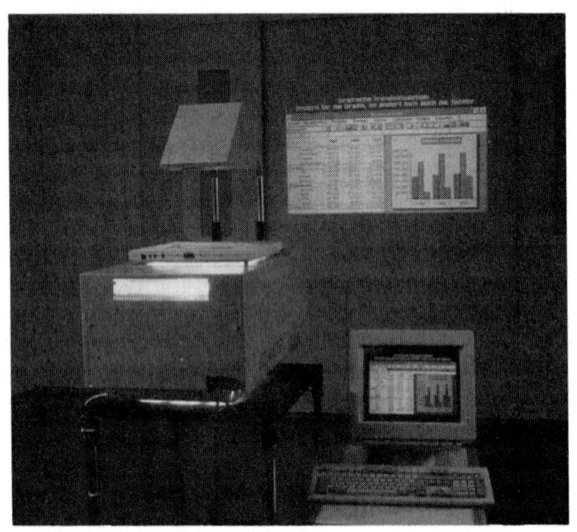

Figure 20.1 The LCD-Master from Audio Visual Inc of Switzerland is suitable for use with a wide range of LCD panels that allow the display output of a personal computer to be shown to groups. With its 1200 W metal halogen lamp, and optics optimized for LCD panel use, it gives a much brighter picture than a standard OHP

LCD overlay panels now represent the most common method of presenting computer displays to large groups. When they were first introduced they were monochrome and of very poor resolution. Now color VGA displays are reasonably priced, and high-resolution panels are available for those who need them. As mentioned in Chapter 8, it is important to use a high-specification OHP as the light source.

LCD projectors with VGA capability are a more elegant method of presenting the image, and are more suitable when video is also required because they have a better contrast ratio. LCD projectors are easy to set up because they are single lens devices.

While it can be expected that LCD projectors will become the standard for video and VGA group presentation, it will remain the case for some time that CRT projectors are best for high-resolution graphics. Modern graphics projectors use 9-inch CRTs, magnetic focusing, large lenses and sophisticated automatic convergence systems to deliver images of outstanding quality.

Scan conversion

While the ideal is always to show the computer image at its original resolution, there are often reasons why this cannot be done. For example, when very big images are needed, the only big auditorium projector capable of directly showing 1280 × 1024 images is a version of the Hughes-JVC ILA projector. When such big images are needed it is usually the case that viewing distances are long, and there is no need for very high resolution – the priority is brightness and size.

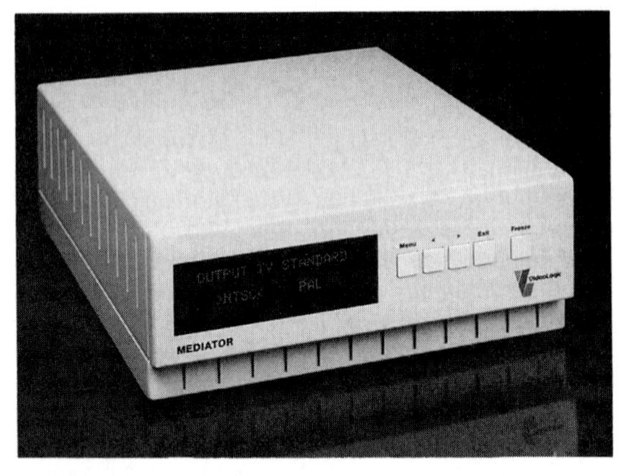

Figure 20.2 The 'Mediator' from Videologic is an inexpensive scan converter which allows VGA data to be seen on ordinary TV sets or recorded to videotape

It is possible to 'down-convert' computer images. Sometimes this is done in the projector. For example, the Barco 8100 'light-cannon' LCD projector can accept signals up to 1152 × 882, but its LCD panels are actually only equivalent to a PAL matrix. Internal 'convolving' circuitry remaps the source image to the available pixels. In practice, the results are excellent, and for large audiences the method is entirely appropriate.

Stand-alone scan converters are available which can down-convert computer images for showing on standard video projectors. Those which simply convert VGA to NTSC are inexpensive, but those which can convert graphics images down to VGA or NTSC/PAL are complex and expensive. Clearly either or both of spatial and temporal resolution are lost in the down-conversion process, but for many applications not involving close-up viewing, the results are quite acceptable.

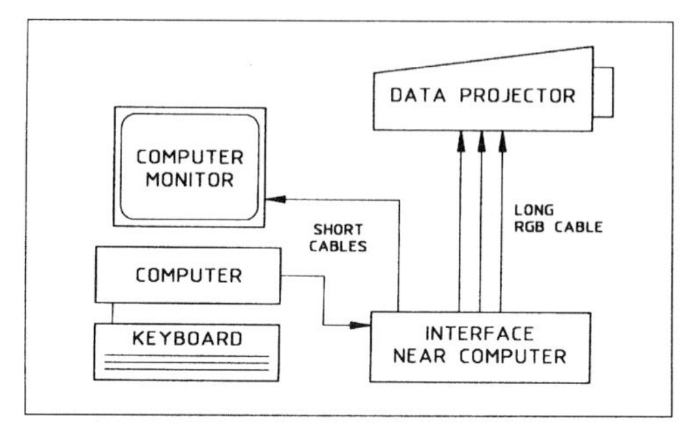

Figure 20.3 Connecting an 'interface' to allow a computer screen to be seen simultaneously on a local monitor and a more distant projector

Computer interfaces

Often there is a requirement for a computer to have its own monitor, for the computer operator, and at the same time for its display output to be shown on a large screen using a projector. It is also the case that the high-frequency scan signals which emerge from the computer are normally intended only to travel the very short distance from the computer to the monitor. For both these reasons it is usual to have an 'interface' sited next to the computer. These devices are of varying complexity, and their function includes:

- allowing the simultaneous use of a local monitor and a projector, without picture degradation on either;
- providing some amplification to allow the use of comparatively long cables to the projector;
- matching the synchronization signals of the source computer to the capability of the display.

The signals coming from a computer destined for a monitor or projector are always separate red, blue and green. The synchronization signals, which time the instant of horizontal and vertical retrace, are sent in one of three ways:

- as a combined sync signal superimposed on the green signal ('sync on green');
- as a combined sync signal carried separately;
- as two separate sync signals, one vertical, one horizontal.

The three methods require, respectively, three, four or five separate coaxial cables.

The switching and routing of computer images requires equipment of considerably greater bandwidth than that used for standard video. For example, distribution amplifiers suitable for standard video are no use at all for computer graphics. The comparison of video bandwidths (typically around 5 MHz for normal video, 20–30 MHz for HDTV) with those needed for graphics signals is not meaningful. The significant figure for graphics signals is the pixel clock rate which is as high as 135 MHz for 1280 × 1024 images. Amplifiers and switchers for this purpose need electrical bandwidths of 200 MHz and higher to ensure that resolution is retained and that images remain sharp. This is an important factor in installations where one projector is served by several different computer sources.

Computer graphics walls

The highest resolution which a single graphics projector can achieve is approximately 1600 × 1200. 1280 × 1024 is more practical and affordable. Special monitors exist to display 2048 × 2048 but none of these devices achieve the resolution of some of the data being processed by the computer, which can only normally be revealed by using alternative output means such as film scanning or graphics plotting.

For example, when film special effects are done by computer, source film may be scanned in at around 4000 × 2000. All computer processing is done at this resolution, but the process is observed on monitors which, for example, are only working at 1280 × 1024. The 'output' of the process can be scaled as required but can also be laser scanned back on to film again without loss of the original resolution.

But what about live computer graphic display? Is it possible, using multiple projectors, to achieve higher resolution than that obtained from a single projector? The answer is a qualified 'yes'; qualified because compromises must be made at the divisions between image sections.

A real need for higher resolution arises in the presentation of complex diagrams representing real-time display of such things as electricity distribution networks, telecommunication networks, vehicle fleet management and process control. A common arrangement is that controlling staff use workstations on a network, often with each member of the control team only observing a portion of an overall diagram.

X-Windows, an operating system for UNIX allows graphics to be defined within a notional 64,000 × 64,000 matrix. On any one workstation it is possible to select any part of the imaginary image, and to scale

Figure 20.4 A computer interface in use

it to fit the available amount of display space. Furthermore it is possible to look at other applications residing on the same network, so images from different applications can, if necessary, be 'windowed' into each other.

A number of vendors including, but not limited to, Barco, M3i and Jupiter, offer software/hardware combinations which allow an X-terminal to be on a network, and to have outputs for multiple projectors, each one at 1280 × 1024. In theory displays of resolutions up to 64,000 × 64,000 could be created in this way if enough (over 3000!) projectors were connected.

A more realistic requirement is shown in Figure 20.5. This is the control room of Scottish Hydro-Electric. Its central display uses M3i X-Wall software in a special X-terminal to achieve an overall resolution of 3840 × 2048 using six graphics projectors, and is sufficient to show the largest single image they need. The high resolution permits close viewing of detail which would not be legible at any significant distance. This raises the point that some of the CAD programs which form the basis of these kind of diagrams have inadequate fonts for distance viewing.

In Stockholm, Barco have installed no less than fifty of their 'Retrographics' 67-inch displays arranged in a 2 × 25 matrix to show a complete rail network. It might be asked, is the dividing line between the screen sections necessary? For most applications of this kind it is. Each back-projection screen is itself an optical system, and cannot be overlapped – quite apart from the practical difficulties of alignment.

However, in the specialist simulator market, some installations use multiple projectors and front projection screens with low gain. In this case an electronic equivalent of the soft-edge technique (described in Chapter 16) can be used. Specialist suppliers such as SEOS in the UK and Panoram Technologies in the USA provide the hardware needed, the technique still being something of a black art. It is really only suitable for rendered images (such as those produced by the high-speed image generators used in simulators) and is not recommended for the line type drawings more commonly encountered in control room situations.

AV program distribution by computer

The success of digital video and the phenomenal increase in the storage capacity of even the smallest computer is opening up completely new possibilities.

Figure 20.5 The Pitlochry control room of Scottish Hydro-Electric uses six 67-inch rear projection screens to show a very high-resolution display

Quite apart from the impact which the new technology is having in the professional market, and from the use of new storage media like CD-video, the computer is now an entirely practical basis for AV program storage and distribution.

While many users will be satisfied with programs delivered in some physical form, for example, a VHS cassette or a CD-ROM, there are advantages in distributing programs via some kind of network, for example, the telephone network or an office computer network. This is especially the case when:

- ☐ programs must be kept up-to-date on a frequent basis;
- ☐ the computer infrastructure is already in place;
- ☐ there is a need to monitor the use of the programs.

A few examples of what is possible now follow. Modern offices work on the 'computer on every desk' principle, and the majority of these have moved or are moving to networked systems. The computer network allows computers to share data and to make more efficient use of software. Now that computers can easily be, or are already, equipped with audio facilities and high-quality color monitors, the obvious question is, 'can AV programs be distributed by an office computer network?' Clearly there can be advantages in doing this, both for keeping staff informed and for ensuring specialist individual training is available at all times.

In principle the answer to the question is yes. In practice, there are some practical limitations which are easily overcome. Computer networks operate on a 'client-server' principle where common data is held on a server and is accessed as required by the 'clients'. For a given number of clients, the carrying capacity of the network must be as great as is necessary to ensure that work is not held up by transmission delays. For example, a client may ask to have word-processing software downloaded on to his computer. He might then work on a document, and then transmit the document as internal e-mail to another client. In this case his use of the network is intermittent, and he would barely notice if his software took one third of a second longer to load because the network was too busy to maintain top speed to him.

But if live video is to be transmitted over a network, there can be no delay. Compressed video, with audio, needs a sustained rate of about 1.2 Mb/s. If there are any interruptions to the datastream, the client will no longer see real-time video. He may see jerky programs if lucky, or he may see and hear a mess. Thus any server intended for use as a video server must be able to maintain the required data rate, and, of course, if

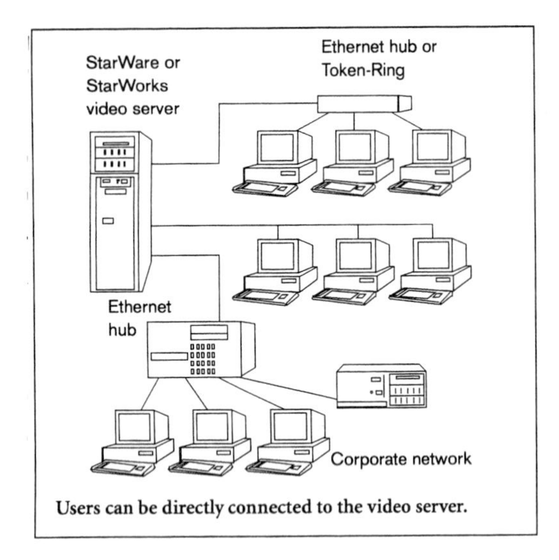

Figure 20.6 Distributing AV programs by computer. The diagram shows three possibilities. A dedicated remote network (top); direct connection to the server (middle); and integration into existing corporate network (bottom). In the last case the network is served simultaneously by the video server and the standard 'business' server. (Diagram from Starlight Networks Inc)

several clients want video simultaneously, both the network and the server must be able to handle the required number of simultaneous datastreams.

Starlight Networks is one company which has addressed the problem of ensuring that video datastreams take priority over other network activities to ensure their continuity. Their software has the following characteristics:

- ☐ It supports most of the popular video compression formats, including Motion JPEG, MPEG-1, Indeo, Cinepak (Quicktime), and DVI's RTV and PLV.
- ☐ It works on Ethernet, Token-ring and FDDI networks.
- ☐ It is available in versions to support different numbers of users. From a server capacity of 6 Mb/sec, which would support four simultaneous users, up to 50 Mb/sec for 40 simultaneous users.
- ☐ It is available as a module for NetwareTM servers, or to run on a UNIX-based server, compatible with Netware™, LAN Manager™ and other networking environments.

One of the first users to take advantage of networked AV programs was CSX Corporation, previously the Baltimore & Ohio Railroad. They have 30,000 employees widely scattered across the Eastern part of the USA. They have set up 30 training areas located at the most convenient locations for their employees, and each area has a network of workstations connected to a video server running Starlight Networks software. Each network can, in turn, access the company's mainframe in Jacksonville, Florida. This facility allows not only the programs to be kept completely up to date, but also allows updating of the 'student' records in CSX's human resources database.

The overall system was a considerable investment, but CSX claim a fast payback, because it has cut down wasted time – a claimed 50 percent reduction in training time – reduced actual training costs by cutting down on the need for peripatetic training staff, and greatly improved the certainty that training has been effective.

Of course, any system such as this is only as effective as the training programs themselves. The training programs would still be developed by expert AV producers. The user now simply has a wider choice of methods of distribution. He can distribute a program either for single user or single site, in which case CD-ROM or laserdisc could be the most practical, or can deliver the same material by computer network.

World-wide web

The distribution of multi-media material by computer is not limited to private networks. The world-wide web on Internet, which connects together thousands of computer networks, allows individuals, institutions and corporations to access computer data from all over the world.

The Internet evolved from computer networks which existed for many years in academic institutions. The big step was to then link these networks into super-networks. From then it was only a matter of time before one huge network evolved. Until recently most users of Internet could only work at telephone modem speeds (typically 14.4 kb/s). This is sufficient for text and simple graphics, but not for multimedia.

Now many of the servers on Internet are able to communicate at very high speed. Users can access the network through ISDN lines or fiber-optic connections. When this is the case it becomes easy to deliver complete AV programs, including full motion video. Thus it is already possible for a publisher of training videos to offer these on the Internet, and for a customer to 'collect' the programs when needed without the need for the transfer of physical inventory.

The problems associated with this method of distribution have now more to do with standardizing on methods of charging for the information and on retaining security than with the difficulties of transmission.

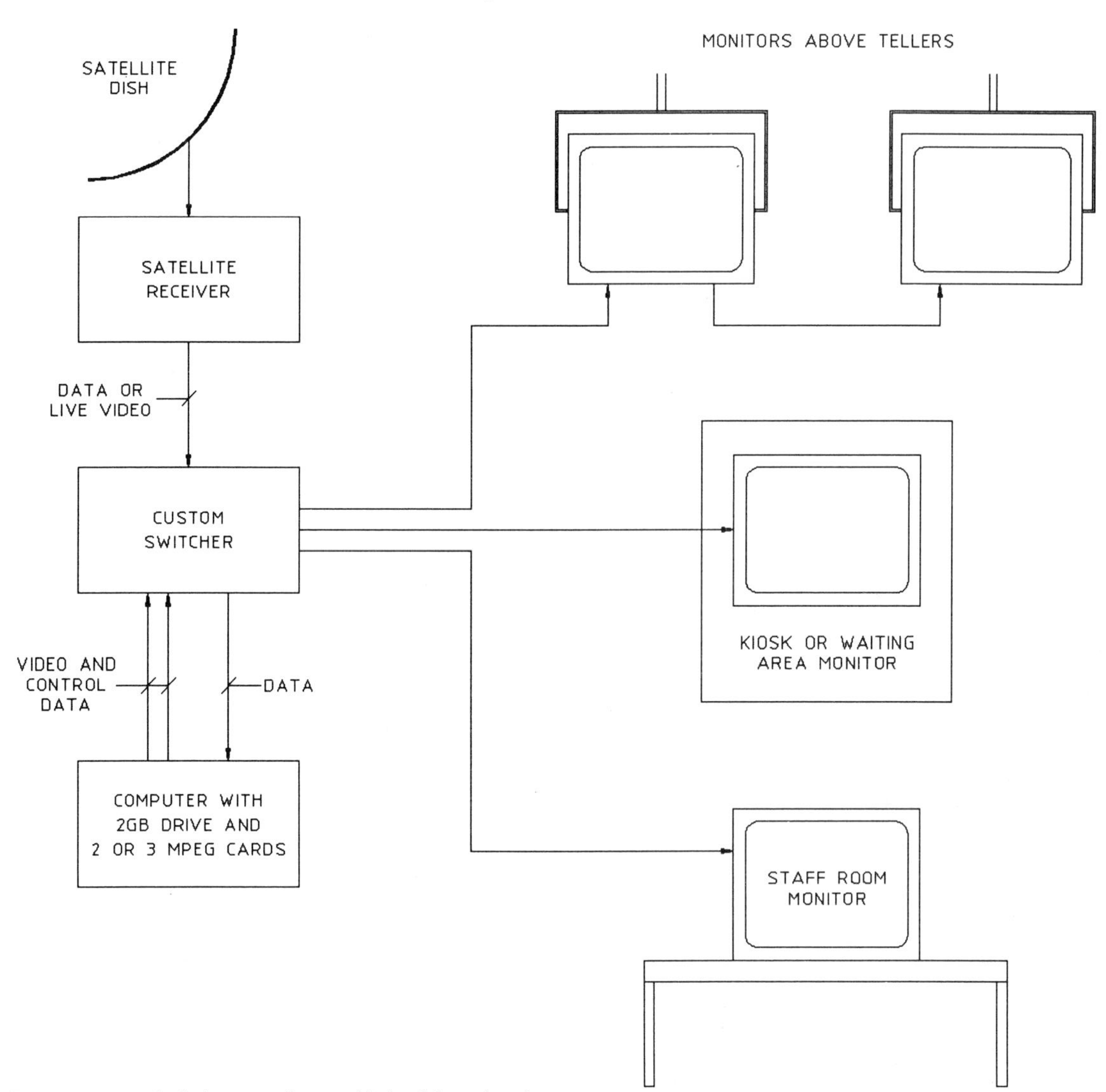

Figure 20.7 Block diagram of a possible bank branch video system

Video on demand

The cost of computer hard disc drives has been dropping at about 50 per cent per year, and at the same time the density has been doubling. Gigabyte drives, once way out of reach of the personal computer, now cost no more than yesterday's 20 Mb drives and occupy no more space. One gigabyte (1024 Mb) can store about $1\frac{1}{2}$ hours of compressed video.

The hard disc drive manufacturer Micropolis is just one company which is specializing in the needs of the emerging video server market. They have optimized their drives for the application, guaranteeing the required continuous data transfer speeds. For video or media server applications, the drives are supplied in a RAID (redundant array of inexpensive discs) configuration which ensures that the failure of any single drive does not result in the loss of data.

A typical use of Micropolis drives is found in hotel TV systems. Until recently these were always based on tape, with programs being changed monthly. This meant that movies could only start at set times, and that often, by the end of the month, the quality of presentation was very poor. New systems based on video servers have, typically, the following characteristics:

□ Ability to store the equivalent of 20 full length movies.

□ Anyone watching a movie has full 'VCR-like' control, being able to start, pause, and 'fast-wind' as required.

□ Up to 32 (or multiples of 32) separate users can be using the system at any time. They can all be looking at any of the films available.

□ The system operators no longer have to deliver physical inventory, and there are no tapes to wear out. Films can be changed by sending new data to the server, either by ISDN line or by satellite.

Systems of this kind are obviously also of interest to anyone with a 'library' of programs which are to be seen at multiple viewing points on a random-access basis. Museums with archive film material, media resource centers at schools and universities, and corporate training facilities are all potential users.

Satellite and ISDN

The ability to distribute AV media to remote sites at will is of great interest to retailers and corporate users. The computer is the key to doing this because in practice it is best to send the program material at a 'non-real-time' data rate, for storage by the remote computer. Often the data is transmitted at night to take advantage of 'off-peak' tariffs, and the data transmitted can also include the scheduling information for showing the programs. In theory the ordinary phone line could be used, but this takes a very long time to send any significant amount of video, so normally ISDN lines or satellite data transmission are used.

Two examples of the successful application of these techniques follow. In Germany, Scheiner Interaktive Medien have developed Show Manager for use with Electrosonic videowalls which are installed in the foyers of cinemas. From a base in Langenfeld, near Düsseldorf, videowall displays in Munich, Berlin and other cities throughout Germany can be controlled and updated. An abridged specification of the system is:

□ It can store up to 2 hours of video, 15 hours of audio, 5000 graphic pages or any combination of these.

□ The unit gets its real-time clock information from the Frankfurt radio clock, thus scheduling is guaranteed to be accurate.

□ The system schedules the showing of program material as specified. For example, it is possible to specify that a given 'spot' will be run 100 times over specified days, and that it will run only between specified times (for example, tobacco and liquor advertisements may only be shown after 6 p.m.).

□ The system can integrate the playing of locally-sourced material (e.g. from laserdisc player) into the schedule if required.

□ The system reports back on what it has actually shown, so advertisers can be sure that their spots have been played.

□ In the single processor version of Show Manager, the unit only looks for program and scheduling updates when it is not playing video. In the dual processor version, it is possible to send new information at any time.

□ Data transmission to and from the remote unit can be either by ISDN line or by modem on the ordinary telephone line.

This kind of approach is already making in-roads into all kinds of point-of-sale video as users realize that for video to really work in support of merchandizing, it is essential to have precise scheduling and to be able to remotely control what is being shown.

Another, much larger, example is given by the service offered by the John Ryan company to several banks in the USA. A typical customer has several hundred, or even several thousand, separate branches, and the problem is how to deliver high-quality AV material to all of them in a way which guarantees its effective use.

The company developed a system for this application which has a surprisingly low capital cost for each outlet, and which has low running costs. Figure 20.7 shows in diagram form a typical branch installation. This shows that each branch has three different uses for video programs. One is for short loops visible to those waiting to see a teller, another is for longer programs which can be viewed in a lounge area where one might wait to see a manager, and the third is the branch's staff room.

The whole system operates fully automatically, even to the extent of switching the power to the video monitors. The satellite link is mainly used for data transmission at off-peak rates, and this includes all scheduling information and all recorded program material. Scheduling can be highly targeted, with different branches having different program material according to their known customer base. If it is known that a particular customer group always comes in their lunch break, advertising can be tailored accordingly.

Staff training programs are usually run before the branch opens its doors to the public. On occasions the satellite link is used for live video, for example, there might be a monthly live program from headquarters featuring the chief executive. Staff at the outlying branches can phone in questions while he is 'on air'.

Systems such as these depend entirely on the availability of low-cost computers, and on the success of the new methods of video data compression. Both the examples given here use MPEG compression for video.

Chapter 21

Lenses

All audio-visual systems using projection finally rely on an objective lens to focus the required image on the screen. The variety of lenses available is what makes projection so flexible. The same principles apply to projection of movie, slides or video. However, it is in slide projection that the versatility of different lenses is best demonstrated. This chapter first describes the possibilities for slide projection, then shows how movie and video projection differ.

Basic lens properties

It is best first to explain some technical terms relating to lenses, so that their importance can be understood when choosing a lens.

The *focal length* of a lens is the distance from the plane in which the lens forms an image of objects at infinity to the node of emission. This textbook description disguises the information that matters; long focal length projection lenses are needed for long-distance projection, and short focal length lenses for short-distance projection. The formula relating projection distance to image size is easy to remember. A 35 mm lens projecting a 35 mm slide gives a picture as wide as the projection distance, e.g. a 2 m (6.5 ft) picture at 2 m projection distance. Multiply the focal length by two and the projection distance doubles, i.e. a 70 mm lens gives a 2 m (6.5 ft) picture at 4 m (13 ft) throw; a 140 mm lens gives it at 8 m (26 ft), and so on. The practical range of projection lenses for 35 mm slide projection is 25 mm to 400 mm, although there is a very expensive 15 mm lens available, and special applications can use focal lengths up to 1000 mm. This is summarized in Table 21.1.

Table 21.1 The standard lens formula

The full formula is:

$$W = \frac{D}{f - 1} \times w$$

where W is the screen width
w is the film aperture width
D is the projection distance measured from the image nodal point of the lens to the screen
f is the focal length of the lens

However, for all practical purposes a much simpler formula can be used:

$$\frac{\text{Screen dimension}}{\text{Film frame dimension}} = \frac{\text{Screen distance}}{\text{Lens focal length}}$$

For slides, the film aperture width is 35 mm. Provided the focal length of the lens is also expressed in mm, the picture width of the projection distance can be easily calculated. It does not matter whether this is done in metres or feet, provided the same units are used for both.

The *relative aperture* of a lens is an expression of its ability to collect, and therefore project, light. Technically, it is the ratio between the effective diameter of the lens and its focal length. The smaller the ratio, the more light passed. The same applies to lenses used for picture taking, and for this reason lenses with low relative apertures of *f*-numbers are called faster lenses. It is difficult to make long focal length lenses with low *f*-numbers because the diameter must increase and they get very big. In slide projection short focal length lenses should be *f*/2.8. With the optical system of standard slide projectors there is no real benefit in having longer focal length lenses with better than *f*/3.5, because the cone of light from the light source is not big enough to take advantage of faster lenses. With lenses of focal length greater than 300 mm it is difficult to get lenses as fast as *f*/3.5.

Modern high-quality projection lenses give an image that is sharp to the corners, that is free of *chromatic aberration* (color fringing) and of *spherical aberration* (the tendency to 'pin-cushion'). The lens includes at least three, and more usually five or six or as many as nine elements. At each glass/air surface there is some reflection, and therefore wasting, of light. For example, if 10 percent of the light is lost at each element of a nine-element lens, the lens loses more than 60 percent of the incident light. A sophisticated

Figure 21.1 The flexibility of slide projection demands a very wide range of lenses for different situations. This range is produced by Docter of Wetzlar, Germany

technique of vacuum coating the lens elements reduces this problem. As much as 25 percent extra light is given this way. A coated lens can be distinguished by a bluish bloom on the lens. Care should be taken not to touch the lens surface, and to clean it as carefully as an expensive camera lens.

Lenses for slide projection

Table 21.2 shows a list of standard slide-projection lenses available from a variety of manufacturers, and gives a good idea of the wide range available.

In general, a permanent installation with a fixed projection distance is best served by use of the appropriate fixed focal length lens because these are less expensive than zoom lenses and usually give more light. However, some modern zoom projection lenses are so good that they can be used in permanent installations when there is no fixed lens suitable. Although the lens manufacturers make special lenses to order, it is not an economic proposition unless several hundred lenses are required.

Table 21.3 draws attention to the fact that there can be slight differences between slide mounts. This affects the image size and must be taken into account when designing precision installations. The differences arise because the actual image on the film occupies a standard area which the mount slightly masks deliberately ensuring a perfectly projected picture. Early standards for masking were somewhat pessimistic, so

Table 21.2 Slide-projection lenses

Fixed focal length, front projection	
90 mm f/2.4	100 mm f/2.8 or f/3.5
150 mm f/3.5	120 mm f/2.8 or f/3.5
180 mm f/3.5	150 mm f/2.8
250 mm f/4.3	200 mm f/3.8
	250 mm f/3.8
	300 mm f/4.5
	350 mm f/5.2
	400 mm f/3.8
Zoom lenses	
70–120 mm f/3.5	
70–125 mm f/3.5	
85–210 mm f/3.5	
110–200 mm f/3.4	
100–300 mm f/3.5	
200–300 mm f/3.5	
Rear projection lenses	
25 mm f/2.8	15 mm f/2.8 (very special)
35 mm f/2.8	37 mm f/2.8
45 mm f/2.8	50 mm f/2.8
60 mm f/2.8	
Perspective control lenses	
35 mm f/2.8	
45 mm f/2.8	60 mm f/2.8 (Extra offset)
60 mm f/2.8	
90 mm f/2.8	
70–125 mm f/3.5	

Table 21.3 Projected apertures

Image	*Width* (mm)	*Height* (mm)
Super 8 movie	5.46	4.01
16 mm movie	9.65	7.26
35 mm movie Academy	21.11	15.29
35 mm movie widescreen 1.85:1	21.11	11.41
35 mm movie anamorphic	21.29	18.29
70 mm movie	48.59	22.10
Superslide	37.3	37.3
35 mm slide ANSI standard	34.2	22.9
35 mm slide WESS No 2	34.8	23.4
35 mm filmstrip	22.3	16.7

slide mount manufacturers increased the area to the maximum practicable.

Zoom lenses

Zoom lenses are lenses with an adjustable focal length. They are invaluable for travelling audio-visual shows as they allow precise filling of the screen from the available projection distance. It is important that these lenses do not introduce distortion, especially in multi-image applications. For economic and practical reasons the zoom range is not usually greater than 2:1. However, as can be seen from Table 21.2, just three lenses can allow zooming all the way from 70 mm to 300 mm. Most AV presentation work using front projection can be based on a 110 to 200 mm or similar lens.

Professional-quality zoom lenses are all supplied in metal barrels, and have a zoom lock that locks the zoom action. This is an essential feature for multi-image work.

Figure 21.2 This range of zoom lenses zoom from 70 mm to 300 mm. The 200–300 mm lens must have a large diameter to achieve its relative aperture of *f*/3.5

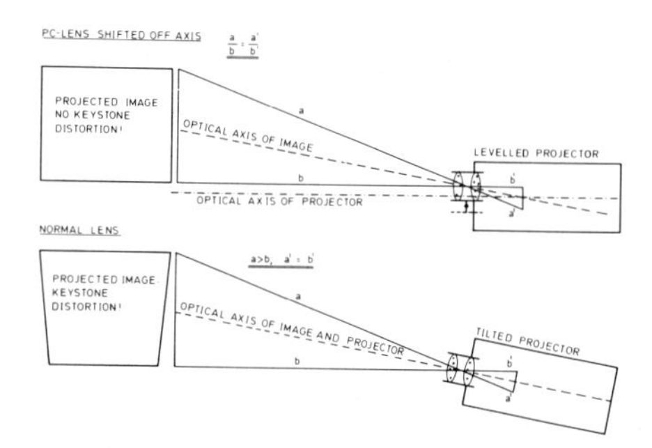

Figure 21.3 The principle of perspective control

It is technically possible, but also rather expensive, to have motorized zoom lenses that correctly maintain focus while zooming. The effect is of limited value except to specialist users like planetaria. Some of these lenses are based on camera lenses, and either have limited light output, or only cover half a slide frame, or both. Some special lens systems have been made with a 20:1 zoom ratio, but their light output is very low.

Back-projection lenses

Although the best picture on back projection is achieved by using a lens with as long a focal length as possible, normally users want to use as little space as possible. A useful range of lenses is available to meet applications varying from compact single-projector display cabinets to complex board room or presentation room systems using several projectors.

A problem arises when more than one projector is needed to serve the same screen area. This always happens when dissolve or multi-image technique is used. Inevitably, only one projector can be perfectly lined up with the screen; more usually, a compromise set-up is used so that none of the projectors is exactly lined up. This results in keystone distortion, where the image is not properly rectilinear. It shows up particularly badly in graphics sequences.

The solution to the problem is the perspective control lens, in which the optical axis of the lens is adjustable. This can effectively eliminate the problem for all pictures greater than 1.8 m (6 ft) wide, and considerably reduce it for smaller pictures. There are also special 'X/Y'-adjustable lens mounts that can eliminate the problem on images as small as 1 m (3.3 ft) wide.

Fixed-focal-length perspective control lenses are often supplied with their exact focal length marked. This means that in a multi-image presentation it is possible to select lenses that match exactly for each image area. Professional lenses are usually guaranteed

Table 21.4 Examples of slide-projection distances using different lenses

Focal length	*Picture width,* m (ft)											
	1 m	(3.3 ft)	1.5 m	(5 ft)	2 m	(6.6 ft)	2.5 m	(8.2 ft)	3 m	(10 ft)	3.5 m	(11.5 ft)
25 mm	0.7	(2.3)	1.1	(3.6)	1.4	(4.6)	1.8	(5.9)	2.1	(6.9)	2.5	(8.2)
35 mm	1.0	(3.3)	1.5	(5.0)	2.0	(6.6)	2.5	(8.2)	3.0	(9.8)	3.5	(11.5)
45 mm	1.3	(4.3)	1.9	(6.2)	2.6	(8.5)	3.2	(10.5)	3.9	(12.8)	4.5	(14.8)
60 mm	1.7	(5.6)	2.6	(8.5)	3.4	(11.2)	4.3	(14.1)	5.1	(16.7)	6.0	(19.7)
90 mm	2.6	(8.5)	3.9	(12.8)	5.1	(16.7)	6.4	(21.0)	7.7	(25.3)	9.0	(29.5)
110 mm	3.1	(10.2)	4.7	(19.4)	6.3	(20.7)	7.9	(25.9)	9.4	(30.8)	11.0	(36.1)
150 mm	4.3	(14.1)	6.9	(21.0)	8.6	(28.2)	10.7	(35.1)	12.9	(42.3)	15.0	(49.2)
180 mm	5.1	(16.7)	7.7	(25.3)	10.3	(33.8)	12.9	(42.3)	15.4	(50.5)	18.0	(59.1)
200 mm	5.7	(18.7)	8.6	(28.2)	11.4	(37.4)	14.3	(46.9)	17.1	(56.1)	20.0	(65.6)
250 mm	7.1	(23.3)	10.7	(35.1)	14.3	(46.9)	17.9	(58.7)	21.4	(70.2)	25.0	(82.0)
300 mm	8.6	(28.2)	12.6	(41.3)	17.1	(56.1)	21.4	(70.2)	25.7	(84.3)	30.0	(98.4)

Figure 21.4 Perspective control lenses eliminate keystone distortion. They are available in focal lengths from 35 mm to 125 mm

Figure 21.5 Schneider make an X/Y lens mount that allows easy setting up of perspective control

to be within 0.5 per cent of their effective focal length.

Anamorphic lenses

An anamorphic lens is one which, when used as a camera lens, squeezes a wider than normal picture onto the film. When used as a projection lens it 'unsqueezes' the picture.

The technique is usually used for movie film. The first major commercial exploitation was Cinema-scope™ Typically, the anamorphic system on movie projection is able to give a picture ratio of 2.4:1, as opposed to the normal wide-screen ratio of about 1.8:1.

The same technique can be used on slides. In this case, the stretching is to about 2:1 from the normal 1.5:1. In both cases the anamorphic lens is in addition to the prime lens.

The anamorphic lens system for use with slides is Iscorama™ made by Isco Optik of Germany. They

Figure 21.6 A short focal length lens for rear projection of 16 mm movie. It uses a built-in reversing mirror. (Photo Buhl Optical)

Figure 21.7 An anamorphic lens as used in professional cinema projection. (Photo Isco Optic)

can also provide professional copying lenses to allow anamorphic slide production on a rostrum camera.

Movie lenses

All rules which apply to slide projection lenses also apply, where relevant, to movie projection. Notice the film aperture is different, so that the focal length needed to produce a given picture size is different from that which applies to slides.

For example, on 16 mm film the width of the actual film aperture is 9.65 mm. A 50 mm lens gives a 3 m (10 ft) wide picture at a projection distance of about 15.5 m (50 ft). Most 16 mm projection is done with fixed focal length lenses although zooms are available. If movie film is back projected, it is necessary to use a reversing mirror; although it is possible to turn a slide round for back projection, this cannot be done with a movie film because of the position of the sound track. The reversing mirror is either external to the lens, or actually built-into it. Lenses down to 9.5 mm are available, to give a picture width the same as the projection distance.

Lenses for 35 mm movie projection are usually made with a small relative aperture, to ensure maximum possible light on the screen and to match professional illumination systems. In principle, it is possible to put more light through a frame of movie film (compared to a slide) because an individual frame is only in the gate for a fraction of a second. Against this, the shutter system is working to reduce the light available.

35 mm movie projection lenses are available in a wide range of focal lengths, in as close as 5 mm increments. Fixed-focal-length lenses are always used, and final matching of the image to a particular screen is done by slight adjustments to the aperture plate. This is a metal mask in the projection gate that is normally the standard projection size, but can be obtained slightly undersize, then enlarged to match the particular installation.

Lenses for 35 mm movie projection are designed with auditorium applications in mind. Lenses for special display applications which require short projection distances must be specially made, or be adapted from slide-projection lenses with a considerable loss in light output.

Projection of 70 mm movies is normally only done in premier cinemas, but it is also used in special displays, especially visitor entertainment in theme parks. A full range of lenses is available for conventional applications, as well as a few high-cost lenses to produce special effects. Some examples include a 50 mm lens, to give a very big image at short projection distance, and very short focal length lenses to fill a whole dome with a single picture. When these lenses are used, there is a premium on picture steadiness and not all projectors are suitable (see Figure 17.8).

Standard video projection lenses

Most commercial video and data projection is based on the use of standard cathode-ray tubes, and this is much less flexible then slide or movie projection. The normal arrangement is three tubes, red, green and blue. The problem is to get light efficiently from the surface of these tubes to the screen.

So little light is available in these systems that it is really only feasible to use fixed-focal-length lenses, and to move the projector to the position that gives the required picture size. The lenses must be about $f/1$ and must deal with an input aperture equivalent to tubes in the range 5–9 ins (12–23 cm) diagonal. This results in a short projection distance, typically 1.5 times the screen width. Fortunately, it is practicable to mount the projector off-axis as keystone distortion can be corrected electronically.

Projection lenses in video projectors are usually made of plastic. However, projectors intended for projecting computer data and graphics benefit from the use of (considerably more expensive) glass or hybrid glass and plastic lenses.

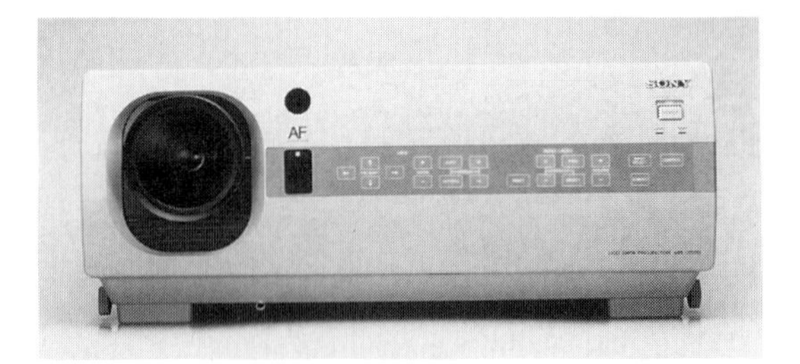

Figure 21.8 This video projector has only a single lens, so the convenience of zoom lenses comes to video and data projection. This Sony VPL-350Q projector also has a perspective control feature to allow off-axis projection

Other types of video projection lenses

There are other methods of video projection which use arc lamps as illumination source. In these cases it is possible to treat the projector much like a slide or movie projector, with a similar flexibility in the choice of lenses.

Liquid crystal video projectors have only a single lens. A typical liquid crystal video projector has a zoom lens giving a choice of image sizes and projection distances, see Figure 21.8. Specialist lens manufacturers now offer short focal length lenses for this type of projector for back projection installations.

Video projectors used for auditorium applications based on LCD, like the Barco 8000 family, have a wide range of standard lenses, allowing projection from distances as short as 1.2 times screenwidth and as long as 7 time screenwidth.

The demands of high-definition television and computer graphics displays are forcing the development of new types of video projection lenses, both to give the required resolution and to accommodate the 16:9 HDTV format.

Chapter 22

Screens

Most AV methods described in this book involve the projected image. In the end the image must be viewed on a screen. This chapter reviews the different types of screen available, and also the reasons for choosing different projection methods.

The concept of screen gain

Some projection screens seem brighter than others, and although it might seem best to use the brightest possible screen for all applications, there could be some disadvantages. To understand why this is, it is necessary to understand what is meant by 'screen gain' and how it is achieved.

The unit of luminous flux, or light output from the projector, is the *lumen*. The 'illumination' of a surface can be described as the luminous flux per unit area; it is measured in lumens per square foot (the foot candle) or lumens per square meter (the lux).

The intensity of light reflected from a surface is referred to as its surface brightness or 'luminance'. One unit of luminance is the foot-lambert, which is the luminance of a uniform diffuser emitting one lumen per square foot.

When a perfectly diffuse reflector is illuminated by one foot candle and has a luminance of one foot lambert, its gain is said to be unity in all directions. Unfortunately, there is no such thing as a perfect reflector, so normally gain is less than one.

However, sometimes a screen has a gain of more than one. An individual point on a diffuse reflector radiates light in all directions; but if a surface reflects more light in one direction than another the gain is greater than one, *when measured in the specified direction.* For example, a screen made with a glass-beaded surface illuminated with 10 lumens per square foot might produce a luminance of 50 foot lamberts when viewed on the axis of projection, i.e. it has a gain of 5. However, when viewed at an angle of 30 degrees off-axis, luminance is only five foot lamberts, i.e. gain is only 0.5.

High-gain screens, therefore, should only be viewed within a specified viewing angle. There are other restrictions on their use, however, which are described later.

Figure 22.1 The screen delivers the final image to the eye, and screen performance has a significant impact on image quality. Modern movie theaters use seamless perforated screens of medium gain. (Photo courtesy United Cinemas International and Harkness Screens Ltd)

Front-projection surfaces

There are various types of surface; some of the common ones are described here.

Matt white

This surface is the nearest approach to the perfectly diffuse reflector. The simplest matt white screen is a wall painted with matt white paint. More usually, however, a matt white plastic or PVC base is used. These screen surfaces have a gain of up to 0.9 on-axis, with only a small fall-off at wide viewing angles (e.g. a gain of 0.75 at 35 degrees off-axis).

Glass beaded

This was one of the first methods of achieving high-gain screens. Beads of glass are embedded in the surface; by internal reflection these ensure that most of the incident light is reflected back on or near the axis of projection. The tendency for the beads to discolor and even fall off, as well as difficulties with transportation means that this type of screen is not now used.

Lenticular

Here the screen is a ridged silver surface, able to produce similar results to the beaded screen but without the problems. It is not quite so high-gain as the beaded, but is a very evenly illuminated screen within a specified viewing angle, beyond which brightness drops off sharply.

Perlux/Ultramatte

These are trade names of Harkness Screens and Stewart Filmscreen Corporation, and are representative of professional coated screens. Here, a matt white surface is sprayed with a special coating to give a seamless medium-gain screen (gain is about two on-axis) which can be viewed at all angles likely to be encountered in a real movie theater auditorium.

Daylight viewing

These are screens of extremely high-gain but also with a narrow viewing angle. They use embossed aluminum surfaces, both to give high-gain and, just as important, to reject light incident from other directions (e.g. light coming from the side or from above) from falling on the screen.

Choice of front-projection screen

Anyone planning a teaching area with only a single projection source, and able to ensure a narrow viewing angle, can consider the use of screens for viewing in high ambient light. Most other users must ensure proper control of lighting conditions.

Teaching and training areas with reasonable lighting control and single-source projection can use lenticular screens provided that the audience are within the optimum viewing angle. If this cannot be assured, then Perlux/Ultramatte or matt white screens must be used.

Auditoria designed for projection (i.e. cinemas, preview theaters, presentation rooms etc.) should use Perlux/Ultramatte, except in those cases where matt white is indicated.

Multi-image projection presents a special problem. Normally matt white must be used to ensure even picture illumination. This is because, when a wide screen is used with many projection sources, any tendency to high gain on one part of the screen necessarily means low gain on another part. For example, someone on the left of the auditorium sees a very bright picture for the part of the image directly in front of him, but the image to his right looks progressively dimmer; what is worse, the screen has a blotchy appearance. Matt white completely solves the problem. In permanent installations with reasonably long projection distances Perlux/Ultramatte may also be used.

In permanent auditoria it is usually necessary for accompanying sound to appear to be coming from the screen. A loudspeaker behind the screen sounds muffled because the screen surface severely attenuates the high frequencies. However, if the screen is *perforated* the sound is fine. In practice, a regular pattern of holes approximately 1 mm diameter is used, with a total perforation of about five per cent of the screen area. This is not suitable for very small auditoria where the audience are close to the screen. Perforated screens can be obtained in both Perlux/Ultramatte and matt white.

Special front-projection screens

Apart from video-projection screens, which are described separately later, there are three other special types of screen that may be of interest:

Deep curve

The curved screens used for some movie presentations can suffer from a lack of contrast due to light from one part of the screen scattering across to another part. Special coatings are available to minimize this problem.

3D

Projection of 3D movies is based on the projection of polarized images. Unfortunately, some conventional screens tend to destroy the polarization. Also, the polarizing process takes away a lot of the light, so a high-gain screen is needed. Special 3D screens, with multiple aluminum reflectors sprayed on to the surface, have an on-axis gain of about four. They cannot be viewed more than 25 degrees off axis.

Very high gain

Front projection screens with a viewing angle of only a few degrees are used in film and TV production. For example, if a film is made which involves action in some exotic location, it can save time and money to use a photograph as the background scene. One way is to use a projected backdrop while action takes place in front of it. No image is seen by the camera on any foreground objects. This is achieved by the projector and camera lens both being on exactly the same axis using half-silvered mirrors, which are actually specially coated glass.

Why back projection?

Although the optimum way of presenting a film or other projected AV program is by front projection in a darkened auditorium, there are many occasions when this is not possible. Very often back projection can solve the problem. Back projection is also the key to several applications of AV that are not auditorium-based.

The majority of back-projection screens reject most light falling on the front of the screen. This means the image does not get washed out by ambient light. Back projection is, therefore, the key to those projection applications where the image must be viewed in high ambient light.

Back projection is essential in applications such as desktop slide/sound units and filmstrip units, microfilm viewers, etc. It also helps in small presentation rooms where AV material must be viewed with lights on, to allow note-taking and discussion.

Another problem in some presentation rooms is the limited height of the room. Back projection can allow a greater image height than would be possible in a room where a front-projection beam might be obstructed by the audience.

Back projection is more flexible to design into exhibition displays. It is also widely used in the theater and other live presentation applications where action takes place in front of a screen.

Back-projection screen material

There is a choice for different applications:

Glass

Ground or sandblasted glass is emphatically of *no use* as a back-projection material. However, glass rear-projection screens where the diffusing surface is either applied as a coating or cast on are available. Glass screens are the most expensive, but also most permanent. Big screens are difficult to transport. In presentation rooms they are best because they cut down sound from projection equipment and are unaffected by differences of air pressure on either side. A compromise is to use a plain piece of glass and a flexible plastic screen in front of it separated by less than an inch. CAT glass is a very fine-grain rear projection glass screen, available in small sizes only, which is used for 'copy and title' work, and even for video transfer where the film or video camera is directed at the screen.

Acrylic

A wide range of acrylic plastic screens is available, from small fine-grain screens for microfilm viewers up to large screens for presentation rooms. Quality of large screens varies from excellent to dreadful, so a demonstration is recommended before purchase. Most acrylic screens consist of a transparent acrylic plastic base with an applied coating. However, the best screens have both the diffusing layer and the pigment cast with the acrylic itself. This gives both better results and a more durable screen. A typical material of this kind is available in sheets up to 1500 mm by 2250 mm (5 ft by 7 ft), and has pigmentation to reject front incident light, a matt front surface to avoid specular reflections, a diffusing layer to give a useful viewing angle of up to 40 degrees off-axis and a transmitted gain of 1.2.

The acrylic lens screens specially developed for video projection should definitely be considered for exhibition work where bright images are required and where their size limitations are acceptable.

Flexible translucent PVC

This is the most widely used material, especially for AV shows given to group audiences and in decor, etc. It is very easy to install; small pieces can be directly stapled into a wood frame. More usually the material is supplied with a strong plastic webbing around the edges, with grommet holes to allow lacing to a frame or with press studs for use with transportable screens. It is available in several different types, some suitable for folding without creasing, some not.

There is a trade-off between amount of light transmitted and degree of dispersion. For AV it is better to sacrifice some light on-axis in order to ensure wide viewing angle and minimum 'hot spot'.

The material is usually dark tinted to help with the rejection of surface incident light. However, white material is available, both for back projection in the dark and for the odd occasions when a surface is needed that serves simultaneously as a back *and* a front projection screen.

It is possible to obtain flexible rear-projection material in seamless rolls up to 3 m (10 ft) wide. Bigger screens can be made by welding the material. The weld seams are virtually invisible when a picture is viewed from a proper viewing distance.

A widely available general-purpose flexible rear-projection material has the following outline specifications:

☐ Vinyl latex base approximately 0.3 mm thick, and suitable for folding. It should be elastic, fungus resistant and fire retardant.

☐ Available off-the-roll in 3 m (10 ft) widths or made up to size, with or without edging, and with either grommets or press studs.

☐ Dark tint; transmitted gain on axis approximately 2.5; low reflectance (about 15 percent at 30 degrees); image resolution 20 line pairs per millimetre.

☐ Maximum recommended viewing angle 45 degrees off-axis; suitable for short focal length lenses.

A rear projection screen images the light from the projector. If wide-angle lenses are used, not all light hits the screen at right angles. There is, therefore, a tendency to hot spot on back projection so that the center of the image appears significantly brighter than the edges. The problem is greatly reduced if *long* focal length lenses are used, but in back projection most users want to use short projection distances. In that case it is essential to use a material which is highly dispersive to minimize or eliminate the problem. In slide-projection terms, the highly dispersive material is suitable for 35 mm lenses, but 60 mm lenses should be used if space is available.

Screen construction

Screens are available in a wide range of executions to meet the needs of different uses and budgets.

Permanent installation should have fixed screens. Where one screen is to serve several different projection needs or formats, it should be fitted with motorized masking. Usually this is only side masking, but in some cases it is also necessary to adjust the top and bottom masking. Masking is done with a non-reflective fabric, usually a black wool serge. Many permanent installations also need a set of motorized curtains or drapes to cover the screen when not in use.

Lecture theaters and teaching environments may use fixed screens, sometimes covered by sliding panels, which may themselves be part of the teaching wall. Very often, however, the multi-purpose nature of the room demands the use of ceiling or wall-mounted roller screens. Ideally, these are motorized, but hand-operated screens are also available.

Travelling AV presentations need some form of portable screen. Within one building the most economical approach is the use of tripod screens for screen sizes up to about 2.5 m (8 ft) wide and box screens up to 3.6 m (12 ft) wide. Neither is particularly elegant, however, and neither is suitable for transporting by car. For these reasons most professionally made and presented AV shows shown on the road use fold-up screens with a folding aluminum frame and screen surfaces that snap on with press studs. These screens are available in a wide range of sizes from about 1.8 m by 1.2 m (6 ft by 4 ft) up to 9 m by 3 m (30 ft by 10 ft) or even bigger. They are available with both front and back projection surfaces, and can also be supplied with dress-up kits to give a draped screen surround. Smaller sizes use a single-tube frame construction; larger ones, especially for wide-screen back projection, use a truss construction.

Figure 22.2 Motorized roll-down screens are suitable for training rooms

Figure 22.3 Fold-up screens are best for travelling AV presentations. They pack into small boxes that easily fit in the back of a car

Screens for video projection

Video projection presents some special problems. The light valve system of video projection is similar to a professional movie projector in its requirements, so a

Figure 22.4 Temporarily installed screens can look better if they are fitted with a dress-up kit. (Photo The Screen Works Inc)

standard Perlux/Ultramatte screen is fine. All the affordable projectors, however, give very little light, so every effort is made to extract the last lumen.

Some video projector units come complete with their own built-in screen system, either back or front projection. These are not described because it must be assumed that manufacturers do their best to optimize screen arrangement. These systems are characterized by a fixed lens-to-screen distance and screens that have as high a gain as possible. The front-projection screens are usually curved, and have a restricted viewing angle.

There are now several of these free-standing video projectors. Most use three cathode ray tubes (one each for red, green and blue), but these have recently been joined by single-lens liquid crystal projectors. With the exception of very expensive units the total equivalent white light output is between 100 and 600 lumens, i.e. at best the output is half that of a standard slide projector.

Where it is possible to install a fixed screen that will *only* be used for the video projector, and where a relatively narrow viewing angle is permissible, it is preferable to install a high-gain screen. The best are rigid screen units with a curved aluminized surface. They are available in sizes up to 3 m (10 ft) wide.

Figure 22.5 The giant Imax screen at the National Musuem of Photography, Bradford, England. At 18 m (60 ft) wide this is a relatively small Imax installation! (Photo courtesy of the museum and Harkness Screens Ltd)

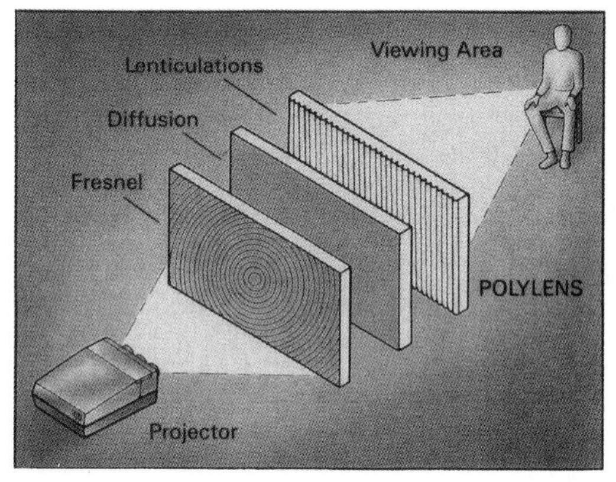

Figure 22.6 The Optixx 'polylens' screen is an example of a high-gain rear projection screen that uses a fresnel lens to concentrate the light, and a front lenticular surface to give it a wide horizontal distribution. (Diagram from Optixx Screen Systems Inc)

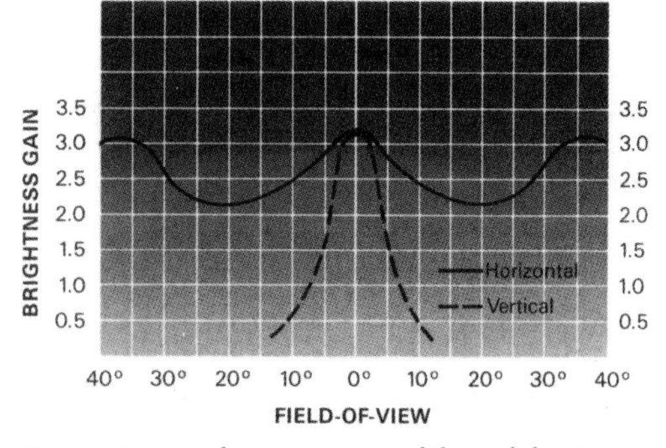

Figure 22.7 The gain pattern of the 'polylens' screen. (Diagram from Optixx Screen Systems Inc)

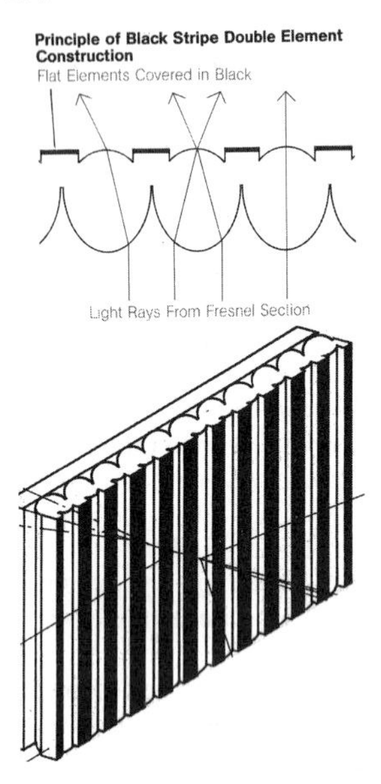

Figure 22.8 Some high-gain rear-projection screens have a black stripe front to improve contrast

Figure 22.10 At Canada's 123rd Birthday Party held at Parliament Hill, Ottawa, a videowall supported the event. For a projected videowall to compete with high ambient light high-gain fresnel lens screens with black stripe front are used. (Photo from Multivision Electrosonic Ltd)

Figure 22.9 Barco offer rear-projection video and data displays that depend for their operation on high-gain lens screens

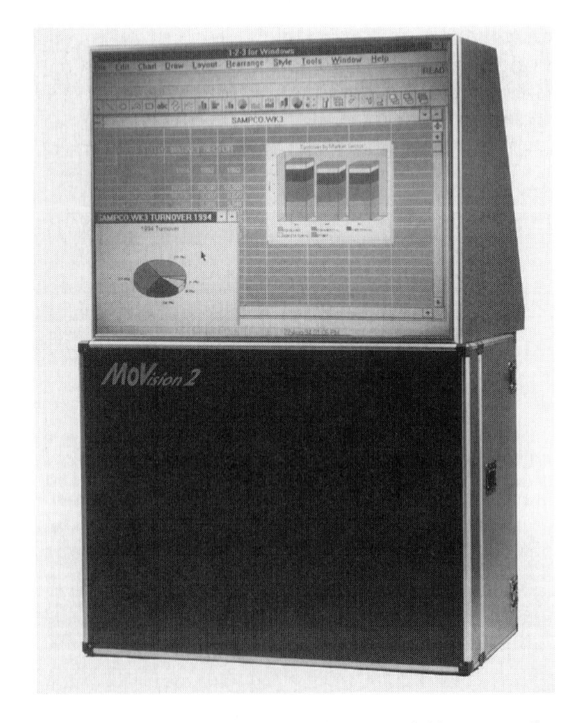

Figure 22.11 MarataVision offer a neat fold-up 50-inch rear projection unit, the MoVision 2, the screen of which is optimized for LCD data projectors. It packs up into a compact case for transport

Often this is not possible, perhaps because a wide viewing angle is required, or the screen system must also serve the needs of other AV shows, or the rigid screen is impractical. Then there is no choice but to use a standard screen. Both Stewart Filmscreen of the USA (Ultramatte 300) and Harkness of the UK (Video Perlux) have special grades of material that may be useful.

Video projectors can also be used for back projection, although in principle only those giving 350 lumens or more should be considered for this purpose, except where a self-contained set is being used. Standard back-projection screen material can be used, and gives reasonable results in the dark.

Special 'lens screens' are available that work on a fresnel lens principle, and have dark non-reflective striped front surfaces. These give significantly better results when high ambient light levels are present; but the difference is not so noticeable in the dark. They are used in self-contained back-projection TV sets, but are also available in a wide range of sizes for other applications.

The smaller lens screens are constructed in two parts. The back of the screen is a large fresnel lens, which bends incoming diverging light rays so that they all come through the screen at right angles to the screen plane.

The front surface of the first element of the screen is another refraction system made up of parallel elements that give efficient dispersal in the horizontal plane, to give good viewing results up to 45 degrees off-axis. The second part of the screen consists of a series of black stripes, arranged so that the light from the lens elements emerges from the transparent slits. The result is a screen that is nearly 50 percent black, giving excellent contrast and immunity to the effects of ambient light.

Lens screens are expensive, but may be essential in exhibition applications and are strongly recommended for presentation rooms. They are the secret to the effectiveness of the projected videowall described in Chapter 19. Double element screens are available in sizes up to 84-inch diagonal, and single element screens (without the black stripe) up to 150-inch. Gain is as high as 5 on-axis, dropping to 2.5 at 35 degrees off-axis in the horizontal plane. Lens screens larger than 150-inch can be made using a different type of lens construction; building up the screen from a number of panels.

Lens screens designed for video projection with three-tube CRT projectors are usually color-corrected to eliminate any trace of 'three-color hotspot' which can be particularly noticeable when low-cost PVC screens are used. Lens screens are also suitable for use with liquid crystal video projectors, but in this case it is important to fit the projector with a lens of the correct focal length to match the screen.

It is not recommended to use small lens screens with low-resolution LCD projectors (less than 300,000 pixels per color). This is because the projected dimension of the LCD pixel is of the same order as the stripe pitch of the screen, resulting in a Moiré patterning effect. In case of doubt, a viewing test should be conducted before committing to a particular projector/screen combination. It is possible to obtain 'depixellation' filters that fit to the lens which reduce or eliminate the problem.

Chapter 23

The audio in audio-visual

High-fidelity sound in the home has made audiences aware of good sound, and they expect similar quality when they hear an AV show. Unfortunately the conditions under which AV shows are presented are not the same as those that apply in the living room, so results are not always as good as expected. This chapter reviews a few topics that can help get the best out of AV sound, and describes some of the changes that are taking place in the methods of sound recording and playback.

Absorption

The average living room is quite small. Even a company presentation room is significantly bigger, and a hotel banqueting suite much bigger still. The result of this is that a sound system that sounds fine at home sounds very disappointing when played to an audience of any size. The sound seems to have lost its bite and words become difficult to understand. This is because the room furnishings, and in particular the audience, are absorbing the high and mid frequencies which give the intelligibility.

The first priority, as far as the audio in audio-visual is concerned, is that the message is heard. This usually means the use of loudspeakers that are larger than may be convenient. If it is not possible to take large loudspeakers to the presentation, then ones that are good on speech should be used.

There are special loudspeakers available that have been designed for auditorium and performance applications, and these should be used for all presentation work. These are of high efficiency and can project the sound to the back row.

Loudspeakers must be installed up high, for example, on either side of the screen. If people can see the loudspeakers, they can probably hear them. If the speakers are on the floor, the front row may be deafened but the back row may not hear a thing. Many performance loudspeakers are available with lightweight telescopic stands that help position the loudspeakers correctly.

Figure 23.1 Travelling AV shows should use performance loudspeakers at screen height

Speech reinforcement

It is surprising how even a small auditorium or meeting room needs a speech reinforcement or public address system. This is because of sound absorption by furnishings and people, and is very noticeable in modern office-type environments with low absorbent ceilings and soft furnishings. Listener fatigue is reduced by always using a simple sound system.

For meetings that take place in rooms on an irregular basis, the portable lectern with built-in amplification can be helpful. However, because the lectern does not permit the loudspeaker to be sited very high, and because the proximity of the loudspeaker to the microphone limits the available sound level, these devices

Figure 23.2 The lectern with built-in amplifier and speaker system is helpful in the small temporary venue. (Photo of the Anchor Ensign from Raper and Wayman)

are best for small audiences in temporary meeting rooms.

Permanent sound reinforcement systems in presentation or meeting rooms with low ceilings often use ceiling loudspeakers. While this has the disadvantage that the sound no longer seems to come from the presenter, it has the advantage of giving high speech intelligibility while allowing the system to be operated at an unobtrusive level.

Larger presentation rooms can be treated more like theaters, with loudspeaker systems designed to give even coverage to all the audience while maintaining the illusion that the sound is coming from the stage. In deep auditoria or large presentation rooms it may be necessary to have additional loudspeakers placed further down the auditorium to get good coverage at the back without the sound being too loud at the front. This is especially the case when trying to reach people in seats under a balcony. Until relatively recently the problem with this was that people sitting at the back not only had the illusion that the sound was coming from very near them, but also often heard the front loudspeakers as an echo because the sound from the front loudspeakers was reaching them appreciably later than that from the nearby loudspeakers. The problem is now completely solved by the use of electronic time-delay units. These delay the sound signal being fed to the back loudspeakers by the necessary few milliseconds to ensure that the sound from the distant loudspeakers arrives at the listeners' ears at exactly the same time as that from the near loudspeakers. This not only eliminates the 'echo' problem, but also helps maintain the illusion that the sound is coming only from the stage.

Another special sound reinforcement problem is common in round-table conferences and large boardrooms. Often the number of participants is so great that a person at one end of the table can have real difficulty hearing what is being said by someone at the other end. The problem can be solved in two ways. The lowest-cost method, and one which is quite suitable for temporary installations, is to equip each delegate with a combined microphone/loudspeaker unit. These units are arranged so that if a particular microphone is switched on, its associated loudspeaker is switched off to prevent acoustic feedback. Otherwise all loudspeakers receive the same signal from whichever microphone is in use.

Switching is sometimes manual, arranged so that

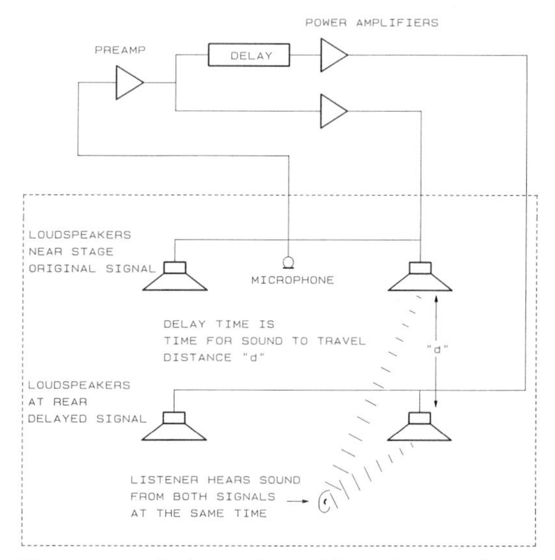

Figure 23.3 In-deep auditoria delay units are used to ensure that the sound from both near and far loudspeakers reaches the ear at the same time

Figure 23.4 The Shure 809 pressure zone microphone is used for boardoom sound systems

the action of switching on any one microphone automatically switches off the one previously in use. Generally, the chairman has an arrangement whereby he cannot be overridden. More recently automatic switching, where the presence of a voice input signal automatically switches on the microphone, has been introduced. This arrangement is a standard feature of the second, more expensive, system in which the automatic sound reinforcement system is built-in to the boardroom table. The design goal is a system which is totally unobtrusive in use, requires no operating by either an operator or the participants, and ensures that everything that is said by any participant is clearly heard by all the others.

These systems make use of 'pressure zone' microphones, which look little different from a flat plate. They are particularly suitable for this application and are quite unobtrusive.

AV show playback

Straightforward commercial AV shows need no more than a good quality mono sound track. Slide and multi-image shows given to large audiences, especially those designed for motivation, promotion or entertainment, benefit from stereo sound. There are some shows that use more than two sound tracks.

For convenience many small shows are run from standard compact cassettes. The track format of these cassettes was originally designed to meet the needs of the domestic user, and they have two pairs of tracks each intended for a stereo audio program. For this use it is important that there is negligible crosstalk between the two programs, but it does not really matter if there is crosstalk within the stereo pair.

As soon as cassettes started being used for AV it was found to be impractical to use one half of the stereo pair for audio, and the other for control signals. When level of the control signal is loud enough to reliably operate the programming device there is objectionable interference on the audio track. Therefore, cassettes used for AV are only used in one direction and the control signal is placed where the second stereo pair is normally situated.

Unfortunately, there is some confusion as to the precise placing of control track information. As far as international standards are concerned, tape cassettes run at 4.75 cm/s, and tracks are placed as shown in Figure 23.5. However, to meet the needs of the musician market some manufacturers have introduced cassette machines with four equally-spaced tracks and a speed option of 9.5 cm/s. There is therefore a tendency to produce multiimage show copies with the control signal in the position of track 4 on the equally-spaced system, rather than in the standard mono position covering both tracks 3 and 4 of the two-pair system. Care must therefore be taken to ensure that show-copy tapes are prepared in a manner that is suitable for the intended replay equipment, because control signals recorded on the 'track 4' basis may be only marginal when played back on equipment intended for 'AV standard' replay.

Tape cassettes are normally only used for small show replay or in the production of very simple slide/sound sequences where copies are not required. Only tape cassettes of the highest quality using standard play tape should be used, and it is wise to always have a spare copy cassette available. Show copy cassettes must *never* be made on high-speed cassette duplicators.

Prestige multi-image or mixed media show playback systems need multi-track capability.

Wide-screen shows benefit from three tracks of audio to ensure a good center-sound channel; conventional stereo can result in a 'hole in the middle' sound. Putting the main commentary on a separate center

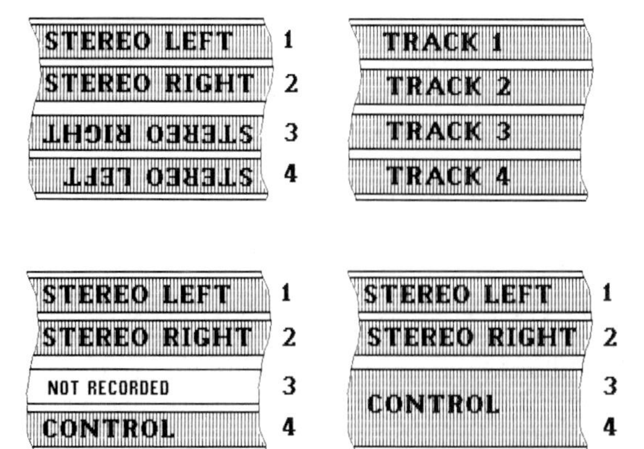

Figure 23.5 Different track standards on the compact audio cassette

Figure 23.6 The Tascam 133 AV cassette recorder has become an industry standard for AV presentation work

channel enables the sound to be rebalanced to ensure commentary is heard properly if there are any acoustic difficulties.

Big shows also use effects tracks played through loudspeakers at the back of the auditorium, and perhaps from other sources to locate the sound. Sometimes programmed sound-switching is used as well as the multi-track technique. A feature of some entertainment shows is the introduction of enhanced bass to improve the effect of full-range orchestral music and sound effects such as trains, explosions etc. This bass can either be derived from all the sound tracks via a special filter, or can be carried on a track of its own. The loudspeakers used are very large and not really suitable for the travelling show! Most loudspeaker systems can give a reasonable performance down to about 50 Hz, but to achieve the visceral impact of these special effects requires use of loudspeakers able to deliver a strong acoustic output at frequencies down to as low as 18 Hz. To do this special drive

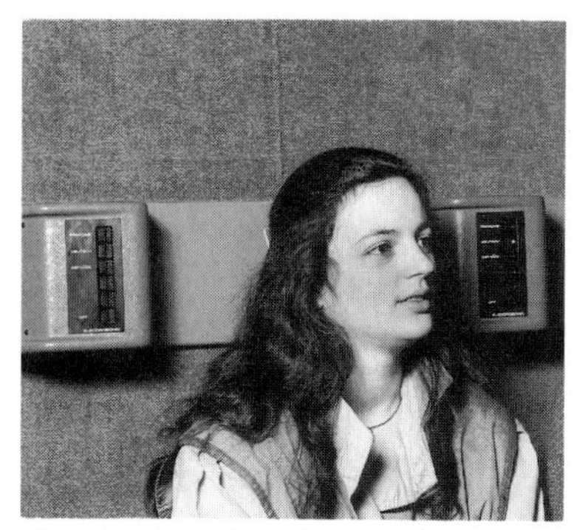

Figure 23.7 In auditorium shows one way of providing multi-language replay is by individual loudspeakers installed in the back row of seats

units are used. In the case of the Intersonics Servo-Drive™ system, for example, electric motor technology is used to move the large volumes of air.

Another obvious use of multi-track sound is to carry different languages. These can be played through the main system as alternatives to the first language, or can be fed via headsets to those members of the audience who need the alternative. This means that either the seats must be wired with headphone sockets and appropriate controls, or visitors must be given suitable wireless receivers with transmission by induction loop or infra-red.

If only a proportion of the audience needs to hear the other languages, an alternative is to fit the back row of seats with headrests with built-in low-level loudspeakers. The system should then be designed so that at the end of the show all of these loudspeakers that have been in use are automatically switched off.

Until recently all big AV shows using multi-track sound would have used multi-track analog tape as the replay medium. Now there is a very wide choice, including:

- ☐ multi-track analog tape with noise reduction, still a cost effective method, even if the show has been made in the digital domain;
- ☐ multi-track digital tape;
- ☐ multi-track digital audio workstation/computer;
- ☐ one or more compact discs;
- ☐ dedicated digital show playback unit using hard-disc;
- ☐ dedicated digital show playback unit using non-volatile flash-EPROM memory.

If a show is only being given for a limited number of performances, for example, a product launch, the playback method is likely to match that of the production and be one of the first three methods listed. In a permanent installation where the show must run all day, every day, then one of the last three methods is more likely to be used.

Often the medium carrying the audio must also carry either or both of timecode and control data (see Chapter 16). This is easily done with multi-track tape, and in each of the other cases, there are ways in which the timing and control information can be incorporated.

Super sound systems

Audio is still to some extent a black art beset with a lot of jargon and riddled with personal opinion. What *is* clear is that the price of sound equipment goes up almost exponentially: to get a significant improvement over a £6,500 ($10,000) system, you might have to spend £25,000 ($40,000).

When designing a sound system for AV presentations a responsible sound contractor starts by defining the basic requirements:

- ☐ How many sound tracks?
- ☐ What are the other sound sources?
- ☐ What is the size and shape of auditorium?
- ☐ How big is the audience?
- ☐ How often does the show run?
- ☐ Is automatic or manual control needed?
- ☐ What is the required sound pressure level?

There is now a good choice of equipment to fulfil the requirements of each part of the system, and it is relatively easy to produce a specification that gives good results for a reasonable sum of money in most circumstances. It is not uncommon to encounter specifications calling for sound pressure levels so high that if the system were actually run at the full level the audience would suffer permanent damage to their hearing. The problem about such specifications is that the user ends up unnecessarily spending money on equipment that might have gone towards improving the show!

The actual sound level needed depends on the expected program content and the nature of the audience. The requirements of a public entertainment show with a full symphony orchestra sound track are quite different from those of ambient sound in a museum. There must then be some headroom to allow the system to deal with transient peak levels without distortion.

Some loudspeaker manufacturers have specialized computer software that allow an audio contractor to verify sound pressure levels that will be achieved in a given space, using a particular loudspeaker/amplifier combination.

Most sound systems require some form of equalization. Equalization is the expert's word for tone control. It might be asked why, if an AV program is recorded professionally for a particular project and state-of-the-art equipment is used for its playback, any modification to the signal is needed at all. It is needed to compensate for the acoustic conditions of the listening space. Therefore the controls used are more sophisticated than the traditional bass and treble controls. The first step is to use parametric equalizers. These are tone controls where the user can select the frequency band being affected.

The next step is to take this approach further and divide the audio spectrum into a number of bands, each of which can be cut or boosted separately. The device that does this is called a graphic equalizer. Professional graphic equalizers allow adjustment of, say, 31 different frequency bands with their center points at one-third octave spacing. The graphic equalizer can ensure that room resonances and unexpected absorption peaks, as well as non-linear loudspeaker response are compensated for electronically.

The first aim of graphic equalization should be to end up with a system that gives a totally predictable

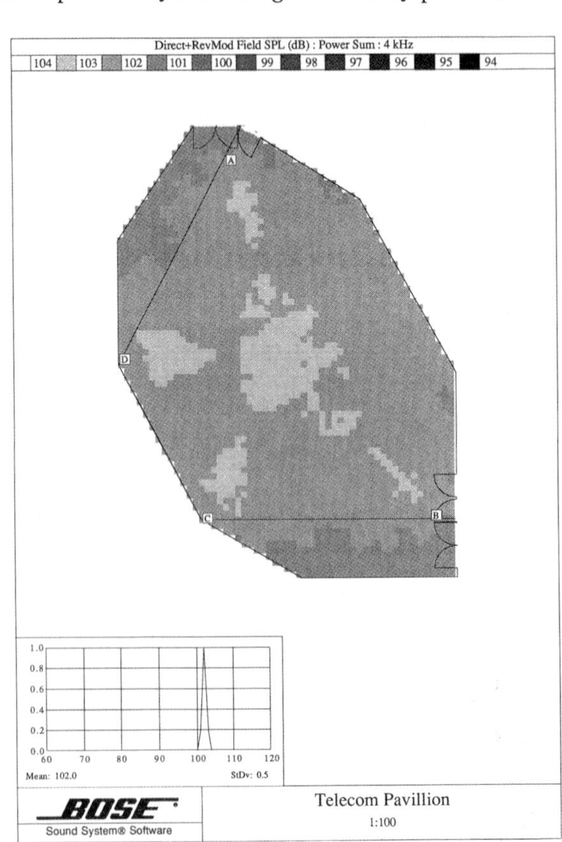

Figure 23.8 The loudspeaker manufacturer Bose provides a computer program that will calculate sound pressure levels within a given space

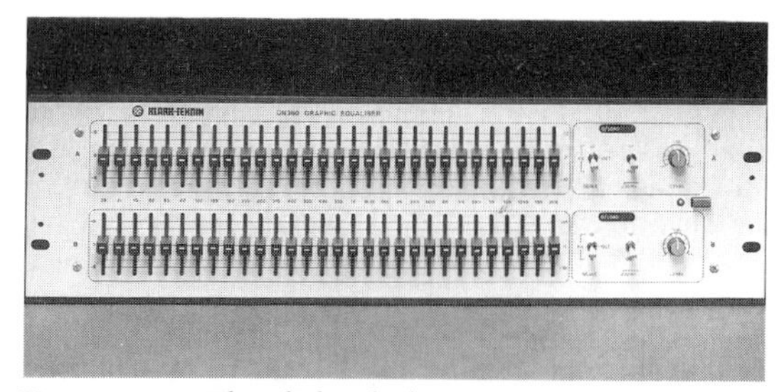

Figure 23.9 The Klark Teknik DN360 stereo one-third octave graphic equalizer can compensate for difficult acoustics in an auditorium

performance. Any program material fed into it should be reproduced without any peaks or troughs in the frequency response. This is achieved by feeding into the complete sound chain a 'pink noise' signal which contains all the frequencies required to be reproduced at the same level. A microphone at the intended listening position picks up the signal that comes out of the loudspeaker and feeds it to a spectrum analyzer. This device displays how much of each frequency is being received. In the ideal system the display is a straight line; in practice, it looks like a small mountain range. The graphic equalizer is the device in the sound chain that is able to selectively boost or attenuate narrow bands in the audio spectrum. It flattens the mountains and lifts up the valleys.

A badly set-up equalizer can do more harm than good. In simple playback systems it is best to have the system professionally set-up, and then to have the controls concealed by a security cover. Modern digital equalizers can have their settings put into permanent memory so that no adjustments can be made by unauthorized users.

Digital equalizers bring other benefits. Some can be programmed by computer, so the installing contractor can keep a record of the settings, and can replace a unit with one already loaded with the correct settings for a particular site. Of more importance is the fact that in many installations there can be a need for more than one setting.

For example, there may be a need within an auditorium to have a different equalizer setting for conference and microphone use, and for program playback. Here the equalizer is used to modify sound deliberately by, for example, ensuring that program replay is flat and full spectrum, but that microphone signals are subject to bass cut, but with some boost at the frequencies that give speech intelligibility.

In other applications acoustic conditions may vary substantially according to the number of people present. Therefore a museum gallery may benefit from having equalization settings that are changed accord-

Figure 23.10 A small part of a theme park sound system. This one is using digital graphic equalizers which have no external controls, and are therefore tamper proof. All equalization settings are entered by computer and stored in non-volatile memory

ing to whether it is a busy or light day. Digital equalizers can have several 'memories' that can be selected according to the current use of a system.

Digital sound

Most audio tapes are analog, that is, they record an electrical analog of the original sound. In digital sound the electrical signal is instantaneously measured and turned into a digital signal consisting only of 0 or 1 values, and coded to represent the instantaneous level. All subsequent processing is done digitally until the final replay when the digital signal is decoded, using a digital-to-analog converter. The method is technically very complex, but made possible by recent advances in electronics. Its advantage is that noise resulting from copying or limitations of the recording or transmission medium is virtually eliminated. The digitization process can introduce quantization and error noise of its own, but in properly designed equipment it is completely filtered out.

Quality of the digitally encoded sound depends on two factors. The frequency response depends on the number of samples taken per second. Mathematically, it can be shown that the sample rate needs to be double that of the maximum frequency required. Thus, if a bandwidth of 10 kHz is required, it is necessary to take 20,000 samples per second. The dynamic range of the sound depends on the number of bits per sample. If there are only eight possible values of sound level, which is all that can be achieved with three-bit sampling, the resulting sound will either be monotonous or will jump in level. A minimum practical number is 8-bit sampling (256 sound levels).

Professional audio recording requires a wide frequency response and extended dynamic range. The industry standard sampling frequencies are 44.1 kHz and 48 kHz that ensure a usable 20 kHz frequency response. Recording is done using at least 16-bit PCM (pulse code modulation). The quantization of each sample is linear, in that the 16 bits represent 65,536 equally-spaced levels of sound.

Once sound has been converted to the digital domain, it is possible to treat it mathematically. As far as the chips are concerned, sound numbers are no different from payroll numbers. The details of how this is done are beyond the scope of this book, but elegant mathematics show that it is possible to create digital filters and to achieve digital data compression.

This is resulting in remarkable products based on various kinds of DSP (digital signal processing) chips. For example, digital graphic equalizers referred to earlier, and chips that are able to reduce the number of bits per sample to only four bits while still maintaining high fidelity sound. Processes such as ADPCM (adaptive delta pulse code modulation) reduce the number of bits needed for signal storage.

Audio professionals like to acquire source material to the highest possible fidelity to the original. This means that they prefer to use methods which do not compress the sound signal in any way. Professional DAT systems (referred to later) and regular audio CDs are examples of products where the digital audio signals are uncompressed, i.e. the original sampling structure is maintained in the recording and no attempt is made to take advantage of redundancy in the signal.

Compression methods work on the basis that where information is repeated, it can be recoded to use less data space. Audio compression can also take advantage of psycho-acoustics to eliminate signals which the human ear simply cannot hear. There is a masking effect where, for example, a loud sound at one frequency will completely mask a softer sound at another frequency. Audio compression algorithms take advantage of this and some of the products and services doing so are:

□ **Apt-x**, used to reduce the bit count in digital transmission systems between studios, etc. Also used in the DTS cinema sound system based on CD-ROM and used first for the movie *Jurassic Park*.

Figure 23.11 The Studer D740 provides direct recording of single CDs. This type of equipment makes CD an accessible playback medium for AV shows

□ **Musicam** used in Europe's Eureka Digital Audio Broadcasting project and the similar, but more complex Perceptual Audio Coding (PAC) from Bell Labs.

□ **Minidisc** from Sony, a re-recordable disc format which uses compression to get CD quality and duration on to a 3'-inch disc.

□ **Dolby AC-3**, a multichannel sound system for providing 'surround sound' in such applications as home theater and the planned 'Grand Alliance' HDTV system.

□ **MPEG**, while primarily a *video* compression system, must, of course, also carry sound. It uses a derivative of the Musicam method. Products such as the solid-state digital sound stores used for the maintenance-free playback of audio in museums, permanent exhibitions, etc., can, and in the case of the equipment shown in Figure 11.11, do use the audio part of MPEG compression.

The compact disc

The most accessible form of digital audio for the consumer is the compact disc. This remarkable device stores up to 74 minutes of stereo sound with a signal to noise ratio of 98 dB and a frequency response that is flat from 20 Hz to 20 kHz. It uses a 44.1 kHz sampling frequency, and data is recovered from the disc at over 2 million bits per second.

Until recently the CD was of limited interest to the AV user because of the difficulty and high cost of having discs made, and because CD players available were all designed for the consumer market. While some consumer players are suitable for AV use, most are not because of their lack of suitable remote control.

There is now a large number of CD manufacturers, some of whom specialize in short-run work. It is quite practical for a visitor's center or museum to commission their own CD, and one CD can carry material to serve many exhibits or shows. Relatively low-cost disc recorders, that record discs compatible with standard CDs, are being introduced to the professional recording and broadcasting industries, so it is even possible to make a single CD for a special application.

CD players are now available with useful remote-control facilities, that allow remote track selection from a computer or system controller. They can therefore be used for long program replay and for applications requiring program selection. For example the Tower Hill Pageant, next to the Tower of London, England, has a 14-minute dark ride requiring stereo sound replay within the car, with a choice of five languages. This just fits in to the CD capacity.

One problem with standard CD players is that there is an indeterminate delay between giving a 'start' signal and the actual start of the program material. This arises because the CD is a constant linear velocity device (see Chapter 13). It presents difficulties when it is required either to synchronize several players together, for multi-track sound replay, or when it is required to synchronize a CD to a sound and light or multi-media sequence. Therefore for many permanent installation applications it can be easier to use standard video laserdiscs as the audio source when precise synchronization is required. Some professional players are available which have buffer memories to standardize the start delay. CD has therefore arrived as a good option for a high quality audio source in prestige permanent AV installations.

Figure 23.12 A show playback system for a large multi-image show. This one is using two CD players as the audio source. The players are fitted with buffer memories to ensure synchronized start

Chapter 14 discussed the many different formats emerging for interactive programs. The compact disc as a physical item is a device for storing large quantities of digital data, but the way in which the data is used is up to the user. The CD represents the current ultimate in accessible and affordable audio storage, but for many applications the sound is actually much better than needed. The CD-I allows sound of lesser, but still very acceptable, quality to be stored in larger quantities. Table 14.3 shows some of the possibilities. Level A audio is excellent for AV work and it is also used in the background music industry. Levels B and C are suitable for various kinds of sound and voice archives.

Figure 23.13 This background music player from AEI-Rediffusion uses a CD-I audio format to allow the storage of 8 hours of high-quality music. The CD is held in a 'caddy'

Digital tapes

Recording digital audio signals on to magnetic tape presents considerable difficulty because of the very high data transfer rate required. This implies a high tape-to-head speed. One satisfactory way of recording digital audio is to use a videocassette recorder and, in fact, the U-matic videocassette has been the *de facto* standard master medium for making CDs. A special encoder allows transcription of a stereo digital audio track on to the space normally occupied by the picture signal.

Some AV producers use the videocassette as an audio source in prestige presentations. This method is attractive because, like using the NTSC videodisc as a digital source, playback equipment is relatively inexpensive.

Digital audio tape recorders are of two kinds. S-DAT recorders use a high linear tape speed and a stationary tape head. The machines are relatively large and are the logical successors to the analog multi-track machines. Like their analog counter-parts they are available as two-track mastering machines, and multi-track production machines.

R-DAT machines use a slow linear tape speed and a rotary head system to achieve the high read/write speed, exactly as used in videotape recorders. The tape is carried in a very small cassette, resulting in compact players and recorders. When people refer to DAT, they are usually referring to this type of digital audio tape. It is a medium that has been a long time reaching the market because of concerns expressed by the record companies about the ease of copying.

Figure 23.14 The Sony TCD-D10 PRO II DAT recorder is widely used as a portable acquisition recorder by professionals. (Photo from Raper and Wayman)

DAT is being used professionally as a mastering medium. Quality is excellent, and recording equipment is compact and very easy to use. It is also an option as an AV-showing medium because it is now possible to obtain DAT recorders that can synchronize to SMPTE timecode.

The use of video-style cassettes as a storage medium has been carried over to audio. The Tascam DA-88 recorder uses a standard Hi-8 $\frac{1}{4}$-inch videocassette for 73 minutes of 8-track digital audio recording and machines like these, and the rival Alesis ADAT format are taking over the traditional role of compact analog multi-track audio recorders. They have the big advantage that, being digital, they do not suffer from the crosstalk between tracks and the limited signal-to-noise performance of their analog forbears.

Consumer confusion

There has been considerable confusion in the consumer markets, largely brought about by manufacturers misjudging the market. Originally DAT tape was expected to be a consumer format, but in fact it only

Figure 23.15 For DAT to be used in AV production it must be synchronizable to timecode. Fostex were the first company to offer 'timecode DAT'. (Photo of the Fostex D-20 recorder from Raper and Wayman)

succeeded in the professional market. Another attempt at introducing digital tapes to consumers was Philips DCC, which uses an ingenious fixed head multi-track arrangement in a low-cost transport. It now seems unlikely that digital tapes will have any significant place in the consumer market, and that digital audio will, for the consumer, mainly mean CD (and its successor multi-media products) and audio in the computer environment. What goes for the consumer also applies to the business and educational user of AV.

Audio production in the digital age

Digital techniques are transforming the way audio programs are produced. For many years the method of making an AV program or video soundtrack was standardized. Professional work was normally carried out in studio premises where often a considerable investment had been made in mixing desks, multi-track tape machines, etc.

Now there are still some kinds of audio production which use traditional set-ups (although the multi-track machines may have gone digital), but in many cases tape has ceased to be used in the editing process. Tape will continue to be used as a means of field acquisition and archiving, but less as the essential means of production.

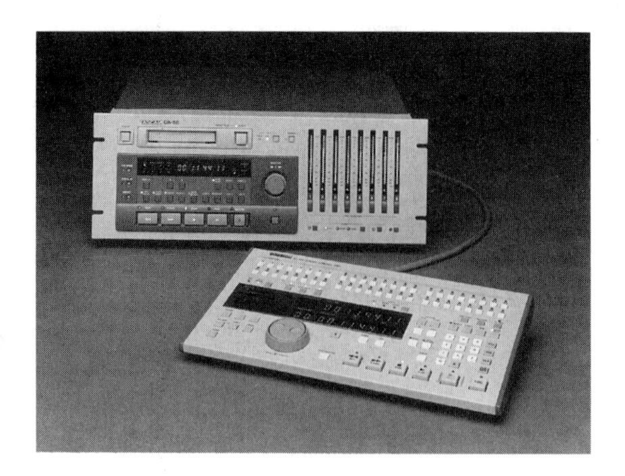

Figure 23.16 The Tascam DA-88 digital multi-track recorder uses Hi-8 videocassettes to record eight tracks of digital audio. Machines like these are replacing their analog predecessors.

Figure 23.17 The Akai DD1500, based on magneto-optical disc technology, brings great flexibility to AV sound track production and is an example of how digital methods are displacing analog tape

This has happened as a result of the arrival of the digital audio workstation. At its simplest this can be a suitably equipped personal computer with audio-editing software, but the professional version is a device better configured to the work in hand. By using hard disc drives (or in some cases magneto-optical drives) as multi-track audio recorders, the user gets a system which is much faster than tape (no 'rewind' time) and can more easily allow multiple versions of mixes to be tried out.

As an example of what has happened, the managing director of one London studio, Sound Developments Ltd, reported that the 24-track analog tape recorder which was once the main item of equipment in one studio dedicated to the post-production of the audio for TV programs had not been switched on once since the arrival of a DAR Sound Station™. This device can be thought of as the equivalent of a 24- (or more) track tape recorder with instant access to any part of any track, and with the ability to compile new mixed tracks from the content of any others. For those requiring more modest facilities, there are programs available giving similar but smaller-scale facilities on both Macintosh and PC computer platforms.

As discussed in Chapter 14 it is now the case that anyone with a personal computer can move into the digital audio business. However, the talent for making worthwhile AV sound tracks is likely to remain with a limited number of AV professionals and the benefits of the new techniques are seen by the sensible AV user as better sound tracks produced more quickly at no more cost, rather than as any temptation to 'do it yourself'.

This point applies, of course, to the entire range of audio-visual activities. Computers have made professional production techniques available at the desk-top; but this does not mean that all programs should be made there. AV users in education, commerce and industry now have an enviable choice of methods, and there are many activities that can and should be done 'in house'. However, when a well-structured AV program is required, it is usually best to work with a team which has professional experience.

Glossary

The brief glossary that follows includes most of the technical terms likely to be encountered on first acquaintance with industrial AV. Many are explained in much more detail in relevant chapters. Definitions and descriptions are those appropriate to industrial AV. For those who require a more comprehensive glossary the BKSTS *Dictionary of Image Technology*, published by Focal Press, is recommended.

Aerial image An optical image formed in space rather than on a screen. It is used in creating film special effects and in slide or film-to-video transfer. In slide-to-video transfer a group of projectors works through an *optical multiplexer* to form a combined *aerial image* that is received by the video camera.

aliasing Appears in video and graphics images as 'stepped' or 'jaggie' diagonal lines, and as stroboscopic effects like wagon wheels going backwards. Arises because the sampling frequency is too low to accurately show image detail. Anti-aliasing is the process in computer graphics of smoothing out these errors.

anamorphic An anamorphic lens squeezes a picture to obtain a wider aspect ratio. See Chapter 21.

answer print The first print of a complete motion picture, including sound track. It is normally used as the basis for making decisions about detailed color correction on the subsequent prints.

aperture In projection terms the aperture is the size of the original image being projected, e.g. 35 by 23 mm for 35 mm slides.

ASCII An acronym for *American standard code for information interchange*. There is a simple digital code for all upper and lower case letters, numbers, punctuation marks etc. This code is used by computers to communicate with printers and other peripherals.

aspect ratio Ratio of the width to the height of an image. Slides are usually 3:2. See Chapter 17 for different film aspect ratios.

assemble-edit A method of video production where sections of the program are re-recorded sequentially to assemble the complete production.

back projection This is when the projector is placed behind the screen, and the audience views the image by *transmission* of light, as opposed to *reflection* when front projection is used. See Chapter 22.

bandwidth A term defining the frequency response of an electrical system. A bandwidth of 20 kHz (20,000 cycles per second) can carry excellent audio, covering the range of hearing. For a video image a bandwith of 5 MHz, (5 million cycles per second) is needed.

BASIC An acronym for *beginner's all purpose symbolic instruction code.* A computer language for writing computer programs, mainly used by beginners but also by experts when programs must be written quickly, but where speed of execution is not important.

baud The rate at which computer data is transmitted, usually in bits per second.

Betacam A trade name of Sony, also licensed to other manufacturers. A professional method of analog video recording which records the video signal in *component* format. A competitive format from Panasonic is M-II. There is now also a *digital Betacam.*

bit The fundamental unit of computer information. It can only take the values of 0 or 1; all calculations within a computer are carried out using this binary notation.

byte A group of bits. An 8-bit byte can represent all numbers up to 256. A kilobyte is 1024 bytes, a megabyte is 1024 kilobytes, a gigabyte is 1024 megabytes, and a terabyte is 1024 gigabytes.

C A widely-used computer programming language.

camcorder A combined video camera and videocassette recorder. In broadcast use the format is usually Betacam or equivalent component system; for AV and consumer use it is VHS, S-VHS, Video 8 or Hi-8.

Carousel A registered trademark of Kodak. It describes their range of automatic slide projectors using a circular slide tray and gravity slide feed.

cartridge In AV it is likely to be encountered (1) as the pick-up on a phonograph or record player; (2) as a device to hold an endless loop of $\frac{1}{4}$-inch tape in continuous sound playing devices; (3) as a device to hold both an endless audiotape and a film strip in desktop AV filmstrip units; (4) as a device to hold an endless loop of Super 8 mm movie film.

cassette The tape cassette was devised to make tape handling easier. The tape is enclosed with its supply and take-up reels. The standard audio cassette is the CC cassette, and is available with different lengths and qualities of tape. See also *videocassette.*

CCD charged-coupled device. A solid-state device used as the image sensor in modern video cameras. A professional camera uses three of them, one each for red, blue and green components.

CD compact disc. Now the standard for storage of high fidelity audio. Stores 74 minutes of stereo audio with a 20 kHz bandwidth in a digital format on a 12 cm disc.

CD-I, CD-ROM, CDTV The use of compact disc medium for audio, data, video and graphics. See Table 14.1 for full list of acronyms.

chip The popular name for an integrated circuit, a single component which itself contains up to many thousands of transistors and other components that have been produced on one substrate by a miniature lithographic process. Components are produced by selective etching and diffusion with trace elements.

chromakey Used in video production to allow one image to overlay another. Typically an actor stands in front of a blue screen, but is seen to be speaking from some exotic location derived from a separate image source. Here the blue is used to 'key' the second image.

chrominance One of the components in component video. In professional systems there are two color difference chrominance signals, Cb = B–Y and Cr = R–Y. Where R, G and B are the red, green and blue signals and Y is the luminance signal, then Y = R + G + B.

component video Electrical signals carrying the information for a color television program, divided into luminance and chrominance components (see Chapter 19).

composite video An electrical signal carrying the information for an entire color television program, but not including the sound. See also RGB and RF.

Compression Methods of reducing the amount of data to be stored or transmitted, as applied to digital signals. See Chapter 14.

D1 A professional digital video recording format. It records uncompressed *component* digital video on to a videocassette using $\frac{3}{4}$-inch tape.

D2 This records *composite* digital video on to a cassette the same size as the D1, but using a different type of tape. A D2 cassette gives more than twice the duration of its D1 counterpart.

D3 A *composite* digital format using the same sampling format as D2, but requiring only a $\frac{1}{2}$-inch tape.

D5 A *component* digital format recording an uncompressed signal on to a $\frac{1}{2}$-inch cassette. The cassette is identical to the D3 cassette, but the tape runs at twice the speed to accommodate the higher data rate. There is no D4 standard because four is an unlucky number in Japan.

D6 A new digital tape format designed by BTS and Toshiba for the recording of high-definition video.

DAT Digital audio tape. A method of recording digital audio signals onto tape (see Chapter 23).

data projector The output of a computer that would normally be shown on a monitor can be fed to a video projector. However, the resolution of such projectors is not always good enough for such applications. Data projectors are video projectors of high bandwidth. Most, but not all, are also suitable for other color video programs. They use an RGB signal.

DCC Digital compact cassette. A compact audiocassette format promoted by Philips. Uses special tape, multiple tracks and compression techniques to record near-CD quality digital audio on a cassette physically the same size as the standard analog audiocassette.

DCT Discrete cosine transform. A method of digital *compression,* used, for example, in the *MPEG* method. DCT is also a digital video recording format promoted by Ampex which records *component* digital video on to $\frac{3}{4}$-inch D2 size cassettes, using approximately 2:1 compression.

digital In electrical terms signals can be digital or analog. Most audio recordings are analog, i.e. the signal on the tape varies in the same way as the sound to be reproduced. In a digital signal there are only two values, on or off, 1 or 0. Digital audio recordings take the instantaneous level and code it as a binary number. This type of system is greedy in the amount of data space it needs, but theoretically, it gives almost noise-free recording with no degradation on copying. All computer signals are digital, so the word often appears in references to computer and programming equipment. Digital processing is also applied to video signals.

Digital Betacam A *component* professional digital video recording format developed by Sony. It uses the same size cassettes as analog Betacam, but to do so it uses approximately 2:1 compression.

dissolve unit Controls two automatic slide projectors to dissolve one image to another using electronic lamp control (see Chapter 15).

DMX A serial digital protocol for the control of lighting. Originally specified for lighting level control through dimmers, but now also used for the control of moving lights, color scrollers and similar devices. Electrically it uses RS485 practice to transmit data at 256 kb/s. In its basic form it controls 512 'channels' each with 8-bit resolution.

DVCPRO A new professional digital videocassette format especially suitable for news gathering and similar applications. The very small cassettes carry one or two hours recording on $\frac{1}{4}$-inch tape. Recording is to the 4:1:1 component standard, and 5:1 intra-frame DCT compression is used.

D-VHS A system from JVC which allows a VHS videocassette transport to be used for recording digital data. Depending on the speed of the tape this could be used either for recording very long 'programs' with a high degree of digital compression (e.g. surveillance or direct recording of satellite delivered digital programs) or for high-quality recordings. It could also be used for non-video data.

DVI Digital video interactive (see Chapter 14).

Eidophor Trade name of Gretag AG to describe its powerful video projector that uses a light-valve technique (See Chapter 19).

episcope Otherwise known as an opaque projector. It is a device for projecting directly from, for example, the pages of a book. Not used much now since it is easier to make an overhead transparency or slide instead. An **epidiascope** is a device that combines the facilities of an episcope with those of a slide projector.

equalizer A device for deliberately modifying the frequency response of an audio system, in other words a sophisticated tone control. *Graphic equalizers* allow individual frequency bands to be boosted or attenuated; there are typically as many as 32 bands in a one-third octave graphic equalizer. This device is used to match an audio system to an auditorium by reducing frequencies that have a resonance and boosting frequencies with exceptional absorption.

feedback The feeding of an output signal back into the input. Some kinds of feedback are beneficial, but the one most AV users are likely to encounter definitely is not! Acoustic feedback is when a microphone picks up its own amplified signal from the loudspeaker, causing a howl. The cure is to turn the volume down, but also to ensure that the loudspeakers are forward of the microphone and pointing away from it.

field A normal TV image frame is made up of two fields. Each field carries half the picture information, and lines of each field are interlaced with each other. There are 60 fields per second in NTSC and 50 in PAL.

filmstrip Individual images intended for still projection printed on a continuous length of film, usually 35 mm. It is cheaper than slides, especially when a lot of copies are needed, but not as flexible. It was used in industrial AV in individual communication devices, but is now obsolete in commercial AV. Still used in education.

flipchart A very large pad of paper, usually supported on an easel. The presenter uses it as a disposable writing board, or prepares a number of pages with information and flips them over as they are described. Writing is by colored marker pens. Flipcharts are useful for meetings and informal presentation, especially where facilities are limited.

floppy disk A flexible magnetic disk for storing computer data and computer programs. It was originally 8 inches in diameter, but now all microcomputers use 5.25-inch disks or protected 3.5-inch disks.

focal length A length, usually expressed in millimeters, that relates the image size produced by a lens to the distance of projection (see Chapter 21).

frame An individual TV or film image. There are 24 frames per second in standard movie film, 30 frames per second in NTSC, and 25 frames per second in PAL.

framestore A solid-state device for storing a video frame. Used in video production, in graphics systems and in other image processing devices, for example videowalls.

frequency Cycles per second or hertz (Hz) as applied to electrical and audio signals. Most audio information is in the range 20–15,000 Hz. Note also that 1000 Hz equals 1 kilohertz (kHz) 1,000,000 Hz equals 1 megahertz (MHz) and 1000 MHz is 1 gigahertz (GHz) A TV signal on videotape needs about 3.5 MHz bandwidth. VHF broadcasts are at around 100 MHz.

genlock All video devices require a source of *sync pulses* that determine the precise moment at which a video frame starts. When several devices, e.g. a video recorder and several cameras, are being used together it is vital that their sync pulse generators work in unison. This is achieved by genlock.

hard disk The standard bulk memory in a computer. While floppy discs are exchangeable, most hard discs are not. However, they hold much more data and can be accessed much quicker.

HDTV High-definition television. Television display systems with about four times the amount of information than carried in a normal TV picture. Now taken to mean at least 1000 TV lines, and with an aspect ratio of 16:9 (see Chapter 19).

HDVS A trade name of Sony. High-definition video system. Covers their group of products working on the 1125 line, 60 field per second analog HDTV system.

Hi-8 A trademark of Sony for a development of the Video 8, 8 mm cassette system that records chrominance and luminance separately.

holography A means of describing an image in terms of the way it interferes with light, an abstruse concept that is not easily compressed into a few words! The *hologram* is the holographic equivalent of a photograph. Holograms are interesting because they store *three-dimensional* images and because if a hologram is broken each part of it can still reconstruct the whole image, albeit at lower definition. Holograms require laser light for their creation and usually for their viewing. They are suitable for special exhibition displays but not for projection, so holography is unlikely to be relevant to mainstream AV for some time yet. There are some cases where an audience believe they have seen projected holograms, when actually they have seen conventional film or video displays viewed indirectly via partially reflecting glass. This is the 'Pepper's ghost' technique, which is very effectively used by, for example, displays in Walt Disney World.

insert-edit The professional method of video editing using timecode, where video sequences are inserted into exact lengths of tape.

instant slide Presentations have a habit of needing visuals prepared at the last minute. Polaroid has an instant slide film requiring no special facilities, and with a development time of a few minutes.

interface A computer cannot be directly connected to the outside world. For example, if it was required to switch on a light bulb it has no circuit element strong enough for the task. Even if it could, the connection would introduce unwanted interference into the computer. To protect it from the outside world a computer uses interfaces to isolate both inputs and outputs. Interfaces allow computers to become an active part of a control system, and are often themselves highly sophisticated devices.

keystone When a projector of any kind is not able to project on-axis the image projected is not rectilinear, and suffers from keystone distortion. With overhead projectors the problem is eliminated by tilting the screen. With multiple-slide projectors the problem is reduced by keeping the projectors as close together as possible, or by the use of special lenses.

JPEG Joint Photographic Experts Group. The ISO standard method of digital image compression for individual images.

laser An acronym for light amplification by stimulated emission of radiation. A device that produces an intense narrow beam of coherent light (i.e. the light waves are all in step) usually of only one or two wavelengths (colors). Very little use in AV except as a gimmick at product launches and similar events. In those cases the system must be under the control of an experienced operator. Baby laser displays are disappointing. However, one useful low-power device is the laser pointer, a battery-operated device ideal for lecturers in big halls. Lasers are also used in creating and viewing holograms. Compact lasers form the basis of precision scanning devices, for example in videodisc players. Experiments are proceeding in using lasers as a means of directly projecting large video images.

LCD Liquid crystal display. A method of modulating light based on the fact that an electrical field can change the plane of polarization of light travelling through the liquid crystal material. Forms the basis of some flat screen displays and video projectors (see Chapter 19) and of overlay panels for displaying computer data on overhead projectors.

lumen The measure of light emitted by a light source, e.g. a 250 watt projector lamp emits about 6000 lumens in all, but in a slide projector only about 1000 lumens reach the screen because the projector optics are only able to collect a proportion of the light. The lux is a unit of illumination, lumens per square meter.

luminance See *chrominance*.

MAC Multiplexed analog component. A method of transmitting TV signals in analog component form. D-MAC is a variant with digital audio.

machine code High-speed computer programs are written in machine code, the actual internal language of the computer. Each instruction accomplishes a tiny task and relates to the manipulation of individual bits and bytes.

magnetic sound Here sound is recorded in the form of magnetic patterns in a magnetic powder applied to some base material. Most AV sound is magnetic, whether it is on cassette or open-reel tape. Magnetic sound can also be applied to movies, in which case one or more magnetic stripes are bonded to the edges of the movie film.

memory The power of a computer is measured not only by its speed but also by the size of its memory. This is divided into high-speed working memory or RAM (random-access memory), typically from 640 kilobytes to 16 megabytes, and the back-up store which is on floppy or hard disc. This can be anything from 600 kilobytes to many hundreds of megabytes.

microcomputer The first computers were big mainframe machines; then came minicomputers, widely used in industry, most recently microcomputers have exploited the power of the microprocessor. The problem is that present-day micros are more powerful than yesterday's minis or even than the day before yesterday's mainframes. Let's just call them computers!

microprocessor An integrated circuit or chip that embodies all the main attributes of a computer. In fact, there must be several additional components around the microprocessor to interface with the outside world. There are many different kinds of microprocessor, some powerful 16- or 32-bit ones suitable for business computers, others with lots of interface capability and their own internal memory suitable for industrial control applications.

MIDI Musical instrument digital interface. A serial control protocol, originally developed for synchronizing and controlling electronic musical instruments, but now also used for show control, using the MSC 'MIDI Show Control' variant.

Minidisc A trade name of Sony describing a recordable digital audio disc. The system uses compression to get over an hour's playing time on to a 3 inch diameter disc.

monitor A precision TV set for monitoring picture quality. The word now tends to be applied to any TV set with a video (as opposed to *RF*) input. A receiver-monitor is a set that is both a conventional TV set for receiving broadcast TV, and has a video input.

MPEG Moving Picture Experts Group. The ISO standard method of digital image compression for moving images. Now with two versions MPEG-1 and MPEG-2, see Chapter 14.

multi-image The programmed activities of many slide projectors on one or several screens.

multimedia Shows using more than one medium, for example mixed multi-image and video, perhaps with synchronized lighting control. The word is now also used to describe computer-based interactive systems combining data, audio, graphics and video.

multivision Another word for multi-image, previously favored in Europe.

noise reduction Most sound recording systems introduce noise (hiss) to the signal. This can be subjectively reduced by using a noise reduction system that effectively boosts the quiet part of a program for recording purposes, but restores the signal to the correct relative level on replay. The best-known system is the Dolby™ system; others are dbx and High Com. Professional AV shows should use noise reduction.

Non-linear video editing Video editing done in the computer domain. All the program material is stored on computer discs, providing extremely fast random access.

NTSC See *PAL*.

off-line Professional video editing equipment is expensive. Most video productions use the technique of off-line editing, where the master material is copied to a lower-quality standard such as U-matic or S-VHS. All editing decisions are taken in this format, and an edit list prepared based on timecode. The final edit is then done almost automatically on the expensive professional format.

optical sound Most 16 mm and 35 mm movies have optical sound tracks, where sound is recorded photographically and replayed by a photocell. Modern 35 mm sound tracks are good, especially when used with a noise reduction system, but 16 mm sound tracks are less so, because they have a restricted frequency response. However, for most applications they give acceptable results and they represent the most reliable method of sound synchronization for movie.

overhead projector A device for projecting large transparencies of page size. It consists of a lamp surmounted by a fresnel lens stage on which the transparency, or transparent foil, is placed. A lens with an angled mirror above the stage focuses the image on a screen which is usually above and behind the presenter. The size of the original image allows very bright images on the screen.

PAL Unfortunately TV signals are not standardized throughout the world, and there are even variations within each main standard. PAL is the standard in Europe, except in France where they have their own SECAM system. NTSC is used in North America and Japan.

pixel Picture-element. Video and graphic images are divided into pixels for processing. The CCIR601 standard for TV production uses 720 pixels on each displayed video line. Graphics systems use many more.

presentation unit Combines the facilities of a dissolve unit with those of an audio visual tape recorder.

programmer Device to program the activities of one of more slide projectors, usually in synchronization with a sound track (see Chapters 15 and 16).

PROM An acronym for programmable read-only memory. When a microprocessor is used as part of a control product, e.g. a dissolve unit, its program is stored in a solid-state read-only memory. These can be specially made for the job in hand, but for many applications it is easier to buy a blank memory from the chip maker, and program it as required using a PROM programmer. There are various kinds of PROM, all of which are 'non-volatile' i.e. they do not lose their memory when the power fails. 'One-time' PROM can only be programmed once. Standard EPROM can be erased by ultraviolet light and then re-recorded. Flash EPROM can be electrically rewritten as often as required.

pulse The control signal on magnetic tape used to control slide projectors. In simple slide/sound systems a 1 kHz pulse of about half-a-second duration is recorded to advance the slide. Multivision systems use a continuous stream of digital pulses of great complexity.

random access The ability to locate rapidly and show a visual aid or AV program (see Chapters 13, 14 and 15).

reel-to-reel Conventional magnetic audio tape is supplied on reels, the tape being threaded through the record/play mechanism to an empty take-up reel.

Most master audio tapes are made reel-to-reel as it is easiest for editing and gives the best sound quality.

relative aperture The ratio of a lens' diameter to its focal length. Also known as the *f* number (see Chapter 21).

RF Broadcast television uses signals at radio frequency or RF. This is the kind of signal picked up by a TV aerial (antenna). It usually includes audio information as well. It is created by using a composite video signal to modulate a very high frequency signal suitable for transmission.

RGB Color television signals are originated as three separate pictures: red, green and blue. Usually they are merged together as a composite video signal but, for highest quality and for specialist applications like data projection, the signals are kept separate.

rostrum camera Copying camera used in making professional AV shows (see Chapters 2 and 5).

RS232 A code of practice which defines the electrical parameters for sending serial digital data between one device and another. RS232 is used, for example, at the serial port of personal computers. It is intended for short distances (a few metres) and comparatively low data rates. RS422 and RS485 use a more robust balanced line method which allows transmission at higher rates over longer distances (1000 m and more).

SECAM See *PAL.*

slide/sound The French use the word diaporama, which sounds much more grand. Here it is taken to mean single-screen AV shows based on slides with a synchronized sound track. It can grow into multivision.

SMPTE Society of Motion Picture and Television Engineers (USA). Best known in the AV context as the sponsors of SMPTE timecode used in video, film and AV production.

soft-edge mask Used in multivision to allow projection of panorama pictures from several projectors, without any obvious picture join showing (see Chapter 16).

stereo The recording of the sound signal on to two tracks of tape to give a sense of space and direction to the sound. Prestige AV shows use stereo and some use three or four tracks for special effects or to give a good center sound.

S-VHS A development of the popular VHS videocassette format that records the luminance and chrominance separately, resulting in much better resolution.

teaching wall The well-designed training room uses a teaching wall that neatly integrates different communication methods – sometimes side-by-side, sometimes using a system of sliding panels. Thus, in one permanent assembly the facilities of flipchart, white board, slide/video screen and overhead projection screen can be combined.

telecine The equipment for transferring movie film to video. Best known is the Rank Cintel equipment that uses a flying spot film scanner. Equipment using CCD devices is also available.

teletext The generic term for public network systems that present alpha-numeric and graphic data as still video images (see Chapter 19).

three-dimensional projection At present the only practical way to provide projected three-dimensional (3D) images is by the long-established method of taking two photos of the subject, with the two camera lenses separated by approximately the spacing of human eyes. The images are then projected by two projectors onto the same screen. Each projector is fitted with a different polarizing filter, and the viewer wears a pair of special glasses with a different polarizing filter for each eye. Each eye then sees only the image intended for it. The principle works equally well for movie film and for slides, and if care is taken with production the results are spectacular. This technique is of rather limited value in the AV field, because few industrial and commercial subjects can benefit from 3D, other than as an occasional gimmick. Experimental systems have been demonstrated that use special lens screens to allow 3D images to be seen without glasses.

time base correction A method of ensuring that a number of separate video devices all work correctly in picture sync (see Chapter 19). It is especially useful when sources such as VHS tape must be copied or mixed with other material, and is sometimes the only way by which a stable picture can be recovered.

timecode A means of accurately identifying time position within a program. Usually recorded on a separate track of a multi-track tape, or within the vertical interval of a TV picture signal. Best known timecode is the SMPTE code, and its european derivative, EBU (European Broadcasting Union) timecode.

track Referred to magnetic tape. A tape can have more than one independent track on it, thus giving the possibility of stereo sound, or sound plus control. The normal audio cassette has two pairs of tracks. Typical reel-to-reel recorders have four tracks on $\frac{1}{4}$-inch tape or eight tracks on $\frac{1}{2}$-inch tape. Studio recorders have up to 24 tracks on 2-inch tape. Typical videocassette recorders have two conventional audio tracks and the helically-scanned video track.

transparency Slides are transparencies, as are color photographs taken on reversal films. In the AV field, however, the word normally refers to the transparency used on overhead projectors, which is nearly page size.

U-matic A type of *videocassette* used for professional production and for industrial video shows. It uses $\frac{3}{4}$ inch tape.

Uni-Hi. A videocassette format, physically similar to a U-matic cassette (but with quite different tape characteristics) used for recording 1125 line high-definition video.

vertical interval The moment when a TV raster scan reaches the bottom of the scan, and jumps back to start scanning a new field or frame. Video switching must be done in the vertical interval to get a clean cut. The vertical interval is long enough to allow other signals, for example timecode, to be recorded within it.

VGA Video Graphics Array. One of many 'standards' for displaying computer data and images on a monitor, VGA shows 640 pixels on each of 480 lines using progressive scan. See Table 3.1.

VHS VHS is a trademark of JVC it stands for *video home system*. It has become the videocassette standard for consumer use, and is based on $\frac{1}{2}$-inch tape.

video-8 The 8 mm tape format for videocassettes, championed by Sony, but subscribed to by many manufacturers. It has not been able to dislodge the supremacy of VHS, except in the area of portable camcorders where it is a level competitor.

video projector Usually video TV images are viewed directly on a cathode-ray tube, but there is a practical limit to how big these can be. Bigger images can only be produced by projection (see Chapter 19).

videorecorder Device for recording TV programs. Videorecorders used in broadcasting use 1-inch tape on reels or special $\frac{1}{2}$-inch tape in cassettes. AV use is based on VHS, S-VHS or U-matic videocassettes. Videorecorders use multiple tape heads to scan tape diagonally in order to achieve the high tape-to-head speed necessary to record high-frequency signals.

videodisc A means of storing video programs rather like a phonograph record (see Chapter 13).

writing board The old school blackboard has come a long way. Some applications are still best met by glass chalkboards, especially in teaching rooms. Other training rooms and some presentation rooms use plastic or metal whiteboards with colored marker pens.

W-VHS. A videocassette format from JVC which can record a video bandwidth of 20 MHz. This makes it suitable for recording 1125-line high-definition video.

zoom As applied to lenses: variable focal length. In camera lenses this gives the ability to take close-up and distant shots with the same lens. In projection lenses it allows projectors to be placed conveniently, with the lens adjusted so the picture exactly fills the screen. Some video camera lenses have zoom ranges of in excess of 10:1. Projection lenses usually only have a range of about 1.8:1.

Index